METHODS IN MOLECULAR

CW00434154

Series Editor
John M. Walker
School of Life and Medical Sciences
University of Hertfordshire
Hatfield, Hertfordshire, UK

For further volumes:
http://www.springer.com/series/7651

For over 35 years, biological scientists have come to rely on the research protocols and methodologies in the critically acclaimed *Methods in Molecular Biology* series. The series was the first to introduce the step-by-step protocols approach that has become the standard in all biomedical protocol publishing. Each protocol is provided in readily-reproducible step-by-step fashion, opening with an introductory overview, a list of the materials and reagents needed to complete the experiment, and followed by a detailed procedure that is supported with a helpful notes section offering tips and tricks of the trade as well as troubleshooting advice. These hallmark features were introduced by series editor Dr. John Walker and constitute the key ingredient in each and every volume of the *Methods in Molecular Biology* series. Tested and trusted, comprehensive and reliable, all protocols from the series are indexed in PubMed.

Spatial Genome Organization

Methods and Protocols

Edited by

Tom Sexton

Institut de génétique et de biologie moléculaire et cellulaire (IGBMC), Illkirch, France

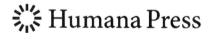 Humana Press

Editor
Tom Sexton
Institut de génétique et de biologie
moléculaire et cellulaire (IGBMC)
Illkirch, France

ISSN 1064-3745   ISSN 1940-6029   (electronic)
Methods in Molecular Biology
ISBN 978-1-0716-2499-9   ISBN 978-1-0716-2497-5   (eBook)
https://doi.org/10.1007/978-1-0716-2497-5

Cover Illustration Caption: Schematic of a chromatin loop bringing a gene and its regulatory element in close proximity. Image courtesy of Angeliki Platania.

Preface

Since Emil Heitz's description of euchromatin and heterochromatin in 1928, it has been appreciated that the spatial arrangement of genetic material within the eukaryotic nucleus is heterogeneous and highly organized. Over the decades, striking correlations with chromosome configurations and the activity of their underlying genes have been uncovered, and identifying any causal and mechanistic links is a highly active area of ongoing research. This book covers the cutting-edge techniques used to interrogate spatial genome organization and is comprised of six main parts.

Studies of genome topology were revolutionized by the advent of the chromosome conformation capture (3C) technique in 2002, identifying pairs of genetic elements in close physical proximity *in vivo*. Part I describes the cutting-edge extensions of this pioneering method, allowing pairwise chromatin interactions to be assessed at ever-increasing coverage. Part II focuses on targeted further adaptations to the 3C-based techniques, maintaining genome-wide views of chromatin interactions (and not necessarily limited to pairwise relationships; Chapter 6) but providing high-resolution views of particular (epi)genetic features of biological interest. Moving beyond spatial associations of genetic elements in relation to each other, Part III covers techniques allowing the relationship of chromatin with non-DNA nuclear structures to be explored genome-wide. A major limitation of all of the above methods is their reliance on obtaining population-average views of large numbers of cells. To better characterize the heterogeneity and dynamics of nuclear architectures, Part IV describes single-cell technologies, and Part V covers cutting-edge approaches to visualize them by microscopy. Finally, to be able to ask functional questions about the role of spatial chromatin organization in genomic control, Part VI gives methods for acute manipulations of chromatin architecture.

As for any complex biological question, no one technique can give the full answer. It is my hope that this book will be a useful resource for you to choose and perform the ones which will best help you with your research question.

Illkirch, France   *Tom Sexton*

Contents

Contributors

POONAM AGARWAL • *Department of Dermatology and Programs in Epithelial and Cancer Biology, Stanford University School of Medicine, Stanford, CA, USA*

EFFIE APOSTOLOU • *Sanford I. Weill Department of Medicine, Division of Hematology/Oncology, Sandra and Edward Meyer Cancer Center, Weill Cornell Medicine, New York, NY, USA*

MARIE-ODILE BAUDEMENT • *Institut de Génétique Moléculaire de Montpellier (IGMM), University of Montpellier, CNRS, Montpellier, France; Faculty of Biosciences, Centre for Integrative Genetics (CIGENE), Norwegian University of Life Sciences, Ås, Norway*

ANDREW S. BELMONT • *Department of Cell and Developmental Biology, University of Illinois at Urbana-Champaign, Urbana, IL, USA; Center for Biophysics and Quantitative Biology, University of Illinois at Urbana-Champaign, Urbana, IL, USA; Carl R. Woese Institute for Genomic Biology, University of Illinois at Urbana-Champaign, Urbana, IL, USA*

EDOUARD BERTRAND • *IGH, University of Montpellier, CNRS, Montpellier, France*

CLAUDIA CATTOGLIO • *Department of Molecular and Cell Biology, University of California, Berkeley, Berkeley, CA, USA; Li Ka Shing Center for Biomedical and Health Sciences, University of California, Berkeley, Berkeley, CA, USA; CIRM Center of Excellence, University of California, Berkeley, Berkeley, CA, USA; Howard Hughes Medical Institute, University of California, Berkeley, Berkeley, CA, USA*

YU CHEN • *Department of Cell and Developmental Biology, University of Illinois at Urbana-Champaign, Urbana, IL, USA; Department of Molecular and Cell Biology, Li Ka Shing Center for Biomedical and Health Sciences, Howard Hughes Medical Institute, University of California, Berkeley, CA, USA*

ELZO DE WIT • *Division Gene Regulation, Oncode Institute, Netherlands Cancer Institute, Amsterdam, The Netherlands*

DAFNE CAMPIGLI DI GIAMMARTINO • *Sanford I. Weill Department of Medicine, Division of Hematology/Oncology, Sandra and Edward Meyer Cancer Center, Weill Cornell Medicine, New York, NY, USA*

DAMIEN J. DOWNES • *MRC Molecular Haematology Unit, Radcliffe Department of Medicine, MRC Weatherall Institute of Molecular Medicine, University of Oxford, Oxford, UK*

ELIZABETH FINN • *Cell Biology of Genomes (CBGE), Center for Cancer Research (CCR), NCI/NIH, Bethesda, MD, USA*

THIERRY FORNÉ • *Institut de Génétique Moléculaire de Montpellier (IGMM), University of Montpellier, CNRS, Montpellier, France*

STEFAN GROB • *Department of Plant and Microbial Biology, University of Zurich, Zurich, Switzerland*

TSUNG-HAN S. HSIEH • *Department of Molecular and Cell Biology, University of California, Berkeley, Berkeley, CA, USA; Li Ka Shing Center for Biomedical and Health Sciences, University of California, Berkeley, Berkeley, CA, USA; CIRM Center of Excellence, University of California, Berkeley, Berkeley, CA, USA; Howard Hughes Medical Institute, University of California, Berkeley, Berkeley, CA, USA; Center for Computational Biology, University of California, Berkeley, Berkeley, CA, USA*

JIM R. HUGHES • *MRC Molecular Haematology Unit, Radcliffe Department of Medicine, MRC Weatherall Institute of Molecular Medicine, University of Oxford, Oxford, UK*

JOP KIND • *Hubrecht Institute, Royal Netherlands Academy of Arts and Sciences (KNAW), University Medical Center Utrecht, Radboud University & Oncode Institute, Utrecht, The Netherlands*

WING LEUNG • *Laboratory for Nuclear Dynamics, Institute for Protein Research, Osaka University, Osaka, Japan; Institute of Medical Science, University of Tokyo, Tokyo, Japan*

NING QING LIU • *Division Gene Regulation, Oncode Institute, Netherlands Cancer Institute, Amsterdam, The Netherlands*

SILKE J. A. LOCHS • *Hubrecht Institute, Royal Netherlands Academy of Arts and Sciences (KNAW), University Medical Center Utrecht, Radboud University & Oncode Institute, Utrecht, The Netherlands*

MICHELA MARESCA • *Division Gene Regulation, Oncode Institute, Netherlands Cancer Institute, Amsterdam, The Netherlands*

MÉLANIE MIRANDA • *Université Paris-Saclay, CEA, CNRS, Institute for Integrative Biology of the Cell (I2BC), Gif-sur-Yvette, France*

TOM MISTELI • *Cell Biology of Genomes (CBGE), Center for Cancer Research (CCR), NCI/ NIH, Bethesda, MD, USA*

BENOIT MOINDROT • *Université Paris-Saclay, CEA, CNRS, Institute for Integrative Biology of the Cell (I2BC), Gif-sur-Yvette, France*

TAKASHI NAGANO • *Laboratory for Nuclear Dynamics, Institute for Protein Research, Osaka University, Osaka, Japan; Institute of Medical Science, University of Tokyo, Tokyo, Japan*

DAAN NOORDERMEER • *Université Paris-Saclay, CEA, CNRS, Institute for Integrative Biology of the Cell (I2BC), Gif-sur-Yvette, France*

A. MARIEKE OUDELAAR • *Max Planck Institute for Multidisciplinary Sciences, Göttingen, Germany*

GIANLUCA PEGORARO • *High-Throughput Imaging Facility (HiTIF), Center for Cancer Research (CCR), NCI/NIH, Bethesda, MD, USA*

ALEXANDER POLYZOS • *Sanford I. Weill Department of Medicine, Division of Hematology/ Oncology, Sandra and Edward Meyer Cancer Center, Weill Cornell Medicine, New York, NY, USA*

COSETTE REBOUISSOU • *Institut de Génétique Moléculaire de Montpellier (IGMM), University of Montpellier, CNRS, Montpellier, France*

THOMAS SABATÉ • *Imaging and Modeling Unit, Institut Pasteur, UMR 3691 CNRS, IP CNRS, Paris, France; IGH, University of Montpellier, CNRS, Montpellier, France; Sorbonne Université, Collège Doctoral, Paris, France*

PELIN SAHLÉN • *Department of Gene Technology, Science for Life Laboratory, Royal Institute of Technology, Stockholm, Sweden*

SÉPHORA SALLIS • *Institut de Génétique Moléculaire de Montpellier (IGMM), University of Montpellier, CNRS, Montpellier, France; Laboratoire de génétique moléculaire du développement, Département des sciences biologiques, Université du Québec à Montréal, Montréal, QC, Canada*

WEI QIANG SEOW • *Department of Dermatology and Programs in Epithelial and Cancer Biology, Stanford University School of Medicine, Stanford, CA, USA*

ELENA SLOBODYANYUK • *Department of Molecular and Cell Biology, University of California, Berkeley, Berkeley, CA, USA; Li Ka Shing Center for Biomedical and Health Sciences, University of California, Berkeley, Berkeley, CA, USA; CIRM Center of Excellence, University of California, Berkeley, Berkeley, CA, USA; Howard Hughes Medical Institute, University of California, Berkeley, Berkeley, CA, USA*

KEVIN C. WANG • *Department of Dermatology and Programs in Epithelial and Cancer Biology, Stanford University School of Medicine, Stanford, CA, USA; Veterans Affairs Palo Alto Healthcare System, Palo Alto, CA, USA*

LIGUO ZHANG • *Department of Cell and Developmental Biology, University of Illinois at Urbana-Champaign, Urbana, IL, USA*

ARTEMY ZHIGULEV • *Department of Gene Technology, Science for Life Laboratory, Royal Institute of Technology, Stockholm, Sweden*

CHRISTOPHE ZIMMER • *Imaging and Modeling Unit, Institut Pasteur, UMR 3691 CNRS, IP CNRS, Paris, France*

Part I

Sailing the '3Cs': Chromosome Conformation Capture and Its Variants

Chapter 1

Quantitative Chromosome Conformation Capture (3C-qPCR)

Cosette Rebouissou, Séphora Sallis, and Thierry Forné

Abstract

Many population-based methods investigating chromatin dynamics and organization in eukaryotes are based on the chromosome conformation capture (3C) method. Here, we provide an updated version of the quantitative 3C (3C-qPCR) protocol for improved and simplified quantitative analyses of intra-chromosomal contacts.

Key words Chromosome conformation capture, Chromatin dynamics and organization, Quantitative PCR, Interaction frequency, Eukaryotic genome

1 Introduction

Chromosome conformation capture (3C) technique [1] and derived technologies (4C, 5C, Hi-C) [2–4] (*see* also Chapters 2–7) have allowed researchers to explore the organization and the dynamics of the eukaryotic genomes with unprecedented resolution and accuracy [5–8]. The creation of a 3C library is therefore a prerequisite to many 3C-based methods. By freezing all chromatin contacts present at a given time in their physiological nuclear context, and then by averaging these events over a population of several million cells, the 3C-qPCR method [9] allows very accurate measurements of the relative interaction frequencies between chromatin segments, in *cis*, on the same chromosome (Fig. 1) [10]. This parameter is a key to investigate the in vivo dynamics of the chromatin because it depends not only on its fundamental biophysical parameters (compaction, rigidity, etc.) [11] but also on important locus-specific factors which control local genomic functions (epigenetic modifications, binding of specific factors, etc.) [12, 13]. Here we provide an updated 3C-qPCR protocol that simplifies previously published protocols [9, 14, 15].

Tom Sexton (ed.), *Spatial Genome Organization: Methods and Protocols*,
Methods in Molecular Biology, vol. 2532, https://doi.org/10.1007/978-1-0716-2497-5_1,

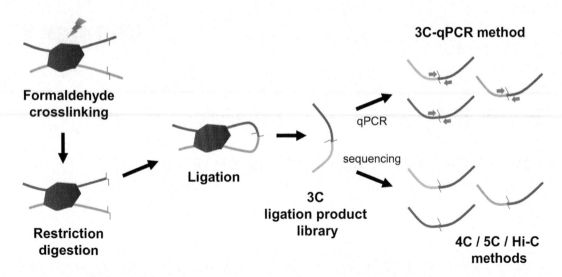

Fig. 1 Principle of the chromosome conformation capture (3C) assay. The principle of the 3C technique relies on three essential steps: formaldehyde cross-linking, restriction digestion and ligation, providing a library composed of 3C ligation products. The relative interaction frequency of each ligated DNA fragments, that reflects their 3D physical proximity, can be accurately determined by quantitative PCR (3C-qPCR method). 3C libraries can also be combined with high-throughput sequencing approaches and used for other 3C based methods, like the 4C, 5C, or Hi-C assays (see text for references)

2   Materials

1. (For tissues) Homogenization buffer: 2.1 M sucrose, 10 mM HEPES pH 7.6, 2 mM EDTA, 15 mM KCl, 10% (v/v) glycerol, 0.15 mM spermine, 0.5 mM DTT, 0.5 mM PMSF, 7 μg/mL aprotinin.

2. (For tissues) Wash buffer: 10 mM Tris–HCl pH 7.4, 15 mM NaCl, 60 mM KCl, 0.15 mM spermine, 0.5 mM spermidine.

3. Phosphate-buffered saline (PBS).

4. Nucleus buffer 1: 0.3 M sucrose, 60 mM KCl, 15 mM NaCl, 5 mM $MgCl_2$, 0.1 mM EGTA, 15 mM Tris–HCl pH 7.5, 0.5 mM DTT, 0.1 mM PMSF, 3.6 ng/mL aprotinin, 5 mM sodium butyrate.

5. Nucleus buffer 2: nucleus buffer 1 with 0.8% (v/v) NP-40.

6. Nucleus buffer 3: 1.2 M sucrose, 60 mM KCl, 15 mM NaCl, 5 mM $MgCl_2$, 0.1 mM EGTA, 15 mM Tris–HCl pH 7.5, 0.5 mM DTT, 0.1 mM PMSF, 3.6 ng/mL aprotinin; 5 mM sodium butyrate.

7. Glycerol buffer: 40% (v/v) glycerol, 50 mM Tris–HCl pH 8.3, 5 mM $MgCl_2$, 0.1 mM EDTA.

8. 3C buffer: 50 mM Tris–HCl pH 8, 50 mM NaCl, 10 mM $MgCl_2$, 1 mM DTT (*see* **Note 1**).

9. 37% (v/v) formaldehyde.

10. 1.25 M glycine.

11. 20% (w/v) SDS.

12. Triton mix: 10% (v/v) Triton X-100 in ligation buffer.

13. High-concentration restriction enzyme (e.g., 40 U/μL HindIII; Fermentas) (*see* **Note 2**).

14. PK buffer: 10 mM Tris–HCl pH 8, 5 mM EDTA, 0.5% (w/v) SDS.

15. 20 mg/mL proteinase K.

16. 1 mg/mL RNase A.

17. Phenol–chloroform–isoamyl alcohol (25:24:1).

18. 2 M sodium acetate pH 5.6.

19. 100% and 70% ethanol.

20. Secondary digestion restriction enzyme (e.g., StyI), supplied with $10\times$ buffer (*see* **Note 3**).

21. 5 M NaCl.

22. Nuclease-free water.

23. PCR primers, designed according to 3C design (see Methods), including control primers for loading controls (Subheading 3.8). For human and mouse genomes the following *Gapdh* primers can be used.

    Forward: acagtccatgccatcactgcc.

    Reverse: gcctgcttcaccaccttcttg.

    Primers used in quantitative PCR are typically 21-23mers with a Tm in the range 55–65 °C with a 2 °C maximum difference between primers used in one experiment. They should be designed close ($\leq$50 bp) to the restriction site used for the 3C assays.

24. qPCR mix: SYBR^R Green PCR Master Mix as described in [16] with modifications described in [11, 12] (*see* **Note 4** for detailed composition).

25. LightCycler 480 II (Roche) or equivalent real-time PCR machine.

26. Ligation buffer: 40 mM Tris–HCl pH 7.8, 10 mM MgCl$_2$, 10 mM DTT, 0.5 mM ATP.

27. 30 Weiss U/μL T4 DNA ligase.

28. 100 mM ATP.

29. 20 mg/mL glycogen.

30. 10 mM Tris–HCl pH 7.5.

31. Genomic DNA, purified from the same species as for the 3C-qPCR experiments, and quantified by A_{260} measurement.

32. BAC(s), spanning the genomic locus of interest.

3 Methods

3.1 Cell Nucleus Preparation

1. Prepare a single-cell suspension from the cell culture or tissue (*see* **Note 5**) of interest. If working with adherent cell cultures, trypsinize cells, wash and filter through a 40 μm cell strainer to make a single-cell suspension. Wash cells in PBS.

2. Resuspend cells in 1.5 mL nucleus buffer 1.

3. Add 0.5 mL nucleus buffer 2, mix slightly and put on ice for 3 min.

4. Put 1 mL in 14 mL tubes containing 4 mL nucleus buffer 3. Save a 4 μL aliquot to count nuclei in a Thoma cell (count used in **step 6** below).

5. Centrifuge for 20 min at 11,300 × g, 4 °C.

6. Remove supernatant and resuspend pellet in glycerol buffer to a final concentration of ~2 × 10^8 cells/mL. Transfer to a 1.5 mL tube, freeze in liquid nitrogen, and store at −80 °C.

3.2 Formaldehyde Cross-Linking

1. Resuspend 5 × 10^6 nuclei in 700 μL 3C buffer and incubate for 5 min at room temperature.

2. Add 19.7 μL 37% formaldehyde, mix by inverting the tube and incubate for 10 min at room temperature.

3. Add 80 μL 1.25 M glycine, mix by inverting the tube and incubate for 2 min at room temperature then place on ice for 5 min.

4. Centrifuge for 3 min at 2300 × g, room temperature, remove supernatant and carefully resuspend nuclei in 1 mL 3C buffer (*see* **Note 6**).

5. Centrifuge for 3 min at 2300 × g, room temperature and remove supernatant.

3.3 Restriction Digestion

1. Resuspend nuclei in 100 μL 3C buffer and transfer to a safe-lock microtube.

2. Add 1 μL 20% SDS and incubate for 1 h at 37 °C, shaking at 350 rpm on a thermomixer.

3. Add 16.8 μL Triton mix and incubate for 1 h at 37 °C, shaking at 350 rpm on a thermomixer.

4. Take a 10 μL aliquot of the sample as the "Undigested control" (do not disturb the mixture) and store at −20 °C until processing at Subheading 3.4.

5. Add 450 U of the selected restriction enzyme (e.g., HindIII) to the remaining sample and incubate overnight at 37 °C, shaking at 350 rpm on a thermomixer (*see* **Note 2**).

6. Take a 10 μL aliquot of the sample as the "Digested control" (do not disturb the mixture) and proceed with this to Subheading 3.4. The remaining sample is processed in Subheading 3.5.

3.4 Determination of Digestion Efficiency

Digestion efficiencies have a significant impact on the assays and should be carefully assessed for each restriction site involved in the analysis. Care should be taken to ensure that digestion efficiencies are in the same range for the sites of interest (*see* **Note 2**).

1. Add 500 μL PK buffer and 1 μL 20 mg/mL proteinase K to the "Undigested" and "Digested" control samples, and incubate for 30 min at 65 °C.

2. Cool samples to 37 °C, add 1 μL 1 mg/mL RNase A and incubate for 2 h at 37 °C.

3. Add 500 μL phenol–chloroform–isoamyl alcohol, mix vigorously and centrifuge for 5 min at 16,100 × *g*, room temperature.

4. Transfer upper aqueous layer to a new microtube. Add 50 μL 2 M sodium acetate and 1.5 mL 100% ethanol and keep at −80 °C for 45 min.

5. Centrifuge for 20 min at 16,100 × *g*, 4 °C, carefully remove supernatant and wash pellet in 500 μL 70% ethanol.

6. Centrifuge for 4 min at 16,100 × *g*, room temperature, remove supernatant and air-dry the pellet.

7. Dissolve DNA in 500 μL 1× secondary digestion restriction enzyme buffer. Add 50 U secondary digestion restriction enzyme and incubate for 2 h at 37 °C (*see* **Note 7**).

8. Add 500 μL phenol–chloroform–isoamyl alcohol, mix vigorously and centrifuge for 5 min at 16,100 × *g*, room temperature.

9. Transfer upper aqueous layer to a new microtube. Add 25 μL 5 M NaCl and 1 mL 100% ethanol and keep overnight at −20 °C.

10. Centrifuge for 20 min at 16,100 × *g*, 4 °C, remove supernatant and wash pellet in 200 μL 70% ethanol.

11. Centrifuge for 4 min at 16,100 × *g*, room temperature, remove supernatant, and air-dry then dissolve pellet in 60 μL nuclease-free water.

12. Perform real-time PCR quantification (*see* **Note 4** and Subheading 3.10 below) on undigested (U) and digested (D) control samples, using primer sets that amplify across each restriction site of interest (R). To correct for any difference in the amounts of templates used in the PCR, also use primer sets to amplify control regions (C) that do not contain the restriction sites of interest. Digestion efficiency is calculated according to the following formula (*see* **Note 8**).

$$\% \text{ restriction} = 100 - 100/2^{\left((Ct_R - Ct_C)_D - (Ct_R - Ct_C)_U\right)}$$

3.5 Ligation

1. Add 12 μL 20% SDS to the digested sample and incubate for 30 min at 37 °C, shaking on a thermomixer at 350 rpm.

2. Carefully transfer digested nuclei to a 12 mL tube (*see* **Note 9**). Add 3.28 mL ligation buffer and 390 μL Triton mix, and incubate for 2 h at 37 °C, shaking in a thermomixer at 450 rpm.

3. Centrifuge for 1 min at 8800 × g, 4 °C and remove 3.27 mL supernatant to leave 500 μL in the tube. Add 6.5 μL 30 U/μL T4 DNA ligase and 3 μL 100 mM ATP and incubate overnight at 16 °C, shaking on a thermomixer at 350 rpm.

3.6 DNA Purification

1. Add 2 mL PK buffer, 1.5 mL nuclease-free water and 5 μL 20 mg/mL proteinase K, and incubate for 1 h at 50 °C.

2. Incubate for 4 h at 65 °C to decrosslink the sample.

3. Add 4 mL phenol–chloroform–isoamyl alcohol, mix vigorously and centrifuge for 15 min at 3900 × g, room temperature (*see* **Note 10**).

4. Transfer upper aqueous layer to a new 12 mL tube and add 200 μL 5 M NaCl, 1 μL glycogen and 8 mL 100% ethanol. Mix and store overnight at −20 °C.

5. Centrifuge for 45 min at 15,700 × g, 4 °C, remove supernatant and wash pellet in 2 mL 70% ethanol.

6. Centrifuge for 15 min at 15,700 × g, 4 °C, remove supernatant, air-dry pellet and dissolve DNA in 50 μL nuclease-free water (*see* **Note 11**).

3.7 Complementary Digestion

1. Add 250 μL 2× secondary digestion restriction enzyme buffer (*see* **Note 3**) and 185 μL nuclease-free water, and transfer sample to a 1.5 mL microtube.

2. Add 5 μL 1 mg/mL RNase A and 100 U secondary digestion restriction enzyme, and incubate for 2 h 30 min at 37 °C.

3. Add 500 μL phenol–chloroform–isoamyl alcohol, mix vigorously and centrifuge for 5 min at 3900 × g, room temperature.

4. Transfer upper aqueous phase to a new 1.5 mL tube and add 25 μL 5 M NaCl, 1 μL glycogen and 1 mL 100% ethanol, and keep overnight at −20 °C.

5. Centrifuge for 20 min at 15,700 × g, 4 °C, remove supernatant and wash pellet in 200 μL 70% ethanol.

6. Centrifuge for 4 min at 15,700 × g, room temperature, air-dry pellet, and dissolve DNA in 150 μL 10 mM Tris–HCl pH 7.5 (*see* **Note 11**).

3.8 Performing Loading Adjustments

1. Determine the DNA concentration of the 3C sample by qPCR (*see* **Notes 4** and **12**), relative to a reference genomic DNA of known concentration. Use a dilution series of 3C material and genomic DNA, and use "internal" primer sets that do not amplify across sites recognized by any of the restriction enzymes used. Set up 10 μL reaction volumes in qPCR plates:

   1 μL DNA (3C or genomic, or nuclease-free water in negative control well);

   7 μL nuclease-free water;

   1 μL 5 μM each primer;

   1 μL qPCR mix (*see* **Note 4**).

2. Perform the following programme in a qPCR thermal cycler (*see* **Note 12**):

   95 °C, 3 min;

   45 × [95 °C, 1 s; 60 °C, 5 s; 72 °C, 15 s].

   Denaturation curve: 45 °C, 30 s; increase to 95 °C at 0.2 °C/s.

3. Use the amplification curve from the known genomic DNA concentrations to estimate the concentration of the 3C DNA samples. Dilute the 3C samples to 25 ng/μL.

4. Repeat **steps 1** and **2** on the new 3C stocks to precisely determine their concentration. These values will be used as the "loading controls" for final quantification (*see* Subheading 3.12).

3.9 Assessment of Sample Purity (Optional)

1. Take a twofold and a fourfold diluted aliquot of the 3C sample (Subheading 3.7, **step 6**) and add genomic DNA to each aliquot to make a final total DNA concentration of 25 ng/μL.

2. Perform qPCR for any 3C primer pair (*see* **Note 4** and Subheading 3.10 below) and verify that the real-time PCR quantifications are reduced according to the dilution factors. If this is not the case, then sample purity is not adequate and the DNA should either be repurified or the sample discarded.

3.10 Real-Time PCR Quantifications of Ligation Products

1. Perform qPCR reactions to obtain the Ct for each assessed ligation product on 25 ng 3C material. Set up 10 μL reaction volumes in qPCR plates:

   1 μL DNA (3C, or nuclease-free water in negative control well);

   7 μL nuclease-free water;

   1 μL 5 μM each primer;

   1 μL qPCR mix (*see* **Note 4**).

2. Perform the following programme in a qPCR thermal cycler (*see* **Note 12**):

95 °C, 10 min;

45 × : [95 °C, 10 s; 69 °C, 8 s; 72 °C, 14 s].

3.11 PCR Control Template Used for Primer Efficiency Control

A control template containing all ligation products in equal amounts is used to optimize real-time quantitative PCR (qPCR) reactions and to establish the minimal amount of ligation product that can still be quantified in a reliable manner. For this qPCR control template, we recommend the use of a single BAC clone covering the genome segment under study. Alternatively, a set of minimally overlapping BAC clones mixed in equimolar amounts can be used. This BAC is then cut with the 3C restriction enzyme of choice (e.g., HindIII) and religated by T4 DNA ligase. A secondary restriction enzyme (e.g., StyI) can be used to linearize DNA circles which may otherwise affect primer hybridization efficiency [17] (*see* **Note 3**). It is then necessary to make serial dilutions of this reaction to obtain standard curves which cover the same range of ligation product concentrations as those that will be obtained in the 3C samples. To mimic 3C sample conditions, the final DNA concentration in these dilutions is adjusted to the amount of DNA used in the 3C samples. Thus, these dilutions are performed in a 25 ng/μL DNA solution made of genomic DNA digested with the second restriction enzyme (e.g., StyI in the present protocol). Using serial dilutions of this control template, a standard curve with specific parameters (slope and intercept) is thus obtained for each of the 3C qPCR primer pairs used. These parameters will be used to correct for potential differences in primer efficiencies.

3.12 3C-qPCR Data Normalization— Primer Efficiency and Loading Controls

To obtain quantification values that are corrected for potential differences in primer efficiencies, the Ct obtained for each ligation product is first normalized using the parameters of the corresponding standard curve (the slope "a" and the intercept "b" obtained in Subheading 3.11). These values are calculated using the following formula: Value $= 10^{(Ct-b)/a}$. For each sample, these values are then normalized to the corresponding "loading control" obtained in Subheading 3.8.

3.13 3C-qPCR Data Normalization to Noise Band

The data should then be normalized to compensate for experimental variations and allow comparison between different 3C-assays. For each biological sample, a Basal Interaction Level (B.I.L.) is calculated and 3C-qPCR data are normalized to this B.I.L.. We first calculate the mean interaction frequency (M) and the mean Standard Deviation (SD) of all the experimental points. Experimental points are selected if their interaction frequency (fx) is both superior to (M − SD) and inferior to (M + SD). The mean fx of the selected experimental points is corresponding to the B.I.L. to which all fx values of the experiment are normalized [14, 15].

4   Notes

1. This buffer is intended to be compatible with the selected restriction enzyme and should thus be adapted accordingly.

2. Digestion efficiency in 3C experimental conditions varies largely depending on the selected restriction enzyme. Some enzymes, like HindIII or EcoRI, have high digestion efficiencies (at least 70 to 80% is recommended for all sites investigated in a given experiment) while others, like BamHI, require sequential addition of the 450 U to reach the same efficiency (add 150U for 2 h, then 150U for 2 h, and finally 150U overnight). Some enzymes, like SacI, do not digest at all in 3C reactions. Enzymes that generate cohesive ends are recommended, as blunted ends do not allow efficient ligation.

3. This enzyme is required to fragment the DNA which helps to improve DNA accessibility and thus PCR amplification efficiencies. Ensure that this enzyme does not digest within the tested amplicons.

4. We report marked improvements over commercial qPCR mixes with a custom quantitative PCR mix composition [16]: 0.24% W1 (polyoxyethylene ether W1), 500 μg/mL BSA, 300 μM dNTP, 50 mM KCl, 30 mM $MgCl_2$, 1/3000 SYB^R Green (10,000× in DMSO, LONZA), 16.24% glycerol, 400 mM 2-amino-2-methyl-1,3-propanediol buffer (adjusted to pH 8.3 using HCl), 0.4 U/μL platinum Taq DNA polymerase (Invitrogen). Use a fresh glycerol stock, as glycerol tends to produce oxidization products that inhibit PCR reactions. W1 can be replaced by a combination of two detergents: 0.09% Brij 56 (Sigma) and 0.15% Brij 58P (Sigma). For better quantification efficiencies, the usual 300 μM dNTP could be replaced by 1500 μM CleanAmp 3'THF dNTP (Tebu Bio). We do not recommend the use of CleanAmp dNTP 3'TBE, as they provide similar quantification efficiencies as classical dNTPs.

5. Nuclei preparation from tissues requires specific adaptations. For mouse liver, homogenize the tissue in homogenization buffer using a Potter. Deposit the homogenate on a cushion of the same buffer and centrifuge at 100,000 × g for 40 min. Wash the pellet with wash buffer and resuspend in glycerol buffer. Freeze in liquid nitrogen and store at −80 °C. For muscle, the same protocol can be used except that the tissue needs to be first disrupted using an Ultra-Turrax before homogenization [18, 19].

6. Rinse the walls of the tube that have been in contact with the reaction mixture to recover all nuclei.

7. Check that the corresponding secondary digestion restriction site is absent from the PCR amplicons used to assess digestion efficiencies.

8. The efficiency of the restriction enzyme digestion should be above 60–70%, but ideally >80% is digested. Samples with lower digestion efficiencies should be discarded (also *see* **Note 2**).

9. Do not rinse the wall of the tube as this may increase significantly the amount of undigested material.

10. If the aqueous phase if very turbid after the first extraction, repeat the phenol–chloroform extraction a second time.

11. If some precipitates do not resuspend, dissolve DNA by gently shaking tubes at 37 °C for up to 30 min. The 3C template may be kept at −20 °C for several months.

12. A_{260} measurements fail to provide an accurate estimate of DNA concentration in 3C samples, probably because of their limited purity. If qPCR reactions are performed in a different thermocycler (than the LightCycler, Roche) the PCR parameters may need to be optimized.

Acknowledgments

Our work is supported by grants from the Agence Nationale de la Recherche (CHRODYT, ANR-16-CE15-0018-04), the AFM-Téléthon (grant N°21024), and the Centre National de la Recherche Scientifique (CNRS).

References

1. Dekker J, Rippe K, Dekker M et al (2002) Capturing chromosome conformation. Science 295:1306–1311

2. Dostie J, Richmond TA, Arnaoult RA et al (2006) Chromosome conformation capture carbon copy (5C): a massively parallel solution for mapping interactions between genomic elements. Genome Res 16:1299–1309

3. Simonis M, Klous P, Splinter E et al (2006) Nuclear organization of active and inactive chromatin domains uncovered by chromosome conformation capture-on-chip (4C). Nat Genet 38:1348–1354

4. Zhao Z, Tavoosidana G, Sjolinder M et al (2006) Circular chromosome conformation capture (4C) uncovers extensive networks of epigenetically regulated intra- and interchromosomal interactions. Nat Genet 38:1341–1347

5. Dixon JR, Selvaraj S, Yue F et al (2012) Topological domains in mammalian genomes identified by analysis of chromatin interactions. Nature 485:376–380

6. Duan Z, Andronescu M, Schutz K et al (2010) A three-dimensional model of the yeast genome. Nature 465:363–367

7. Lieberman-Aiden E, van Berkum NL, Williams L et al (2009) Comprehensive mapping of long-range interactions reveals folding principles of the human genome. Science 326: 289–293

8. Sexton T, Yaffe E, Kenigsberg E et al (2012) Three-dimensional folding and functional organization principles of the *Drosophila* genome. Cell 148:458–472

9. Hagège H, Klous P, Braem C et al (2007) Quantitative analysis of chromosome

conformation capture assays (3C-qPCR). Nat Protoc 2:1722–1733

10. Ea V, Baudement MO, Lesne A et al (2015) Contribution of topological domains and loop formation to 3D chromatin organization. Genes (Basel) 6:734–750

11. Court F, Miro J, Braem C et al (2011a) Modulated contact frequencies at gene-rich loci support a statistical helix model for mammalian chromatin organization. Genome Biol 12:R42

12. Court F et al (2011b) Long-range chromatin interactions at the mouse Igf2/H19 locus reveal a novel paternally expressed long non-coding RNA. Nucleic Acids Res 39: 5893–5906

13. Ea V, Sexton T, Gostan T et al (2015) Distinct polymer physics principles govern chromatin dynamics in mouse and *Drosophila* topological domains. BMC Genomics 16:607. https://doi.org/10.1186/s12864-12015-11786-12868

14. Braem C, Recolin B, Rancourt RC et al (2008) Genomic matrix attachment region and chromosome conformation capture quantitative real time PCR assays identify novel putative regulatory elements at the imprinted Dlk1/Gtl2 locus. J Biol Chem 283:18612–18620

15. Ea V, Court F, Forné T (2017) Quantitative analysis of intra-chromosomal contacts: the 3C-qPCR method. Methods Mol Biol 1589: 75–88

16. Lutfalla G, Uzé G (2006) Performing quantitative reverse-transcribed polymerase chain reaction experiments. (translated from eng). Methods Enzymol 410:386–400. (in eng)

17. Weber M, Hagège H, Lutfalla G et al (2003) A real-time polymerase chain reaction assay for quantification of allele ratios and correction of amplification bias. Anal Biochem 320:252–258

18. Milligan L, Antoine E, Bisbal C et al (2000) H19 gene expression is up-regulated exclusively by stabilization of the RNA during muscle cell differentiation. Oncogene 19: 5810–5816

19. Milligan L, Forné T, Antoine E et al (2002) Turnover of primary transcripts is a major step in the regulation of mouse H19 gene expression. EMBO Rep 3:774–779

Chapter 2

Detection of Allele-Specific 3D Chromatin Interactions Using High-Resolution In-Nucleus 4C-seq

Mélanie Miranda, Daan Noordermeer, and Benoit Moindrot

Abstract

Chromosome conformation capture techniques are a set of methods used to determine 3D genome organization through the capture and identification of physical contacts between pairs of genomic loci. Among them, 4C-seq (circular chromosome conformation capture coupled to high-throughput sequencing) allows for the identification and quantification of the sequences interacting with a preselected locus of interest. 4C-seq has been widely used in the literature, mainly to study chromatin loops between enhancers and promoters or between CTCF binding sites and to identify chromatin domain boundaries. As 3D-contacts may be established in an allele-specific manner, we describe an up-to-date allele-specific 4C-seq protocol, starting from the selection of allele-specific viewpoints to Illumina sequencing. This protocol has mainly been optimized for cultured mammalian cells, but can be adapted for other cell types with relatively minor changes in initial steps.

Key words Circular chromosome conformation capture (4C), Nuclear organization, DNA loops, Topologically associating domains (TADs), High-throughput sequencing, Single-nucleotide polymorphism (SNP)

1 Introduction

The 3D organization of genomes plays important roles in the transcriptional regulation of genes [1]. Over the past two decades, the development of "chromosome conformation capture" assays has deepened our understanding of how chromosomes are folded, and the regulatory consequences of this folding [2, 3]. In particular, the discovery of the relatively insulated topologically associating domains (TADs) sheds light on an important regulatory level [4–8]. The formation of the TADs is largely driven by the coordinated action of CTCF and the Cohesin complex [9–11]. These two molecular actors are found enriched at the TAD boundaries, thereby restricting, according to current models, promoter–enhancer communication to a single TAD [10, 12, 13].

Tom Sexton (ed.), *Spatial Genome Organization: Methods and Protocols*,
Methods in Molecular Biology, vol. 2532, https://doi.org/10.1007/978-1-0716-2497-5_2,
© The Author(s), under exclusive license to Springer Science+Business Media, LLC, part of Springer Nature 2022

High-throughput sequencing-based chromosome conformation capture (3C) assays come in different flavors [3]. Hi-C (Chapter 3) scores for interactions between all possible pairs of genomic loci at a resolution that is largely dictated by the sequencing depth [14]. 4C-seq (circular chromosome conformation capture), on the other hand, queries for the chromatin interactions between a locus of interest (thereafter named the "viewpoint") and essentially any other genomic loci [15, 16]. The resolution of 4C-seq assays is not limited by the sequencing power, but rather determined by the first restriction enzyme used. Capture-C (Chapters 5 and 6) presents an alternative way to determine the interaction frequencies between a relatively large set of genomic loci and the rest of the genome [17, 18]. Yet 4C-seq assays will be more flexible when interested in a few viewpoints (up to around 10 viewpoints), especially if 3D-contacts need to be described in an allele-specific manner.

The central principle of 3C is the detection of 3D interactions through the identification and the quantification of a ligation junction between two restriction fragments as a measure for initial proximity. High-resolution 4C-seq experiments start by the fixation of chromatin contacts using chemical cross-linking, for which formaldehyde is most commonly used. The chromatin is then digested using a restriction enzyme that recognizes a 4-bp sequence. This is followed by proximity ligation to create concatemers of restriction fragments whose linear arrangement directly reflects their initial spatial proximity. To characterize the ligation junctions between fragments, the resulting DNA is purified and subjected to a second round of enzymatic digestion and ligation, which results in circularization of the DNA. The circles containing the region of interest (viewpoint) and potential captured partners are amplified using an inverse PCR that concurrently generates ready-to-sequence libraries (Fig. 1). The 3D partners are finally identified and quantified after Illumina sequencing of the ligation junctions and bioinformatical analysis of the resulting reads.

Although for most genomic intervals, the two alleles of a diploid cell are assumed to adopt a virtually identical average 3D conformation, there are a variety of cases where this may not be the case, for example in the context of X-chromosome inactivation [19], genomic imprinting [20, 21], or cases of allelic exclusion [22–24]. Of particular relevance for this chapter, different strategies have been adopted to set up allelic 4C-seq assays and determine allele-specific chromosomal contacts (Fig. 2). These strategies fall into two distinct approaches. A first approach relies on an allele-specific restriction site, thereby solely amplifying the partners captured by one of the two alleles (Fig. 2a, b). A second approach relies on the sequencing of a single-nucleotide variant located between one of the primers and a digested restriction site (Fig. 2c, d). Both approaches have their own set of advantages and limitations (Fig. 2). For instance, in the first approach, the design of viewpoints

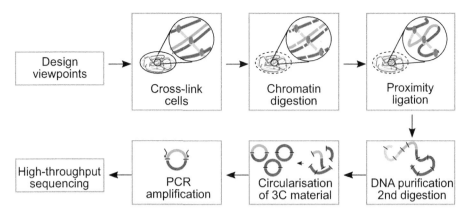

Fig. 1 Workflow for in-nucleus Circular Chromosome Conformation Capture (4C-seq) experiments. Once viewpoints have been designed and the restriction enzymes defined, chromatin contacts are fixed by formaldehyde cross-links, followed by enzymatic chromatin digestion using a first restriction enzyme and enzymatic ligation. All these steps are done in permeabilized nuclei. After cross-link reversal, the recombined DNA is purified, digested by a second restriction enzyme and circularized. An inverse PCR allows for the amplification of the circles containing the viewpoint of interest. Finally, the captured ligation partners are identified by high-throughput sequencing

is less flexible but there is a much-reduced chance that reads are incorrectly attributed to the wrong allele, as restriction enzymes are very efficiently blocked by changes to the genomic sequence. In contrast, the second approach is much easier for viewpoint design but there is a considerable chance that reads are attributed to the incorrect allele because the allelic identification is based on the sequencing of usually a single polymorphism.

In this chapter, we provide a detailed protocol of a high-resolution and allele-specific 4C-seq assay that relies on an allele-specific restriction site. This is a further development from our previous protocol [25], which includes a set of further improvements and adaptations. In particular, this in-nucleus version of the 4C-seq protocol incorporates the technical optimizations introduced for Hi-C experiments to preserve the nucleus prior to ligation and reduce the frequency of random ligations occurring in diluted conditions [13, 26]. Although a special focus is given to the design of allele-specific experiments, the following methods and guidelines will be useful for setting up both allelic and nonallelic 4C-seq assays.

2   Materials

1. 1× phosphate-buffered saline (PBS), pH 7.2.

2. Cell isolation buffer: 10% fetal bovine serum (FBS) in 1× PBS, prepared fresh for each experiment.

3. Cross-link buffer: 2% formaldehyde in cell isolation buffer, prepared fresh for each experiment.

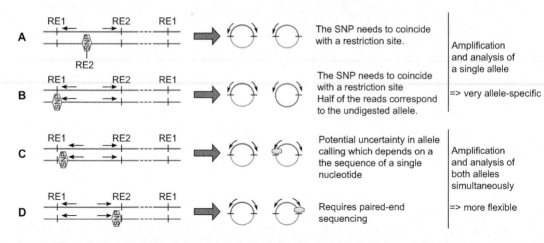

Fig. 2 Different strategies for allele-specific 4C-seq studies. To detect allele-specific 3D contacts, 4C-seq uses single nucleotide polymorphisms (SNP) to distinguish between the alleles. From top to bottom: four different published approaches, with ligation products after circularization of both alleles shown on the right. In (**a**) the SNP creates a new recognition site for the secondary restriction site (RE2), which prevents the inclusion of the binding site for the reverse primer in the circularized ligation product (pink circle) [19, 22]. In (**b**) the SNP perturbs the recognition site for the primary restriction enzyme (RE1), which results in a circularized ligation product that contains the "undigested" restriction fragment directly downstream of the viewpoint (pink circle) [20]. In (**c**) a SNP is located in-between the forward 4C primer and the first cutter [36], which will be included in the sequencing result and thus allows for the distinction between the alleles, and in (**d**) the SNP is located in-between the reverse 4C primer and the second cutter, which will be included in the sequencing result if paired-end sequencing is used [22, 30, 37, 38]. Strategies (**a**) and (**b**) generate 100% allele-specific data (blue circles only), but only one of the two alleles is analyzed at a time. The other allele is analyzed using another set of primers or in the reciprocal cross if working with laboratory strains. Strategies (**c**) and (**d**) are more flexible, but the distinction between the alleles relies on the sequencing of a single polymorphism, which is sensitive to sequencing errors. For strategy (**c**), the SNP needs to be located in close proximity to RE1, as sequencing starts from the 5′-side of the 4C primer facing RE1. This constraint is alleviated in strategy (**d**), but this requires paired-end sequencing. RE1/2: restriction site of the primary and secondary restriction enzyme. The arrows represent the 4C primers amplifying a single (**a**) or both alleles (**b**, **c**, **d**). The yellow hexagons indicate the location of the SNP

4. 1 M glycine.

5. Cell lysis buffer: 10 mM Tris–HCl pH 8, 10 mM NaCl, 0.2% (v/v) NP-40, 1× Complete EDTA-free protease inhibitor cocktail, prepared fresh and kept on ice for each experiment.

6. 20% (w/v) SDS. 0.5% solution also made fresh for each experiment.

7. Molecular biology–grade water.

8. 20% (v/v) Triton X-100.

9. High-concentration primary restriction enzyme (e.g., *Dpn*II, *Mbo*I, or *Nla*III) with compatible 10× restriction buffer (*see* **Note 1**).

10. 20 mg/mL BSA.

11. 5 M NaCl.

12. 1 M Tris–HCl pH 7.5.

13. 0.5 M EDTA pH 8.

14. 20 mg/mL proteinase K.

15. 10 mg/mL RNase A.

16. Phenol–chloroform–isoamyl alcohol (25:24:1).

17. 2 M sodium acetate pH 5.6.

18. 10 mg/mL glycogen.

19. 100%, 80%, and 70% ethanol.

20. 10 mM Tris–HCl pH 8.

21. $1.5\times$ Ligation master mix for the first ligation: $1.5\times$ commercial T4 ligation buffer, 1.5% Triton X-100, 0.16 mg/mL BSA.

22. 20 U/μL (Weiss units) T4 DNA ligase.

23. DNA dye incorporation assay for high and low-concentration DNA (i.e., Qubit fluorimeter with dsDNA BR and HS assays, or equivalent).

24. Secondary restriction enzyme with compatible $10\times$ restriction buffer (*see* **Note 1**).

25. $10\times$ homemade ligation buffer for the second ligation: 660 mM Tris–HCl pH 7.5, 50 mM $MgCl_2$, 50 mM DTT, 10 mM ATP. Aliquot and store at $-20\,°C$ for up to 6 months.

26. Qiagen PCR Purification kit (or equivalent), containing buffers PB and PE and QiaQuick spin columns.

27. Expand Long Template PCR system (Roche).

28. 10 mM dNTPs.

29. 100 μM 4C PCR primers (*see* Subheading 3.1 for design and testing principles, Fig. 3 and **Note 2**).

30. SPRI beads (e.g., AMPure XP).

31. Magnetic rack.

32. Illumina sequencing platform.

3   Methods

See **Notes 1–5** for important guidance regarding the experimental design before starting the experiment.

3.1 Design of Allele-Specific 4C Primers

The construction of 4C-seq primers is indicated in Fig. 3. From the 5′-end to the 3′-end, they are composed of an Illumina P5 or P7 sequence for cluster generation, an Illumina Read Sequencing Primer sequence, and a viewpoint specific sequence that anneals to the 4C template. A variable index for multiplexing strategies can

Primer facing the first cutter restriction site

5'-	Illumina P5 sequence	Illumina Read 1 Seq Primer	Viewpoint-specific	-3'

AATGATACGGCGACCACCGAGATCTACACTCTTTCCCTACACGACGCTCTTCCGATCTNNNNNNNNNNNNNNNNNNNNN

Primer facing the second cutter restriction site

5'-	Illumina P7 sequence	Index	Illumina Read 2 Seq Primer	Viewpoint-specific	-3'

CAAGCAGAAGACGGCATACGAGATCGTGATGTGACTGGAGTTCAGACGTGTGCTCTTCCGATCTNNNNNNNNNNNNNNNNNNNNN
ACATCG
GCCTAA

Fig. 3 Design of the 4C-seq primers with Illumina adapters. The primers are composed of Illumina P5/P7 adapters (for cluster generation), sequencing primers, and the viewpoint specific sequence at the 3'-end of the primers. A 6 bp index can be added in the primer facing the site of the secondary restriction enzyme used. These primers are compatible with Illumina NextSeq flow cells and single index multiplexing (read using i7 index primer)

be added to the primer facing the restriction site of the second cutter (Fig. 3). For the allele-specific 4C-seq described here, the P5-containing primer faces the allele-specific restriction site of the first cutter (Fig. 2b). This requires to know beforehand the location of allelic variants. In this respect, the mouse model represents a convenient system as a comprehensive list of SNPs in numerous mouse strains is available on the Mouse Genomes Project (*see* **Note 3**). This list needs to be intersected with the location of restriction sites to list allele-specific restriction sites. The primer facing the allelic cut site of the first restriction enzyme should anneal as close as possible to restriction site (ideally within 30 bp), as Illumina reads will start from the 5'-end of the viewpoint specific sequence of the 4C-seq primer. Each 4C primer pair should be tested and validated beforehand (*see* **Note 4**). 4C-seq primers with Illumina adapters are long and therefore need to be PAGE-purified to eliminate shorter by-products (*see* **Note 2**).

3.2 Cell Cross-Linking

1. Collect $1–1.5 \times 10^7$ cells and centrifuge (5 min, $300 \times g$, room temperature). Resuspend the cell pellet in 10 mL cell isolation buffer (*see* **Note 5** if working with nonmammalian cells).

2. Centrifuge (5 min, $300 \times g$, 4 °C) and discard supernatant.

3. Resuspend the pellet in 12 mL cross-link buffer. Gently pipet up-and-down to dissociate any aggregates. Incubate at room temperature for exactly 10 min on a tube roller mixer, then immediately place the tube on ice.

4. Immediately quench the cross-link reaction by adding 1.8 mL of a cold 1 M glycine. Mix by inversion and keep the tube on ice.

5. Centrifuge (5 min, $300 \times g$, 4 °C) and discard the supernatant.

6. Resuspend the cell pellet in 10 mL cold 1× PBS. Centrifuge (5 min, 300 × g, 4 °C) and remove most of the supernatant.

7. Resuspend the cell pellet in the remaining liquid and transfer to a 1.5 mL Safe-Lock tube.

8. Centrifuge (1 min, 300 × g, 4 °C) and discard supernatant. The cell pellet can be snap-frozen in liquid nitrogen and stored at −80 °C for up to 2 years.

3.3 Cell Lysis and Nucleus Permeabilization

1. Gently resuspend the cell pellet in 1 mL cold cell lysis buffer. Dissociate any aggregates by pipetting up and down several times.

2. Transfer cells to a 15 mL cold conical tube. Add 9 mL cold cell lysis buffer and incubate for 20 min on ice.

3. Centrifuge the lysed cells (5 min, 2500 × g, 4 °C) and discard the supernatant.

4. Resuspend the pellet in 10 mL cold cell lysis buffer and incubate again for 10 min on ice.

5. Centrifuge the lysed cells (5 min, 2500 × g, 4 °C) and discard the supernatant.

6. Resuspend the pellet with 500 μL cold cell lysis buffer and transfer the suspension to a 1.5 mL Safe-Lock tube.

7. Centrifuge (5 min, 2500 × g, 4 °C) and discard the supernatant.

8. Resuspend nuclei in 100 μL 0.5% SDS solution and incubate for exactly 10 min at 62 °C while shaking in a Thermomixer (900 rpm).

3.4 First Digestion

1. Add x μL molecular biology–grade water and 25 μL 20% Triton X-100. The volume x is calculated as follows: 323 μL *minus* the volume of enzyme used at **step 5**. This allows for a final volume of 500 μL for the digestion.

2. Incubate for 15 min at 37 °C while shaking in a Thermomixer (900 rpm) (*see* **Note 6**).

3. Add 50 μL of the appropriate 10× restriction buffer and 2 μL 20 mg/mL BSA.

4. Save a 5 μL aliquot ("undigested" sample). Keep at 4 °C or at −20 °C until ready for processing in Subheading 3.5.

5. Add 400 U primary restriction enzyme. Mix well and incubate for 4 h at 37 °C while shaking in a Thermomixer (900 rpm).

6. Add another 400 U primary restriction enzyme. Mix well and incubate overnight at 37 °C while shaking in a Thermomixer (900 rpm).

3.5 DNA Digestion Efficiency

1. Save a 10 μL aliquot ("digested" sample) in a 1.5 mL micro tube. Store the rest of the chromatin material at 4 °C until ready for processing in Subheading 3.6. Add 95 μL molecular biology–grade water and 5 μL 5 M NaCl to "undigested" and "digested" aliquot samples.

2. Reverse cross-links by incubating for 90 min at 65 °C while shaking in a Thermomixer (750 rpm).

3. Add 5 μL 1 M Tris–HCl pH 7.5, 2 μL 0.5 M EDTA, and 2 μL 20 mg/mL proteinase K. Incubate for 1 h at 45 °C while shaking in a Thermomixer (750 rpm).

4. Add 2 μL 10 mg/mL RNase A and incubate for 30 min at 37 °C while shaking in a Thermomixer (750 rpm).

5. Add 120 μL phenol–chloroform–isoamyl alcohol. Shake vigorously and centrifuge (15 min, 16,000 × g, room temperature). Transfer aqueous phase to a new 1.5 mL Safe-Lock tube.

6. Precipitate the DNA by adding 11 μL 2 M sodium acetate pH 5.6, 1 μL 10 mg/mL glycogen and 275 μL 100% ethanol. Centrifuge (30 min, 16,000 × g, 4 °C) and carefully discard the supernatant.

7. Wash the precipitate with 70% ethanol and centrifuge again (10 min, 16,000 × g, 4 °C) and carefully discard the supernatant.

8. Air-dry the precipitate and resuspend in 20 μL 10 mM Tris–HCl pH 8.

9. Visualize DNA on 1.5% agarose gel to evaluate digestion efficiency. Well digested material runs as a smear with DNA fragments ranging from 100 bp to a few kb. A light band of poorly digested material can remain, but its size is usually slightly smaller than undigested control (Fig. 4a). If sample is under-digested (Fig. 4a), see **Note 7**.

3.6 Restriction Enzyme Inactivation and Ligation

1. Inactivate the primary restriction enzyme by incubating the nuclei (from Subheading 3.5, **step 1**) at 65 °C for exactly 20 min while shaking in a Thermomixer (600 rpm).

2. Add 990 μL 1.5× ligation master mix. Incubate at 16 °C for 15 min while shaking in a Thermomixer (900 rpm).

3. Add 5 μL (20 U/μL, Weiss units) T4 DNA ligase. Incubate for 4 h at 16 °C while shaking in a Thermomixer (900 rpm).

4. Decrosslink samples by incubating overnight at 65 °C while shaking in a Thermomixer (900 rpm).

3.7 DNA Purification

1. Divide the samples over two (~750 μL each) new Safe-Lock tubes and add 50 μL 20 mg/mL proteinase K and 30 μL 20% SDS to each. Incubate for 4 h at 45 °C in a Thermomixer (900 rpm).

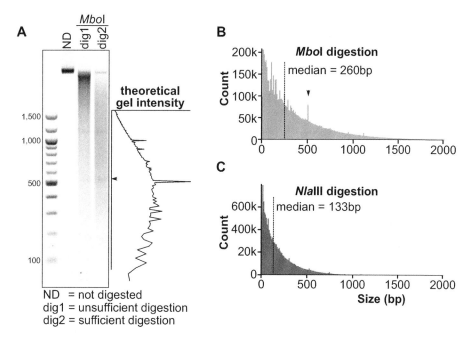

A

MboI

ND dig1 dig2

theoretical
gel intensity

1,500
1,000
500
100

ND = not digested
dig1 = unsufficient digestion
dig2 = sufficient digestion

B

MboI digestion
median = 260bp

C

NlaIII digestion
median = 133bp

Size (bp)

Fig. 4 Validation of chromatin digestion by the primary restriction enzyme. The progression of chromatin digestion needs to be carefully evaluated before ligation. We routinely load an aliquot of digested material after cross-link reversal and DNA purification. (**a**) Migration pattern observed before (ND) and after (dig1, dig2) *MboI* digestion. Dig1 corresponds to insufficiently digested material; dig2 for sufficiently digested material. A theoretical in silico gel intensity profile (mouse genome) is shown alongside the gel. Because digestion of cross-linked chromatin is not 100% efficient, the digested DNA runs at higher molecular weight than in silico predictions. The arrowhead indicates a discrete band around 500 bp observed with *MboI* or *DpnII* that corresponds to the digestion of a specific repeat element. (**b**) and (**c**) Length of the restriction fragments obtained by the in silico digestion of the mouse genome with *MboI* (or *DpnII*) (**b**) and *NlaIII* (**c**). The arrowhead (**b**) indicates the overrepresented bin leading to the discrete band around 500 bp seen in **a**. Bin = 10 bp. On average, the digestion by *MboI* generates larger fragments in the mouse and human genomes than *NlaIII*

2. Add 65 μL 5 M NaCl and incubate for 1 h at 65 °C in a Thermomixer (900 rpm).

3. Add 15 μL 10 mg/mL RNase A and incubate for 1 h at 37 °C in a Thermomixer (900 rpm).

4. Combine both tubes in a 15 mL Falcon tube and add 4.5 mL water and 6 mL phenol–chloroform–isoamyl alcohol. Shake vigorously and centrifuge (15 min, 2200 × *g*, room temperature). Transfer the aqueous phase to a 50 mL conical tube.

5. Precipitate DNA by adding 1.2 mL 2 M sodium acetate pH 5.6, 5 μL 10 mg/mL glycogen and 30 mL 100% ethanol. Mix gently and store at −80 °C for at least 1 h.

6. Centrifuge (45 min, >2200 × *g*, 4 °C) and carefully discard the supernatant.

7. Wash the precipitate with 10 mL cold 70% ethanol and centrifuge (10 min, >2200 × g, 4 °C). Discard the supernatant carefully.

8. Air-dry the precipitate for up to 1 h at 55 °C.

9. Dissolve the precipitate in 200 μL of molecular biology–grade water. Incubate for 1 h at 37 °C.

10. Quantify a 2 μL aliquot of the 3C material using a DNA dye-incorporation assay for high concentration DNA (e.g., Qubit dsDNA BR assay kit; **Note 8**). Visualize 200 ng of 3C material on a 1% agarose gel to verify the ligation efficiency. Correctly ligated material will run at a considerably increased average length compared to digested control (highest average signal from several kb up to >10 kb).

3.8 Secondary Digestion and Ligation

1. Dilute up to 60 μg 3C material DNA to 100 ng/μL in the appropriate 1× restriction buffer for the second restriction enzyme (*see* **Note 1**).

2. Add 1 U of the secondary restriction enzyme per μg of 3C material and incubate overnight at 37 °C (750 rpm).

3. Inactivate the secondary restriction enzyme by incubating for 20 min at 65 °C in a Thermomixer (600 rpm).

4. Take a 5 μL aliquot as "digested control" and visualize on a 2% agarose gel. Digested material should run as a smear ranging from 100 bp to 2 kb (*see* **Note 9**).

5. Transfer DNA to a 50 mL conical tube. Dilute the digested 3C material by adding molecular biology–grade water to a total volume of 12.6 mL, then add 1.4 mL homemade 10× ligation buffer.

6. Cool down to 16 °C in a Thermomixer, then add 10 μL 20 U/μL (Weiss units) T4 DNA ligase and invert the tube a few times. Incubate for 4 h at 16 °C, followed by 30 min at 25 °C in a Thermomixer (600 rpm).

3.9 DNA Purification

1. Add 14 mL phenol–chloroform–isoamyl alcohol, shake vigorously and centrifuge (15 min, 2200 × g, room temperature).

2. Transfer aqueous phase to new 50 mL conical tube and add 14 mL molecular biology–grade water and 2.8 mL of 2 M sodium acetate pH 5.6.

3. Divide the sample into two 50 mL conical tubes and add 35 mL 100% ethanol to each. Mix gently and incubate for at least 2 h at −80 °C.

4. Centrifuge (45 min, >2200 × g, 4 °C). Carefully discard the supernatant.

5. Wash both precipitates with 20 mL 70% ethanol and centrifuge (10 min, >2200 × g, 4 °C). Carefully discard the supernatant.

6. Air-dry the precipitate for up to 1 h at 55 °C. Dissolve both precipitates in 200 μL 10 mM Tris–HCl pH 8 each. Incubate for 1 h at 37 °C.

7. Add 1 mL PB buffer (from QIAquick PCR purification kit, or equivalent buffer) to each tube and mix by pipetting up and down a few times.

8. Load mix on a total of four QIAquick Spin Columns (two columns per conical tube).

9. Centrifuge (1 min, 16,000 × g, room temperature) and discard the flowthrough.

10. Wash the columns twice with 600 μL each PE buffer, centrifuging (1 min, 16,000 × g, room temperature) then discarding flowthrough each time.

11. Transfer the columns to new 1.5 mL Safe-Lock tubes and elute the 4C material by adding 50 μL 10 mM Tris–HCl pH 7.5, then centrifuging (1 min, 16,000 × g, room temperature). Combine the four eluates into a single 1.5 mL microtube.

12. Quantify a 2 μL aliquot of the 4C material using a DNA dye-incorporation assay for high concentration DNA (e.g., QuBit dsDNA BR assay kit; **Note 8**) *see* **Note 10**. Store the 4C material at −20 °C.

3.10 4C-seq
Test PCR

Testing the correct amplification of 4C material should be conducted by a titration PCR using, ideally, a validated pair of 4C primers (*see* **Note 4**). The setup of the four PCR reactions is as follows:

1. Assemble in four individual PCR tubes, with each containing a different amount of 4C material:

   (a) 5 μL 10× Expand Long Buffer 1 (Roche).

   (b) 1 μL 10 mM dNTPs.

   (c) 0.5 μL 100 μM forward 4C primer.

   (d) 0.5 μL 100 μM reverse 4C primer.

   (e) 4C material (25 ng, 50 ng, 75 ng or 100 ng).

   (f) 0.75 μL (3.75U) Expand Long Enzyme mix (Roche).

   (g) Molecular biology–grade water to 50 μL.

2. Run in a thermocycler as follows.

   (a) 94 °C, 2 min.

   (b) [94 °C, 15 s; 55–60 °C, 1 min; 68 °C, 3 min] × 30 cycles.

   (c) 68 °C, 7 min.

3. Load 15 μL of the PCR reaction on a 1.5% Agarose gel (*see* **Note 4**).

3.11 Preparation of 4C-seq Illumina Sequencing Material

1. Do as many PCR reactions as necessary to amplify 1–1.2 µg 4C input material, using the highest concentration of 4C material that still provides linear amplification in the 4C-seq test PCR (*see* **Notes 11** and **12**).

2. Pool the PCR reactions. Mix well and transfer a 100 µL aliquot into a new 1.5 mL tube. Store the remainder of the unpurified pooled PCR reactions at −20 °C as backup.

3. Purify the aliquot on SPRI beads (*see* **Note 13**).

   (a) Allow the SPRI beads to warm up to room temperature (30 min).

   (b) Vortex thoroughly the SPRI beads and immediately transfer precisely 85 µL of the beads to the 100 µL aliquot from the PCR amplification (0.85 volumes beads to 1 volume of PCR reaction). Pipet up-and-down at least 10 times or vortex. Do a short spin to collect all liquid in the solution and incubate for 5 min at room temperature.

   (c) Place the 1.5 mL microcentrifuge tube on a magnetic rack. Remove the supernatant once the solution is clear.

   (d) Gently add 1 mL freshly prepared 80% ethanol while the tube remains on the magnet, without disturbing the beads. Remove the ethanol after 30 s. Repeat once more for a total of two 80% ethanol washes.

   (e) Briefly spin the tube to pellet the remaining drops of ethanol. Place the tube on the magnetic rack and carefully remove the residual ethanol. Air-dry for 30 s.

   (f) Resuspend the beads in 40 µL 10 mM Tris–HCl pH 7.5 by pipetting up-and-down at least 10 times. Incubate for 5 min at room temperature.

   (g) Briefly spin the tube and place the tube on the magnetic rack. Once the solution is clear, transfer most of the solution to a new 1.5 mL microcentrifuge tube.

4. Quantify a 4 µL aliquot of the PCR material using a DNA dye-incorporation assay for high concentration DNA (e.g., QuBit dsDNA BR assay kit).

5. Run 15 µL or up to 400 ng of purified 4C-seq PCR on a 1.5% Agarose gel. Verify the correct amplification (*see* **Note 4**) and check for the absence of primer dimers.

6. The amplified material is ready to be sequenced. Multiplex by combining equal amounts of 4C-seq Illumina sequencing material in a DNA LoBind Tube (*see* **Note 14**). A 3.3 ng/µL DNA concentration is equivalent to 20 nM when DNA fragments have an average length of 250 bp. Confirm the final concentration of the pooled 4C PCR material by quantifying

a 2 μL aliquot using a DNA dye-incorporation assay for low concentration DNA (e.g., QuBit DNA HS assay kit). This material can be used directly for sequencing on an Illumina NextSeq 500 platform in a 75-bp Single-End run (*see* **Note 15**). Paired-end sequencing would be required if primers were designed according to strategy D (Fig. 2d).

3.12 Data Analysis

After Illumina sequencing, the resulting data should be analyzed using bioinformatics tools to create 4C-seq interaction profiles that depict the interaction frequency of each viewpoint with the rest of genome. Different 4C-seq analysis pipelines are available in the literature, which include 4Cseqpipe [16], r3Cseq [27], FourCseq [28], Basic4Cseq [29], fourSig [30], w4Cseq [31], or pipe4C [32]. Another option, which we frequently used, was found in the recently defunct HTSstation [33]. A stand-alone version of this analysis tool for 4C-seq data, combining python, R, and bash scripts, can be downloaded from https://github.com/NoordermeerLab/c4ctus.

All 4C-seq data analysis pipelines roughly follow a similar data analysis strategy, including mapping of data, translation to interaction frequencies and either the identification of genomic regions with significantly enriched signal or the possibility to quantitatively compare between samples. Here, we describe the key steps for processing allele-specific 4C-seq reads obtained by strategy B (Fig. 2b). By design, all reads for which the targeted restriction sites has been digested and ligated, originate from the single allele. Our preferred data analysis approach can be subdivided into five essential steps.

1. Demultiplexing and trimming of the reads. This is done by matching the 5′-end of the reads with the sequence of the 4C primers used for amplification, with some mismatches tolerated. Reads are trimmed at the location of the restriction sites for the following mapping step, with the part complementary to the primer sequence being removed.

2. Mapping of the reads to the reference genome using an alignment tool like Bowtie2 [34].

3. Virtual digestion of the reference genome to generate a list of valid restriction fragments. This list is filtered out for restriction fragments lacking a cut site for the secondary restriction enzyme or restriction fragments containing repeated sequences at both extremities.

4. Assignment of read density to restriction fragment ends, followed by the calculation of a 4C-score per valid restriction fragment, taking into account if reads could be mapped to both restriction fragments ends or only one of the two [33]. The score of the 5 restriction fragments centered on the viewpoint are set to zero. This generates the RAW 4C-seq data.

5. Smoothing and normalization of the RAW 4C-seq data. Given that RAW 4C-seq scores can be influenced, to a significant extent, by the distance between the recognition sites of the primary and secondary restriction enzymes [15], the base composition at its extremities, as well as differences in chromatin accessibility that could affect the digestion and ligation steps, RAW 4C-scores are smoothed using a running mean across several restriction fragments. In our hands, averaging over 11 restriction fragments generates reliable 4C-seq profiles.

To compare different samples, smoothed 4C-seq profiles needs to be normalized. To do so, we consider that the sum of interaction frequencies within a region of several megabases (e.g., a number of TADs) centered on the viewpoint should be identical across different samples. We then multiply the 4C-scores by an appropriate scaling factor so that the sum of all 4C-scores within this region equals one million.

To compare interaction frequencies between samples, a log2 ratio can be calculated between the normalized smoothed 4C-seq scores from both samples. In particular, this is done when comparing interaction frequencies between alleles, using the 4C-score from a F1 hybrid (one allele analyzed) and the one from the F1 of the reciprocal cross (second allele analyzed).

4 Notes

1. Different combinations of restriction enzymes can be used for high-resolution 4C-seq, provided the primary enzyme is able to efficiently digest fixed chromatin. In our hand, the most reliable ones for the first digestion are *Dpn*II (and its isoschizomer *Mbo*I) and *Nla*III. For optimal ligation and circularization of the targeted viewpoint, the distance between restriction sites should ideally be as follows: (a) at least 700 bp between the two consecutive recognition sites of the primary restriction enzyme; (b) at least 250 bp between the recognition sites of the primary and secondary restriction enzyme. Some deviation from this guidance can be tolerated, especially if one of the restriction sites of the first cutter should coincide with a SNP for allelic 4C-seq. However, PCR amplifications of 4C material might be less reliable if distances become too small.

2. Primers containing Illumina adapters should be PAGE-purified to eliminate shorter DNA molecules aborted during oligonucleotide synthesis. To avoid the use of very long primers, a two-step PCR strategy can be considered, whereby the first step is done with viewpoint-specific primers with a noncomplementary 5'-end tail while the second amplification step relies on primers annealing to this tail and containing the Illumina adapters in 3' [32].

3. The mouse Genomes Project [35] catalogs genetic variations among numerous strains of mouse. Although the official release dates from 2015 (REL-1505), two more recent releases can be downloaded from ftp://ftp-mouse.sanger.ac.uk/. The list of available strains can be queried as follows: `bcftools query -l ftp://ftp-mouse.sanger.ac.uk/[DIRECTORY]/[FILE.vcf.gz]`. The variants of some strains are only available in the more recent releases. This is the case for the JF1_MsJ strain.

4. The quality of 4C-seq primers needs to be evaluated. To validate the amplification robustness, we start by testing 4C primers lacking the Illumina adapters by running two replicate PCR reactions on agarose gel. Reliable amplifications are characterized by (1) a reproducible and uniform long smear ranging from 100 bp to 1–2 kb and (2) potentially a few discrete bands, including one that may correspond to the undigested fragment whose length can be predicted. Primers passing this first quality check are ordered with Illumina adapters and tested likewise. Primers will be rejected if the smear is too faint (hard to see on agarose gel), not sufficiently wide-ranging (for instance if centered on 200–300 bp), if many discrete bands are observed (more than 3) or if strong primer dimers are observed (not to be confused with unincorporated primers, which for sequencing material will be removed using SPRI-beads).

5. This protocol has been optimized for mammalian cells grown in tissue culture. When working with organisms with lower DNA content per cells, the amount of starting material should be adjusted. Importantly, cell lysis may need to be optimized for each cell type, especially if starting from tissue samples. Different lysis conditions are proposed in Matelot et al. (2016) [25].

6. Aggregates of nuclei might be visible at this or later stages. To limit their formation, reactions need to be agitated, especially during the overnight digestion with the first restriction enzyme. Using a Thermomixer and shaking at 900 rpm limits formation of aggregates. Pipetting up-and-down the reaction with a 200 μL tip may help to dissociate the aggregates. The presence of large aggregates is often associated with poor cell lysis and tends to inhibit digestion efficiency.

7. If sample is underdigested after the overnight primary digestion (*see* Fig. 4a), add 400 U of the primary restriction enzyme to the nuclei, and incubate overnight at 37 °C (900 rpm). If after two rounds of overnight digestion, the chromatin remains underdigested, it is usually associated with poor cell lysis/permeabilization at initial steps. In this case, the cell lysis should be optimized on a new sample.

8. When oxidized, DTT exhibits a strong absorbance at 280 nm, which alters precise DNA quantifications based on UV-absorbance. For this reason, we strongly advise to rely on fluorescence-based dye-incorporation assays for DNA quantification, especially since the DTT contained in ligation buffer will not be completely removed by the cleanup steps after the second ligation reaction.

9. In case of incomplete digestion with the secondary enzyme, another round of overnight digestion should be done by adding the same amount of the secondary restriction enzyme. Improved digestion efficiency should be checked before continuing.

10. When starting from 10^7 mammalian cells, one should expect around 30 μg of 4C material at this step.

11. It is critical to perform multiple 4C-PCR reactions in parallel, to ensure enough complexity before sequencing. For mammals, we amplify at least 1 μg 4C material, usually by doing 16 PCR reactions in parallel, each containing 75 ng of 4C material ($16 \times 75 = 1.2$ μg). 1.2 μg of DNA corresponds to 4×10^5 copies of haploid mouse/human genome, thereby allowing a theoretical maximum detection of 4×10^5 captured 3D-partners, as only one partner can be captured per viewpoint containing restriction fragment. Because the chromatin digestion is never 100% efficient, and because local 3D-partners will be captured several times, the effective number of unique 3D-partners identified is expected to be smaller. The amount of 4C material used for PCR amplification can be adjusted when working with smaller or bigger genomes.

12. To find the optimal amount for PCR amplification, a titration of novel 4C material is done, ideally using a validated 4C primer pair (see **Note 4**). For mammals, we amplify 25, 50, 75, and 100 ng of 4C material and load the 50 μL PCR reactions on a 1.5% agarose gel [25]. The optimal quantity for amplification corresponds to the highest quantity for which (a) the amplification yield is high, (b) the intensity of the smear linearly increases but (c) the PCR smear does not shift toward the amplification of smaller fragments. In our hands, the optimal quantity of starting 4C material ranges between 50 and 75 ng for mammals.

13. Alternatively, 4C-seq PCR material can be purified with Qiagen PCR purification columns instead of SPRI magnetic beads (e.g., AMPure XP beads). Purification using a 0.85× SPRI beads ratio has the advantage that larger fragments can be removed (~200 bp for SPRI beads versus ~120 bp for PCR purification columns), which more efficiently removes unincorporated 4C-seq primers and primer dimers which tend to be incompletely removed by silica columns.

14. The multiplexing of 4C-seq PCR material originating from multiple viewpoints or samples can be done at two levels before sequencing.

    (a) Multiple 4C-seq PCRs from different viewpoints can be mixed in equimolar quantities and sequenced together, as the first sequenced base pairs will correspond to the primer used for amplification. The demultiplexing is done by comparing the 5′-end of the read with the sequences of the different 4C-seq primers used.

    (b) Multiple 4C-seq PCRs from different samples but with the same primer-set can also be multiplexed as the adapters used here are compatible with Illumina single-indexed sequencing. To do so, an Illumina TruSeq index is included in the P7-containing primer and read using Illumina i7 Index Primer. In this case, demultiplexing is done using the dedicated Illumina software.

    Up to around 20 4C-seq PCR samples, either from different viewpoints or carrying different indexes, can be pooled and sequenced together on a NextSeq 500/550 high output flow cell (aiming for 15–20 million reads per 4C-seq PCR). This number can be adjusted if using a different flow cell/sequencing platform.

15. To ensure sufficient complexity in the first bases during the sequencing, 20% Phi-X balancer DNA should be added prior to loading on the Illumina flow cell.

Acknowledgments

We acknowledge funding from the Agence Nationale de la Recherche (project "IMP-REGULOME"—ANR-18-CE12-0022-02) to D. N.

References

1. Vermunt MW, Zhang D, Blobel GA (2019) The interdependence of gene-regulatory elements and the 3D genome. J Cell Biol 218:12–26. https://doi.org/10.1083/jcb.201809040

2. Dekker J, Rippe K, Dekker M et al (2002) Capturing chromosome conformation. Science (New York, NY) 295:1306–1311. https://doi.org/10.1126/science.1067799

3. Davies JOJ, Oudelaar AM, Higgs DR et al (2017) How best to identify chromosomal interactions: a comparison of approaches. Nat Methods 14:125–134. https://doi.org/10.1038/nmeth.4146

4. Nora EP, Lajoie BR, Schulz EG et al (2012) Spatial partitioning of the regulatory landscape of the X-inactivation Centre. Nature 485:381–385. https://doi.org/10.1038/nature11049

5. Dixon JR, Selvaraj S, Yue F et al (2012) Topological domains in mammalian genomes identified by analysis of chromatin interactions. Nature 485:376–380. https://doi.org/10.1038/nature11082

6. Shen Y, Yue F, McCleary DF et al (2012) A map of the cis-regulatory sequences in the mouse genome. Nature 488:116–120. https://doi.org/10.1038/nature11243

7. Andrey G, Montavon T, Mascrez B et al (2013) A switch between topological domains underlies HoxD genes collinearity in mouse limbs. Science (New York, NY) 340:1234167. https://doi.org/10.1126/science.1234167

8. Lupiáñez DG, Kraft K, Heinrich V et al (2015) Disruptions of topological chromatin domains cause pathogenic rewiring of gene-enhancer interactions. Cell 161:1012–1025. https://doi.org/10.1016/j.cell.2015.04.004

9. Sanborn AL, Rao SSP, Huang S-C et al (2015) Chromatin extrusion explains key features of loop and domain formation in wild-type and engineered genomes. Proc Natl Acad Sci U S A 112:E6456–E6465. https://doi.org/10.1073/pnas.1518552112

10. Fudenberg G, Imakaev M, Lu C et al (2016) Formation of chromosomal domains by loop extrusion. Cell Rep 15:2038–2049. https://doi.org/10.1016/j.celrep.2016.04.085

11. Chang L-H, Ghosh S, Noordermeer D (2020) TADs and their Borders: free movement or building a wall? J Mol Biol 432:643–652. https://doi.org/10.1016/j.jmb.2019.11.025

12. Razin SV, Ulianov SV (2017) Gene functioning and storage within a folded genome. Cell Mol Biol Lett 22:18. https://doi.org/10.1186/s11658-017-0050-4

13. Rao SSP, Huntley MH, Durand NC et al (2014) A 3D map of the human genome at kilobase resolution reveals principles of chromatin looping. Cell 159:1665–1680. https://doi.org/10.1016/j.cell.2014.11.021

14. Lieberman-Aiden E, van Berkum NL, Williams L et al (2009) Comprehensive mapping of long-range interactions reveals folding principles of the human genome. Science (New York, NY) 326:289–293. https://doi.org/10.1126/science.1181369

15. Noordermeer D, Leleu M, Splinter E et al (2011) The dynamic architecture of Hox gene clusters. Science (New York, NY) 334:222–225. https://doi.org/10.1126/science.1207194

16. van de Werken HJG, Landan G, Holwerda SJB et al (2012) Robust 4C-seq data analysis to screen for regulatory DNA interactions. Nat Methods 9:969–972. https://doi.org/10.1038/nmeth.2173

17. Hughes JR, Roberts N, McGowan S et al (2014) Analysis of hundreds of cis-regulatory landscapes at high resolution in a single, high-throughput experiment. Nat Genet 46:205–212. https://doi.org/10.1038/ng.2871

18. Davies JOJ, Telenius JM, McGowan SJ et al (2016) Multiplexed analysis of chromosome conformation at vastly improved sensitivity. Nat Methods 13:74–80. https://doi.org/10.1038/nmeth.3664

19. Splinter E, de Wit E, Nora EP et al (2011) The inactive X chromosome adopts a unique three-dimensional conformation that is dependent on Xist RNA. Genes Dev 25:1371–1383. https://doi.org/10.1101/gad.633311

20. Llères D, Moindrot B, Pathak R et al (2019) CTCF modulates allele-specific sub-TAD organization and imprinted gene activity at the mouse Dlk1-Dio3 and Igf2-H19 domains. Genome Biol 20:272. https://doi.org/10.1186/s13059-019-1896-8

21. Noordermeer D, Feil R (2020) Differential 3D chromatin organization and gene activity in genomic imprinting. Curr Opin Genet Dev 61:17–24. https://doi.org/10.1016/j.gde.2020.03.004

22. Holwerda SJB, van de Werken HJG, de Almeida CR et al (2013) Allelic exclusion of the immunoglobulin heavy chain locus is independent of its nuclear localization in mature B cells. Nucleic Acids Res 41:6905–6916. https://doi.org/10.1093/nar/gkt491

23. Johanson TM, Chan WF, Keenan CR et al (2019) Genome organization in immune cells: unique challenges. Nat Rev Immunol 19:448–456. https://doi.org/10.1038/s41577-019-0155-2

24. Clowney EJ, LeGros MA, Mosley CP et al (2012) Nuclear aggregation of olfactory receptor genes governs their monogenic expression. Cell 151:724–737. https://doi.org/10.1016/j.cell.2012.09.043

25. Matelot M, Noordermeer D (2016) Determination of high-resolution 3D chromatin organization using circular chromosome conformation capture (4C-seq). Methods Mol Biol 1480:223–241. https://doi.org/10.1007/978-1-4939-6380-5_20

26. Nagano T, Várnai C, Schoenfelder S et al (2015) Comparison of hi-C results using in-solution versus in-nucleus ligation. Genome Biol 16:175. https://doi.org/10.1186/s13059-015-0753-7

27. Thongjuea S, Stadhouders R, Grosveld FG et al (2013) r3Cseq: an R/Bioconductor package for the discovery of long-range genomic interactions from chromosome conformation capture and next-generation sequencing data. Nucleic Acids Res 41:e132. https://doi.org/10.1093/nar/gkt373

28. Klein FA, Pakozdi T, Anders S et al (2015) FourCSeq: analysis of 4C sequencing data. Bioinformatics (Oxford, England) 31:3085–3091. https://doi.org/10.1093/bioinformatics/btv335

29. Walter C, Schuetzmann D, Rosenbauer F et al (2014) Basic4Cseq: an R/Bioconductor package for analyzing 4C-seq data. Bioinformatics (Oxford, England) 30:3268–3269. https://doi.org/10.1093/bioinformatics/btu497

30. Williams RL, Starmer J, Mugford JW et al (2014) fourSig: a method for determining chromosomal interactions in 4C-Seq data. Nucleic Acids Res 42:e68. https://doi.org/10.1093/nar/gku156

31. Cai M, Gao F, Lu W et al (2016) w4CSeq: software and web application to analyze 4C-seq data. Bioinformatics (Oxford, England) 32:3333–3335. https://doi.org/10.1093/bioinformatics/btw408

32. Krijger PHL, Geeven G, Bianchi V et al (2020) 4C-seq from beginning to end: a detailed protocol for sample preparation and data analysis. Methods (San Diego, Calif) 170:17–32. https://doi.org/10.1016/j.ymeth.2019.07.014

33. David FPA, Delafontaine J, Carat S et al (2014) HTSstation: a web application and open-access libraries for high-throughput sequencing data analysis. PLoS One 9:e85879. https://doi.org/10.1371/journal.pone.0085879

34. Langmead B, Salzberg SL (2012) Fast gapped-read alignment with bowtie 2. Nat Methods 9:357–359. https://doi.org/10.1038/nmeth.1923

35. Keane TM, Goodstadt L, Danecek P et al (2011) Mouse genomic variation and its effect on phenotypes and gene regulation. Nature 477:289–294. https://doi.org/10.1038/nature10413

36. Mugford JW, Starmer J, Williams RL et al (2014) Evidence for local regulatory control of escape from imprinted X chromosome inactivation. Genetics 197:715–723. https://doi.org/10.1534/genetics.114.162800

37. de Wit E, Bouwman BAM, Zhu Y et al (2013) The pluripotent genome in three dimensions is shaped around pluripotency factors. Nature 501:227–231. https://doi.org/10.1038/nature12420

38. Dixon JR, Jung I, Selvaraj S et al (2015) Chromatin architecture reorganization during stem cell differentiation. Nature 518:331–336. https://doi.org/10.1038/nature14222

Chapter 3

Tough Tissue Hi-C

Stefan Grob

Abstract

The ability to decipher the three-dimensional chromosome folding in many eukaryotes is a major asset in molecular biology. It is not only required to study the biological relevance of chromosome folding in cellular processes but also for the de novo assembly of genomes of nonmodel species. With lowering DNA sequencing costs, the latter has recently become interesting to many scientists, ranging from molecular biologists that aim to establish new model organisms, to evolutionary biologists and ecologists, interested in genome evolution and diversity. Hi-C is regarded as the method of choice to characterize three-dimensional genome folding and, thus, also has been integrated as a standard method in assembly pipelines. However, Hi-C is a demanding molecular biology technique, and its application can be considerably challenged by the tissue used. Hi-C relies on efficient and pure nuclei isolation, which is, especially in many plant species, inhibited by the tough nature of plant tissues and cell walls. The Hi-C protocol presented here has been optimized for such tissues and has been shown to generate Hi-C samples of sufficient quality in various plant and animal tissues.

Key words Hi-C, 3D chromosome folding, Genome assembly, Nonmodel species, Tough tissues

1 Introduction

Since its first introduction in 2009 [1], Hi-C has become a widely applied method to not only study three-dimensional folding of chromosomes but also to aid genome assembly [2–4]. The latter use of Hi-C requires protocols that produce reliable results in many nonmodel species. However, in certain species, especially plant species, standard Hi-C protocols may not be well-suited, as the toughness of the tissue or secondary metabolites may inhibit important steps of the Hi-C protocol, including chromatin fixation, nuclei isolation, and chromatin digestion. The workflow presented here takes these challenges into account and has been validated in several plant species.

The formaldehyde-based fixation (cross-linking) of chromatin can become a primary obstacle, as tough tissues are not necessarily permeable to formaldehyde. Similarly, cross-linking efficiency may

Tom Sexton (ed.), *Spatial Genome Organization: Methods and Protocols*,
Methods in Molecular Biology, vol. 2532, https://doi.org/10.1007/978-1-0716-2497-5_3,
© The Author(s), under exclusive license to Springer Science+Business Media, LLC, part of Springer Nature 2022

vary significantly in different tissues and is difficult to monitor during the experimental procedure. Careful experimental design with sufficient replication and different cross-linking conditions (duration, formaldehyde concentration) can minimize potentially confounding biases.

Nuclei extraction from tough tissues can be rather tedious, as the nuclei have to be physically or chemically separated from the cells and cellular compounds, such as other organelles. Therefore, thorough tissue homogenization is a must for a successful Hi-C experiment.

Inadequate purity of nuclei may inhibit chromatin digestion and ligation reactions. High purity of nuclei can often not be reached; however, a brute-force approach, using large enzyme quantities and prolonged incubation time can usually overcome problems associated with impure nuclei.

Establishing a robust Hi-C protocol in nonmodel species is always challenging and flexibility of the experimenter is required. The protocol presented here is based on previously published plant Hi-C protocols [5, 6] and should be viewed as a rough guide rather than an exact map to a successful Hi-C experiment. Depending on the tissue, the species, and the restriction enzyme used, the protocol will need some adjustments, usually less so in the type of reagents used but in the amounts of single reagents and volumes of the entire sample. The protocol is accompanied by numerous comments, which may help to further optimize and customize the protocol to fit the experiment's needs.

2   Materials

1. Cross-linking buffer: 10 mM potassium phosphate buffer pH 7, 50 mM NaCl, 0.1 M sucrose, 2% formaldehyde, 0.01% Triton X-100 (see **Notes 1** and **2**).

2. Netting material (e.g., nylon mesh, mesh size approx. 2–3 mm).

3. Vacuum desiccator.

4. 2 M glycine.

5. Liquid nitrogen.

6. Mortar and pestle (see **Note 3**).

7. NIB (nuclei isolation buffer): 20 mM Hepes pH 8, 250 mM sucrose, 1 mM $MgCl_2$, 0.5 mM KCl, 40% (v/v) glycerol, 0.25% (v/v) Triton X-100, 0.1 mM phenylmethanesulfonylfluoride (PMSF), 0.1% 2-mercaptoethanol, 1× total protease inhibitors (Roche), prepared fresh for each experiment.

8. 7 mL Dounce homogenizer.

9. Miracloth.

10. Restriction enzyme with supplied 10× buffer (*see* **Note 4**).

11. 10% (w/v) SDS.

12. 20% (v/v) Triton X-100.

13. Fill-in mix: 133 μM biotin-14-dCTP, 133 μM dATP, 133 μM dGTP, 133 μM dTTP, 1 U/μL DNA polymerase I large (Klenow) fragment in 1× restriction enzyme buffer (*see* **Note 5**).

14. Ligation buffer: 50 mM Tris–HCl pH 7.5, 10 mM MgCl$_2$, 10 mM DTT, 1 mM ATP, 0.2 mg/mL BSA. Prepare fresh for each experiment from stock solutions of 100 mM ATP, 20 mg/mL BSA, and a homemade 10× stock (500 mM Tris–HCl pH 7.5, 100 mM MgCl$_2$, 100 mM DTT), stored in aliquots at −20 °C.

15. High-concentration T4 DNA ligase.

16. 10 mg/mL proteinase K.

17. 5 M NaCl.

18. Phenol–chloroform–isoamyl alcohol (25:24:1).

19. Chloroform–isoamyl alcohol (24:1).

20. 3 M sodium acetate pH 5.2.

21. 20 mg/mL glycogen.

22. 100%, 80%, and 70% ethanol (prepare fresh).

23. Qubit fluorometer with DNA HS quantification assay kit, or equivalent.

24. 5× biotin removal mix: 0.5 mM dATP, 0.5 mM dGTP, 1 mg/mL BSA, 0.25 U/μL T4 DNA polymerase in 5× supplied T4 DNA polymerase buffer (e.g., NEBuffer 2), made fresh for each experiment (*see* **Note 6**).

25. 0.5 M EDTA, pH 8.

26. AMPure XP beads, or equivalent SPRI beads.

27. Magnetic stand.

28. 10 mM Tris–HCl pH 7.5.

29. Covaris M220 sonicator.

30. Streptavidin C1 magnetic beads.

31. Tween Wash Buffer (TWB): 5 mM Tris–HCl pH 7.5, 0.5 mM EDTA, 1 M NaCl, 0.05% (v/v) Tween-20.

32. 2× Binding Buffer (BB): 10 mM Tris–HCl pH 7.5, 1 mM EDTA, 2 M NaCl.

33. KAPA Hyper Prep Kit (Roche), containing end repair and A-tailing buffer and enzyme mix, sequencing adapters, ligation buffer, ligase, and PCR primer mix.

34. Kapa HiFi HotStart ready mix (2×).

35. Agilent TapeStation and D1000 ScreenTape, or equivalent.

3 Methods

3.1 Tissue Collection, Formaldehyde Cross-Linking, and Nuclei Isolation

1. Collect 0.5–5 g of tissue, cut into small pieces (~5 × 5 mm) and immediately transfer it to a 50 mL conical tube on ice (*see* **Note 7**).

2. Add 15 mL cross-linking buffer and ensure that the tissue is completely submerged using a ring and nylon mesh netting placed inside a 50 mL conical tube. Place it in a vacuum desiccator and apply vacuum for 1 h at room temperature. Release and reestablish the vacuum a few times during the incubation to increase formaldehyde infiltration (*see* **Note 8**).

3. To quench the residual formaldehyde, add 1.9 mL 2 M glycine and continue applying the vacuum for 5 min.

4. Wash the tissue a few times with deionized water, place on a filter paper and dry between paper towels (*see* **Note 9**).

5. Snap-freeze the tissue in liquid nitrogen, grind it to a very fine powder using a pestle and mortar, and transfer to a 50 mL conical tube on ice with a precooled spatula (*see* **Note 10**).

6. Immediately add 7 mL NIB and gently resuspend by swirling the tube. Incubate on ice for 30 min (*see* **Note 11**).

7. Grind the tissue in a 7 mL Dounce homogenizer, using both pestles A and B (*see* **Note 12**). Transfer suspension to a fresh 50 mL conical tube on ice. Wash the homogenizer with 7 mL NIB to recover as much tissue as possible and transfer the suspension to the same conical tube to get a final of 14 mL of suspension.

8. Filter the suspension twice through one layer of Miracloth into a fresh 50 mL conical tube (*see* **Note 13**). Finally, wash the residual nuclei off the Miracloth into the same tube with additional NIB.

9. Centrifuge for 15 min at 3000 × g, 4 °C and immediately discard the supernatant by inverting the tube.

10. Resuspend the nuclear pellet in 3 mL NIB (*see* **Note 14**) and transfer 1 mL aliquots to fresh 1.5 mL microcentrifuge tubes (*see* **Note 15**).

11. Centrifuge for 5 min at 2000 × g, 4 °C and discard supernatant, obtaining a white pellet of ~30 μL volume (*see* **Notes 16–18**).

12. Resuspend the nuclei in 500 μL 1.2× restriction enzyme buffer, centrifuge for 5 min at 2000 × g, 4 °C and discard supernatant.

3.2 Digestion

1. Resuspend the nuclei in 500 μL 1.2× restriction buffer and add 10 μL 10% SDS. Incubate for 5 min at 65 °C with shaking on a Thermomixer at 900 rpm (*see* **Note 19**).

2. Add 50 μL 20% Triton-X100 and incubate for 1 h at 37 °C with shaking at 900 rpm.

3. Transfer an 80 μL aliquot to a new microtube ("chromatin integrity" control), add 120 μL molecular biology-grade water, and store at 4 °C, ready for processing in Subheading 3.3, **step 2**.

4. To the rest of the nuclei, add 120 μL molecular biology-grade water and 100 U restriction enzyme, and incubate for 3 h at 37 °C with shaking on a Thermomixer at 750 rpm (*see* **Note 4**).

5. Add 200 U restriction enzyme and incubate overnight at 37° with shaking on a Thermomixer at 750 rpm.

6. The next morning, add 100 U restriction enzyme and incubate for 3 h at 37 °C with shaking on a Thermomixer at 750 rpm.

7. If the restriction enzyme can be heat-inactivated at temperatures ≤65 °C, incubate the nuclei for 20 min at 65 °C with shaking on a Thermomixer at 750 rpm (*see* **Note 20**). Transfer a 60 μL aliquot to a new microtube ("digestion" control), add 140 μL molecular biology-grade water, and store at 4 °C, ready for processing in Subheading 3.3, **step 2**.

8. To the rest of the sample, add 60 μL fill-in mix (*see* **Note 21**) and incubate for 1.5 h at 23 °C (*see* **Note 22**).

9. If the restriction enzyme was not heat-inactivated (**step 7**), add 12 μL 10% SDS and incubate for 25 min at 65 °C with shaking on a Thermomixer at 750 rpm. Add 37 μL 20% Triton-X100 and incubate for 1 h at 37 °C with shaking on a Thermomixer at 750 rpm. Transfer a 60 μL aliquot to a new microtube ("digestion" control), add 140 μL molecular biology-grade water, and store at 4 °C, ready for processing in Subheading 3.3, **step 2**.

3.3 Ligation and DNA Purification

1. Add molecular biology-grade water to bring the volume to 616 μL, then add 84 μL ligation buffer (*see* **Note 23**) and 25 Weiss units T4 DNA ligase. Incubate for 2 h at 16 °C, add another 25 Weiss units of T4 DNA ligase and continue incubation for another 2 h at 16 °C. Finally, without adding additional DNA ligase, incubate for 45 min at room temperature.

2. Add 10 μL 10 mg/mL proteinase K and 30 μL 5 M NaCl. To the "chromatin integrity" and "digestion" control tubes, add 5 μL 10 mg/mL proteinase K and 10 μL 5 M NaCl. Incubate all tubes for 1 h at 37 °C, then overnight at 65 °C (*see* **Note 24**).

3. Add 10 μL 10 mg/mL proteinase K (5 μL for control tubes) and incubate for 2 h at 65 °C (*see* **Note 25**), then cool to room temperature.

4. Add 700 µL phenol/chloroform/isoamyl alcohol (200 µL for control tubes) and shake vigorously for >30 s (*see* **Note 26**).

5. Centrifuge for 10 min at 13,000 × *g*, room temperature and carefully transfer the upper aqueous phase to a fresh microtube without disturbing the interphase.

6. Add 700 µL chloroform/isoamyl alcohol (200 µL for control tubes) and shake vigorously for >30 s. Centrifuge for 10 min at 13,000 × *g*, room temperature, and carefully transfer the upper aqueous phase to fresh 5 mL (1.5 mL for controls) reaction tube without disturbing the interphase.

7. To the Hi-C sample, add 800 µL molecular biology-grade water (*see* **Note 27**), 150 µL 3 M sodium acetate, 5 µL of 20 mg/mL glycogen and 3.75 mL 100% ethanol. To the control tubes, add 20 µL 3 M sodium acetate, 2 µL of glycogen and 550 µL 100% ethanol. Mix by inverting the tubes and store for >2 h (can be overnight) at −80 °C.

8. Centrifuge for 1 h at 20,000 × *g*, 4 °C, remove the supernatant without disturbing the pellet and wash with 1 mL 70% ethanol.

9. Centrifuge for 20 min at 20,000 × *g*, 4 °C, carefully remove supernatant, air-dry the pellet and dissolve in 100 µL molecular biology-grade water overnight at 4 °C (*see* **Note 28**).

10. Measure the DNA concentration using a Qubit fluorometer (*see* **Note 29**).

    Assess 180 ng of Hi-C sample, as well as the "chromatin integrity" and "digestion" controls, by gel electrophoresis (*see* Fig. 1). Furthermore, the efficiency of the fill-in reaction can be assessed by exploiting the fact that the original restriction enzyme site sequence has been lost upon blunt-end ligation [5, 7] (*see* Fig. 2).

11. Pool all Hi-C subsamples passing quality control into a single microtube. The following steps assume a total volume of 150 µL; alter the reaction volumes accordingly to keep the same ratios.

12. Add 40 µL 5× biotin removal mix (*see* **Note 6**), adjust the volume to a final 200 µL with nuclease-free water and incubate for 2 h at 12 °C.

13. Stop the reaction by adding 4 µL 0.5 M EDTA.

14. Add 360 µL of AMPure XP beads, mix and incubate for 15 min at room temperature on a rotating wheel (*see* **Note 30**).

15. Reclaim the beads on a magnetic stand and remove supernatant once the solution is cleared.

16. Leaving the tubes on the magnetic stand, add 500 µL of freshly prepared 80% ethanol, incubate for 30 s, and remove supernatant. Repeat this step once.

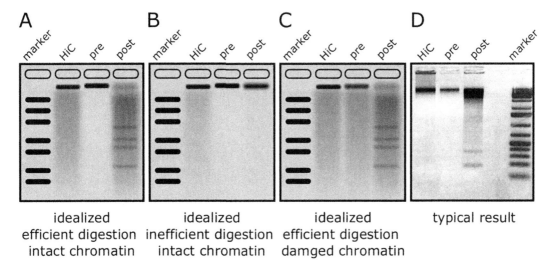

Fig. 1 Chromatin integrity, digestion, and ligation control. The "chromatin integrity control" (pre) sample should form a single high-molecular band, with little or no smear below. Any smear visible is indicative of compromised initial chromatin integrity. In case no high-molecular band is visible but only a smear, the corresponding Hi-C sample should be discarded. The "digest control" (post) should appear as a smear. The size of the center of the smear should correspond to the restriction enzyme used (4-cutter approx.: 256 bp, 6-cutter approx.: 4096 bp). The presence of a high molecular band indicates insufficient digestion efficiency. The Hi-C sample should show a smear, which is shifted upward compared to the "digest control" sample. Depending on the restriction enzyme used and the electrophoresis conditions, only a high molecular band may be visible. However, the band should run lower than that of the "chromatin integrity control" sample (Figure adapted from [5])

17. Remove residual ethanol and air-dry the beads until no liquid is visible (*see* **Note 31**).

18. Elute the purified Hi-C DNA, by resuspending the SPRI beads in 50 µL 10 mM Tris–HCl pH 7.5 and incubate for 5 min at room temperature.

19. Place the reaction tube on a magnetic rack until the suspension is cleared and transfer the eluted Hi-C DNA present in the supernatant to a fresh 1.5 mL reaction tube. Store the Hi-C sample at 4 °C (up to 1 week) or −20 °C (several weeks).

20. Measure the concentration with a Qubit fluorimeter. A total amount of 500 ng of Hi-C DNA is sufficient to perform standard library preparations.

3.4 Library Preparation (See Note 32)

1. Sonicate to desired size using a Covaris M220 sonicator in a 50 µL sonication tube using the following settings: Peak Power 75 W, Duty Factor 20, Cycles per Burst 200, Duration 150 s (*see* **Note 33**). Retain a 5 µL aliquot for later fragment size analysis and replace the missing 5 µL with nuclease-free water. Keep the residual sonicated Hi-C samples on ice until further processing at **step 3**.

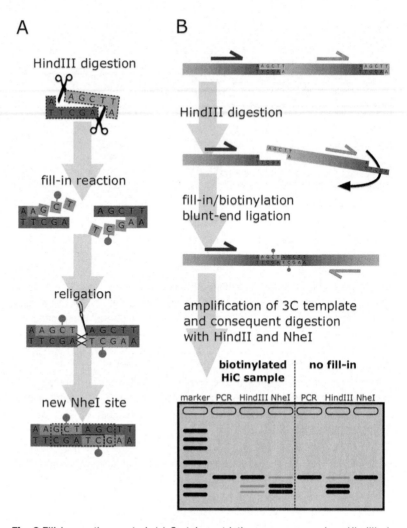

Fig. 2 Fill-in reaction control. (**a**) Certain restriction enzymes, such as HindIII, do not leave an overhang covering the entire restriction enzyme recognition site. Thus, by filling in the overhangs and subsequent blunt-end ligation, the HindIII restriction enzyme recognition site is lost. A novel restriction enzyme recognition site is created (NheI). (**b**) Using tandem-oriented primers close to two restriction sites (neighboring or within a few kb) guarantees that only fragments that were initially digested and religated are amplified. Upon PCR amplification using Hi-C DNA as a template, the PCR product can be digested with either HindIII or NheI. The Hi-C fill-in reaction should have mutated most of the HindIII sites, thus, when HindIII can digest a significant amount of PCR product, the fill-in reaction had low efficiency. Vice-versa, high digestion efficiency with NheI indicates high fill-in reaction efficiency. (Figure adapted from [5])

2. Transfer 60 μL Streptavidin C1 magnetic beads to a low-binding reaction tube and add 400 μL of TWB. Incubate for 5 min at room temperature on a rotating wheel. Reclaim the beads on a magnetic stand and remove supernatant. Repeat TWB wash once. Remove the supernatant.

3. Add 300 μL of 2× BB, 50 μL purified Hi-C DNA, and 250 μL of nuclease-free water. Incubate for 15 min at room temperature on a rotating wheel. Reclaim the beads on a magnetic stand and immediately proceed to end repair and A-tailing.

4. Resuspend the Hi-C DNA binding Streptavidin beads in 50 μL nuclease-free water, 7 μL end repair and A-tailing buffer, and 3 μL end repair and A-tailing enzyme mix. Resuspend the beads by gently pipetting up and down and transfer the suspension to a PCR strip.

5. In a thermal cycler, incubate the reaction for 30 min at 20 °C, followed by 30 min at 65 °C (set lid temperature to 85 °C).

6. To the end-repaired and A-tailed bead-bound Hi-C sample, add 5 μL of 2 μM sequencing adapters, 5 μL nuclease-free water, 30 μL ligation Buffer, and 10 μL DNA ligase. Mix thoroughly and incubate for 20 min at 20 °C.

7. Transfer the end-repaired and A-tailed Hi-C samples to a fresh low-binding 1.5 mL reaction tube and place it on a magnetic stand for 5 min. Remove the supernatant and resuspend the reclaimed beads in 400 μL of TWB. Transfer the suspension to a new 1.5 mL reaction tube. Repeat the TWB wash once.

8. After reclaiming the beads, resuspend in 200 μL of 1× BB. Transfer the suspension to a new 1.5 mL reaction tube.

9. Reclaim the beads on a magnetic stand and finally resuspend in 50 μL 10 mM Tris–HCl pH 7.5.

10. Using KAPA HiFi HotStart Ready Mix (2×), perform a trial library amplification to determine the optimal cycle number (*see* **Notes 34–36**). Set up three reactions (for 8, 11, and 14 cycles) by mixing 2 μL of adapter-ligated Hi-C DNA (on beads), 10 μL 2× PCR mix, 2 μL 10× PCR primer mix, and 6 μL nuclease-free water. Run the following program.

    (a) 98 °C, 45 s.

    (b) [98 °C, 15 s; 60 °C, 30 s; 72 °C, 30 s] × 8, 11 or 14 cycles.

    (c) 72 °C, 5 min.

    (d) 12 °C, hold.

11. Transfer the PCR reaction to a low-bind 1.5 mL reaction tube, place on a magnetic stand, and transfer the supernatant to a fresh tube.

12. Add 30 μL nuclease-free water and 37 μL AMPure XP beads, mix and incubate for 15 min at room temperature on a rotating wheel.

13. Reclaim the beads on a magnetic stand and once the solution is cleared, remove the supernatant.

14. Leaving the tubes on the magnetic stand, add 500 μL of freshly prepared 80% ethanol, incubate for 30 s and remove supernatant. Repeat this step once.

15. Remove residual ethanol and air-dry the beads until no liquid is visible (*see* **Note 31**).

16. Elute the purified DNA by resuspending the beads in 20 μL of 10 mM Tris–HCl pH 7.5 and incubate for 5 min at room temperature. Transfer the supernatant to a fresh low-binding reaction tube.

17. Measure the DNA concentration with a Qubit fluorimeter and determine optimal cycle number allowing linear amplification.

18. Perform final amplification reaction by making up PCR reactions of 10 μL bead-bound Hi-C DNA, 25 μL 2× PCR mix, 5 μL 10× primer mix and 10 μL nuclease-free water per reaction. Run the thermal cycle from **step 10**, using the appropriate cycle number.

19. Transfer the PCR reaction to a low-bind 1.5 mL reaction tube, place on a magnetic stand, and transfer the supernatant to a fresh tube.

20. Add 37 μL AMPure XP beads, mix and incubate for 15 min at room temperature on a rotating wheel.

21. Reclaim the beads on a magnetic stand and once the solution cleared, remove the supernatant.

22. Leaving the tubes on the magnetic stand, add 500 μL of freshly prepared 80% ethanol, incubate for 30 s and remove supernatant. Repeat this step once.

23. Remove residual ethanol and air-dry the beads until no liquid is visible.

24. Elute the purified library by resuspending the beads in 20 μL 10 mM Tris–HCl pH 7.5 and incubate for 5 min at room temperature.

25. Measure DNA concentration using a Qubit fluorimeter and analyze library fragment size distribution on a TapeStation using a D1000 ScreenTape (*see* **Note 37**) comparing it to the initial sonicated but not adapter-ligated aliquot from **step 1**.

26. Perform paired-end sequencing of the Hi-C library.

4   Notes

1. It is crucial to use fresh formaldehyde (not older than six months), as the cross-linking efficiency is rapidly lost with the age of formaldehyde (especially the number of times the bottle has been opened before). Single-use glass vials are the best choice to ensure the quality.

2. This buffer contains fewer toxic and "sticky" compounds and is thus more practical to use. However, if nuclear integrity is a problem, consider replacing with 2% formaldehyde in NIB.

3. Make sure that the mortar has a smooth surface. With age and repeated use mortars tend to have a rather rough surface, which is suboptimal to thoroughly grind tissue to a fine powder.

4. The chosen restriction enzyme must efficiently digest fixed chromatin. The prevalence of its recognition site will determine the maximal theoretical resolution of the Hi-C experiment. As a six-cutter (lower theoretical resolution, but fewer sequences required for equivalent coverage due to lower relative complexity of Hi-C material), HindIII often displays evenly dispersed restriction sites, can be purchased in large quantities at moderate price, and shows high digestion efficiency. For similar reasons, DpnII represents a bona fide four-cutter restriction enzyme (higher resolution; be aware of the inhibition by adenosine methylation). Be sure to perform digestions in the compatible buffer and at the correct temperature for the enzyme.

5. Depending on the sticky end produced, another biotinylated nucleotide (e.g., biotin-14-dATP) may be preferred. The rationale is that the biotinylated nucleotide should be moved further inside the overhang and not be the outermost nucleotide, as the biotin tag at the outermost position may inhibit blunt-end ligation [7]. Also, be aware of the type of overhang produced by the restriction enzyme (5' or 3' overhang).

6. The correct choice of nucleotides added depends on the initial restriction enzyme used. The nucleotides should be complementary to the part of the overhang inside of the biotin tag. It is important to add a low amount of nucleotides and not to provide the full set of complementary nucleotides to prevent complete refilling of the 5' overhang.

7. Try to minimize the time between tissue collection and cross-linking to avoid nuclei degradation.

8. In our experience, cross-linking efficiency can be very variable and depends on the type of tissue used. Thus, when later comparisons between Hi-C samples are intended, it is crucial to keep that in mind. It is difficult to readily assess cross-linking efficiency in the wet lab; however, during later Hi-C data processing, the ratio between intrachromosomal and interchromosomal contacts may constitute a valid proxy for cross-linking efficiency. Generally, higher inter- to intrachromosomal contact ratios indicate higher cross-linking efficiency. It is extremely difficult to determine an ideal cross--linking level that leads to cross-linking of all relevant chromosomal contacts without generating false positive contacts. The experimenter can modify the level of cross-linking by the formaldehyde concentration

and incubation time. In our hands, the latter has a greater influence. Whereas the determination of ideal cross-linking levels is not an exact science, it crucial for comparative studies to minimize this important bias: always try to cross-link samples for comparison simultaneously and try to use a uniform tissue type across the samples.

9. Be aware that residual liquid may be problematic for the subsequent grinding of the tissue, as it can form ice crystals.

10. Tissue grinding may well be the most important step of the protocol concerning final Hi-C DNA yield. Thus, the necessary efforts should be put into the grinding to reach a very fine powder. Ideally no single clumps can be visible anymore. Also, be sure that the tissue never thaws during the grinding step. Be aware that the time needed for thorough grinding can vary significantly, depending on the nature of the tissue and the number of samples. When several samples (≥ 6) are processed in parallel, grinding may take up to 3 h or more.

11. The incubation time of 30 min is sufficient, however, if experimental planning allows, even longer incubation is recommended. Longer incubation helps to disintegrate the tissue and, thus, facilitates later steps of the protocol. Final low Hi-C sample yield is often caused by inefficient initial nuclei isolation. Thus, prolonged incubation of the powder in NIB is advisable in case problems with yield and purity occur.

12. In case the pestle of the Dounce homogenizer cannot be moved, the tissue suspension may be further diluted using NIB. For practical reasons in further steps of the protocol, a total of 30 mL NIB can be used.

13. Filtration may be tedious, as the Miracloth is likely to clog when either too much tissue has been used as input or the tissue has not been ground sufficiently. Before removing the suspension from a clogged Miracloth, one can try to declog by using the upper end of a P1000 pipette tip and gently scraping along the Miracloth.

14. Nuclei are fragile. Be sure to resuspend them throughout the protocol with great care. If resuspension by pipetting up and down is necessary, cut the tip and pipet up-and-down very slowly.

15. It is recommended to split the material at this stage into 2–4 subsamples (resuspend pellet in the appropriate volume of NIB to make the required number of 1 mL aliquots. At a later stage of the protocol, digestion and ligation efficiency of the subsamples will be assessed and only subsamples that reach quality control criteria are further processed. With little experience, only one out of four subsamples may be of sufficient quality.

In experienced hands, most subsamples will fulfill the criteria; thus, less subsamples can be generated to guarantee a successful experiment.

16. The nuclear pellet should be white at this stage. Green or brown color indicates that chloroplasts or other cellular debris are present and requires further nuclear isolation by resuspending in 1 mL NIB, recentrifugation, and removal of supernatant. The exact number of NIB washing steps may vary between experiments; typically, two to four NIB washing steps are required to reach a white or pale green nuclei pellet.

17. In plants, starch granules copelleting with the nuclei may constitute a major contaminant and may inhibit later enzymatic reactions. As previously proposed [6], if applicable concerning the experimental setup, plants can be grown several days in the dark before the experiment, which will significantly reduce the amount of starch in the plant.

18. A frequent mistake is to use too much input material. A final yield of some hundreds of ng of final Hi-C DNA is completely sufficient to produce informative Hi-C sequencing libraries. As a rule of thumb, a pellet greater than 50 μL may be too abundant in contaminants inhibiting the enzymatic reactions. In this case, it is advisable to further split the (sub)sample.

19. At this stage nuclei start to aggregate and form flakes, which can be observed when the reaction tube is held against the light.

20. It is not recommended to incubate at higher temperatures or for a longer period, as this may lead to decrosslinking of the sample and hence loss of contacts.

21. Check that the Klenow reaction is efficient in the restriction enzyme buffer. If it is not, pellet the digested nuclei by centrifugation for 1 min at $10,000 \times g$, room temperature, discard the supernatant, and resuspend the nuclei in 540 μL appropriate buffer (e.g., $1\times$ NEB CutSmart) before adding the fill-in mix. Make sure the fill-in mix is also made in this appropriate buffer, rather than the restriction enzyme buffer.

22. Previous protocols suggest incubation times of up to 4 h [7]. Whereas prolonged incubation likely increases the efficiency of the fill-in reaction, it may not be practical, as it will result in a very long working day. In our hands, 1.5 h incubation can result in fill-in efficiency of approx. 90%.

23. High ATP concentration inhibits the blunt-end ligation reaction. Furthermore, T4 DNA ligase is also inhibited by high NaCl concentration (>200 mM), thus depending on the buffer composition already present after the fill-in reaction (which depends on the restriction enzyme used), the amount of ligation buffer must be adjusted accordingly.

24. If time is a crucial issue, the samples can be directly incubated at 65 °C without prior incubation at 37 °C.

25. Decrosslinking is a crucial step in any Hi-C protocol. Inefficient decrosslinking leads to significant losses in yield; thus, it is important to give the decrosslinking reaction sufficient time. The incubation time in the morning can also be significantly extended (up to 5 h), which may help if low final yield is a problem.

26. It is best to shake by hand while keeping a finger on the lid.

27. It is important to dilute the Hi-C samples before precipitation, as the DTT present in the ligation buffer can inhibit precipitation.

28. After the 70% ethanol wash, the pellet may get loose (even after prolonged centrifugation). Do not overdry the pellet. When no movements of liquid can be observed by flicking the tube, the pellet can be considered sufficiently dry.

29. Qubit fluorometers give the most reliable results. Spectrometers, such as Nanodrop are not recommended in plant tissues, as impurities stemming from plant secondary metabolites can significantly distort the signal. If no Qubit is available, it is rather recommended to estimate the DNA concentration using an agarose gel and DNA samples of known concentration (e.g., DNA ladders).

30. This assumes a SRPI beads to sample ratio of 1.8, which for AMPure XP beads does not lead to size selection. Be aware that SPRI beads from other sources may have different ratios to prevent size selection.

31. Be aware that ethanol evaporates quickly, and that overdrying of the beads may make it difficult to elute the bound DNA. To assess whether any liquid remains, it is recommended to flick tubes and check whether small splashes are formed on the inner wall of the reaction tube. As soon as the layer of SPRI beads start to form cracks, immediately proceed to elution to prevent overdrying.

32. Hi-C sequencing libraries can be generated with a wide range of library preparation protocols, however, these protocols need be customized to allow the enrichment of biotin-labeled fragments. Typically, after sonication-based fragmentation, the libraries are prepared while the biotinylated fragments are bound to streptavidin beads. Thus, end-repair, adenylation, adapter ligation, and library amplification are performed on beads, which may make library preparation less tedious, as the streptavidin beads can play a role similar to SPRI beads to perform washes in-between the major library preparation

steps. As biotin-streptavidin binding cannot easily be reversed, the final free adapter-ligated library molecules are generated during the amplification step. The original biotin-tagged and streptavidin bound Hi-C fragments are discarded after library amplification and non–biotin-labeled copies of these Hi-C fragment are retrieved from the supernatant of the on-bead library amplification reaction. Here we describe a specific protocol, which is based on using the KAPA Hyper Prep Kit (Roche). Please note that all reagents used during library preparation stem from this kit and are not identical with reagents (e.g., DNA ligase) to those used in previous steps of the Hi-C protocol.

33. Other sonication devices, such as Bioruptor are also suitable to reach satisfactory fragmentation. Ideally, Hi-C DNA is fragmented to a size of ~300 bp.

34. The uncertainty about the relevant amount of input material also makes it difficult to define the correct number of cycles for amplification. Typically, the number of cycles must be increased considerably (>10 cycles) to generate sufficient library amounts. It is recommended to try different cycle numbers before the entire library is amplified.

35. Determining the correct cycle number may also work using qPCR, however, as the template is bound to magnetic beads, this may interfere with the detection system of the qPCR machine.

36. Typically, 12 cycles of amplification generate sufficient product and still provides linear amplification.

37. Concentrations above 10 μg/μL are sufficient for sequencing. Due to the adapter ligation, the fragment size distribution of the final library should shift up by approximately 120 bp compared to the initial preligation aliquot.

Acknowledgments

SG thanks Ueli Grossniklaus and the University of Zurich for financial support. Current research of SG is furthermore supported by a project grant by the Swiss National Science Foundation (Project number 310030_200704/1).

References

1. Lieberman-Aiden E, van Berkum NL, Williams L et al (2009) Comprehensive mapping of long-range interactions reveals folding principles of the human genome. Science 326(5950): 289–293

2. Kaplan N, Dekker J (2013) High-throughput genome scaffolding from in vivo DNA interaction frequency. Nat Biotechnol 31(12): 1143–1147

3. Selveraj S, Dixon JR, Bansal V, Ren B (2013) Whole-genome haplotype reconstruction using proximity-ligation and shotgun sequencing. Nat Biotechnol 31(12):111–118

4. Burton JN, Adey A, Patwhardan RP et al (2013) Chromosome-scale scaffolding of de novo genome assemblies based on chromatin interactions. Nat Biotechnol 31(12):1119–1125

5. Grob S, Grossniklaus U (2013) Chromatin conformation capture-based analysis of nuclear architecture. In: Kovalchuk I (ed) Plant epigenetics: methods and protocols. Boston, MA, Springer US p, pp 15–32

6. Liu C (2017) In situ Hi-C library preparation for plants to study their three-dimensional chromatin interactions on a genome-wide scale. In: Kaufmann K, Mueller-Roeber B (eds) Plant gene regulatory networks: methods and protocols. New York, NY, Springer New York, pp 155–166

7. Belaghzal H, Dekker J, Gibcus JH (2017) Hi-C 2.0: an optimized Hi-C procedure for high-resolution genome-wide mapping of chromosome conformation. Methods 123:56–65

Chapter 4

Mapping Mammalian 3D Genomes by Micro-C

Elena Slobodyanyuk, Claudia Cattoglio, and Tsung-Han S. Hsieh

Abstract

3D genome mapping aims at connecting the physics of chromatin folding to the underlying biological events, and applications of various chromosomal conformation capture (3C) assays continue to discover critical roles of genome folding in regulating nuclear functions. To interrogate the full spectrum of chromatin folding ranging from the level of nucleosomes to full chromosomes in mammals, we developed an enhanced 3C-based method called Micro-C. The protocol employs Micrococcal nuclease (MNase) to fragment the genome, which overcomes the resolution limit of restriction enzyme-based methods, enabling the estimation of contact frequencies between proximal nucleosomes. Such improvements successfully resolve the fine-scale level of chromatin folding, including enhancer–promoter or promoter–promoter interactions, genic and nucleosomal folding, and boost the signal-to-noise ratio in detecting loops and substructures underlying TADs. In this chapter, we will thoroughly discuss the details of the Micro-C protocol and critical parameters to consider for generating high-quality Micro-C maps.

Key words Micro-C, Chromosomal conformation capture (3C), Hi-C, 3D genome, Loop, TAD, Enhancer–promoter interaction

1 Introduction

Mapping 3D genome organization progresses our understanding of the role of physical chromatin interactions in the context of a functional genome, including the processes of transcription, DNA replication, and DNA repair. Various genome-wide 3D mapping approaches such as Hi-C (Chapter 3), Capture-C (Chapter 5), and HiChIP (Chapter 7) have tackled a variety of biological questions and unveiled many vital links between genome organization and nuclear functions [1]. For example, two levels of chromosome conformation—compartmentalization and local chromatin folding—greatly contribute to the 3D genome structure of interphase chromosomes and are typically superimposed on each other in Hi-C contact maps, due to heterogeneity in cell populations or

Tom Sexton (ed.), *Spatial Genome Organization: Methods and Protocols*,
Methods in Molecular Biology, vol. 2532, https://doi.org/10.1007/978-1-0716-2497-5_4,
© The Author(s), under exclusive license to Springer Science+Business Media, LLC, part of Springer Nature 2022

distinct folding processes [2]. A and B compartments, or compartmental domains, correspond to active and inactive chromatin segments, respectively, and appear as a plaid-like pattern in Hi-C contact maps. Transcriptional activity and perhaps the chromatin states in the milieu are likely to be involved in compartment segregation. On the other hand, the loop-extrusion model suggests that the ATP-dependent SMC complex, cohesin, mediates the formation of local chromatin structures, including topologically associating domains (TADs) and loops [3] . Cohesin encloses chromatin loci and extrudes chromatin until paused by CTCF. The extruding chromatin spatially organizes into a self-interacting domain, and the CTCF binding site constitutes the domain boundary insulating interdomain interactions. Stabilization of cohesin at CTCF sites sometimes gives rise to sharp corner peaks in contact maps, which are referred to as loops or loop domains. Importantly, mounting evidence suggests that TADs and loops are functionally linked to stem cell differentiation [4], V(D)J recombination [5], and many developmental processes [6].

Various 3D genome mapping methods differ in their resolution (nucleosome- or restriction-size fragments), genome coverage (all-by-all or point-by-point), and the involvement of a target enrichment step (protein or DNA/RNA pull-down), which determine their detection power for different levels of chromatin structures. Which method to choose depends on the experimental design and the scope of 3D structures of interest. We summarize the key features of these methods and their ideal applications in Table 1.

In the Hi-C protocol [7], chromatin is first cross-linked by formaldehyde and subsequently fragmented with a 4- or 6-cutter restriction enzyme. Religation of fragments yields chimeric DNA readouts in the Hi-C data that capture genomic loci that were crosslinked to one another in vivo. While many factors such as sequencing depth and library complexity impact the effective resolution of a Hi-C data set, a fundamental limit to genomic resolution is the size of the fragments generated before physical interactions are captured via ligation. We developed an enhanced Hi-C method, called Micro-C [8], that interrogates chromosome folding at lengths from single nucleosome to the full genome, which successfully uncovered the fine scale of chromatin structures in budding and fission yeast [9–12], as well as in mouse [13, 14] and human cells [15]. Here we highlight the key aspects of the Micro-C protocol that enable the identification of fine-scale chromatin features, namely, (1) Micro-C uses micrococcal nuclease (MNase) instead of restriction enzymes to fragment the genome into mononucleosomes, which allows the capture of nucleosome and protein interactions and nucleosome-resolution analysis of the chromosomes;

Table 1
Choice of chromosomal conformation capture assay (ligation-based)

	Micro-C (This Chapter)	Hi-C (Chapter 3)	4C (Chapter 2)	5C	Capture-C (Chapter 5)	HiChIP/PLAC-seq/ ChIA-PET (Chapter 7)
Resolution	Single nucleosome (~150 bp)	Depends on restriction enzyme	Depends on restriction enzyme	Depends on restriction enzyme	Depends on restriction enzyme and probe density	Depends on restriction enzyme and antibody
Readout	All-by-All	All-by-All	One-by-All	Multiple-by-Multiple	Targets-by-All	Peaks-by-Peaks
Compartment or compartmental domain	+++	+++	−	+++	−	−
TAD	+++	+++	−	+++	−	−
Nested structures within TAD	+++	+/−	−	++	−	−
Loop/Loop domain (anchored by CTCF/cohesin)	+++	+	++	++	+++	+++
E-P link	+++	+/−	++	++	+++	+++
Gene folding	+++	−	−	−	−	−
Nucleosomal folding	+++	−	−	−	−	−

(2) A dual-cross-linking protocol combining formaldehyde (~2 Å) and the protein-protein crosslinker DSG (disuccinimidyl glutarate; 7.7 Å) or EGS (ethylene glycol bis(succinimidyl succinate); 16.1 Å) substantially increases the signal-to-noise ratio to unmask previously indiscernible structures, such as chromosomally interacting domains (CIDs) in yeast [8] and highly abundant chromatin loops in mammals [14, 15]. The higher resolution of Micro-C improves the capture of the enhancer–promoter connectome that is intimately linked to cell type-specific transcriptional regulation [14].

In this chapter, we will focus on the Micro-C protocol, including: (1) Preparation of cross-linked chromatin from cell culture; (2) Micrococcal nuclease digestion and titration; (3) DNA fragment end-repair; (4) Proximity ligation, removal of unligated ends, and reverse cross-linking; (5) Di-nucleosomal DNA purification; and (6) Micro-C library preparation. Finally, we will briefly introduce the computational analysis of the generated Micro-C data sets.

2 Materials

1. Low-retention tubes and pipette tips (*see* **Note 1**).

2.1 Chromatin Cross-Linking

1. Cell culture medium (without fetal bovine serum), according to cell line used.

2. Trypsin, if using adherent cells, according to cell line used.

3. Formaldehyde cross-linking medium: 1% formaldehyde (FA) in cell culture medium (*see* **Notes 2** and **3**).

4. 1 M and 10 mM Tris–HCl pH 7.5.

5. 1× phosphate-buffered saline (PBS).

6. DSG cross-linking solution: 3 mM disuccinimidyl glutamate (DSG; from 300 mM stock solution in DMSO) in 1× PBS (*see* **Notes 2** and **3**).

2.2 Micrococcal Nuclease Digestion and Titration

1. Micro-C Buffer #1 (MB#1): 10 mM Tris–HCl pH 7.5, 50 mM NaCl, 5 mM $MgCl_2$, 1 mM $CaCl_2$, 0.2% (v/v) NP-40, 1× Protease Inhibitor Cocktail (Sigma) (*see* **Note 4**).

2. MNase solution: 20 units/µL micrococcal nuclease stock (Worthington Biochem) in 10 mM Tris–HCl pH 7.5. Stored at −80 °C in 20 µL aliquots; avoid repeated freeze–thaw cycles (*see* **Note 5**). Just before use, thaw an aliquot and prepare a diluted (1 unit/µL) solution in 10 mM Tris–HCl pH 7.5.

3. 500 mM EGTA (*see* **Note 6**).

4. 1× TE buffer: 10 mM Tris–HCl pH 7.5, 1 mM EDTA.

5. Reverse cross-linking solution: 0.5 mg/mL Proteinase K (from a 40 mg/mL stock in 40% glycerol, 1× TE buffer), 1% (w/v) SDS, 0.1 mg/mL RNase A in 1× TE buffer.

6. Phenol–chloroform–isoamyl alcohol (25:24:1, v/v).

7. 100% and 80% ethanol.

8. 3 M sodium acetate pH 5.5.

9. Zymoclean DNA Clean & Concentrator Kit.

10. NanoDrop spectrophotometer.

11. Micro-C Buffer #2 (MB#2): 10 mM Tris–HCl pH 7.5, 50 mM NaCl, 10 mM $MgCl_2$. Filtered through a 0.22 µm filter.

2.3 DNA Fragment End Repair

1. End-chewing master mix: 1.2× NEBuffer 2.1, 2.4 mM ATP, 5.9 mM DTT.

2. 10 units/µL T4 polynucleotide kinase (PNK) (NEB).

3. 5 units/µL DNA polymerase I Klenow fragment (NEB).

4. End-labeling master mix: 1× T4 DNA ligase buffer (NEB), 0.2 mM biotin-14-dATP, 0.2 mM biotin-11-dCTP, 0.2 mM dTTP, 0.2 mM dGTP, 0.1 mg/mL bovine serum albumin (BSA).

5. 500 mM EDTA.

6. Micro-C Buffer #3 (MB#3): 50 mM Tris–HCl pH 7.5, 10 mM MgCl$_2$. Filtered through a 0.22 μm filter.

2.4 Proximity Ligation, Removal of Unligated Ends, and Reverse Cross-Linking

1. Ligation master mix: 1× T4 DNA ligase buffer (NEB), 0.1 mg/mL BSA, 20 units/μL T4 DNA ligase (NEB).

2. Exonuclease III master mix: 1× NEBuffer 1 (NEB), 5 units/μL exonuclease III (NEB).

2.5 Dinucleosomal DNA Purification

1. 3% TBE NuSieve (Lonza) agarose gel (*see* **Note 7**).

2. Zymoclean Gel DNA Recovery Kit.

3. Qubit dsDNA HS Assay Kit (ThermoFisher).

2.6 Library Preparation

1. End polishing master mix, made from components of the End-It DNA End-Repair Kit (Lucigen): 3.125× End-It buffer, 3.125 mM ATP, 781.25 μM each dNTP, 1/16 volume End-It enzyme mix (i.e., 0.5 μL per 8 μL added to one reaction).

2. Dynabeads MyOne Streptavidin C1 (ThermoFisher).

3. 2× BW buffer: 10 mM Tris–HCl pH 7.5, 2 M NaCl, 1 mM EDTA. Also prepare a 1× stock. Both buffers are filtered through a 0.22 μm filter.

4. 1× TBW buffer: 1× BW, 0.1% (v/v) Tween-20.

5. Magnetic stand for 1.5 mL microcentrifuge tubes.

6. NEBNext Ultra II DNA Library Prep Kit for Illumina, including the End Prep enzyme mix, Illumina adapters and primers, ligation master mix and enhancer, USER enzyme and Q5 master mix.

7. End Prep master mix. Per 30 μL reaction: 3.5 μL End Prep reaction buffer, 1.5 μL End Prep enzyme mix, 25 μL nuclease-free water.

8. AMPure XP beads.

9. Next-generation sequencer (Illumina).

3 Methods

Carry out all procedures on ice, unless otherwise specified. Use low-retention tubes and pipette tips for all samples.

3.1 Preparation of Cross-Linked Chromatin from Cell Culture

1. Culture cells in the recommended conditions (*see* **Note 8**).

2. Trypsinize cells if needed and count cells after inactivating trypsin with media. Harvest cells by centrifugation for 5 min at 850 × *g* at room temperature (*see* **Note 9**).

3. Resuspend cells in formaldehyde cross-linking medium at a concentration of 1×10^6 cells/mL, for a maximum of 30 mL cross-linking medium in a 50 mL tube. First resuspend the pellet using a 1 mL pipette tip, and then add the rest of the media with a larger pipette without touching the cells. Incubate for 10 min at room temperature while nutating (*see* **Note 10**).

4. Add 1 M Tris–HCl pH 7.5 to a final concentration of 375 mM to quench the reaction. For example, add 18 mL 1 M Tris–HCl to a 30 mL sample. Incubate for 5 min at room temperature. Centrifuge for 5 min at $850 \times g$ at 4 °C. Aspirate the supernatant.

5. Wash cells twice with cold 1× PBS at a concentration of 1×10^6 cells/mL. For each wash, centrifuge for 5 min at $850 \times g$ at 4 °C and aspirate the supernatant.

6. Resuspend the cell pellet in the DSG cross-linking solution at a concentration of 1×10^6 cells/mL. First resuspend the pellet using a 1 mL pipette tip, and then add the rest of the medium with a larger pipette without touching the cells. Incubate for 45 min at room temperature while nutating (*see* **Note 11**).

7. Add 1 M Tris–HCl pH 7.5 to a final concentration of 375 mM to quench the reaction. For example, add 18 mL 1 M Tris–HCl to a 30 mL sample. Incubate for 5 min at room temperature. Centrifuge for 5 min at $850 \times g$ at 4 °C. Aspirate the supernatant.

8. Wash cells with cold 1× PBS at a concentration of 1×10^6 cells/mL. For each wash, centrifuge for 5 min at $850 \times g$ at 4 °C and aspirate the supernatant.

9. Resuspend the cell pellet in the appropriate volume of 1× PBS to aliquot 1 mL of sample into multiple tubes, at a concentration of 1×10^6 cells/mL (*see* **Note 12**). Centrifuge for 5 min at $850 \times g$ at 4 °C and aspirate the supernatant. Snap freeze cell pellets in liquid nitrogen or proceed to the MNase titration. Store frozen pellets at −80 °C (*see* **Note 13**).

3.2 Micrococcal Nuclease Titration (See Note 14)

1. Thaw a cell pellet of 1×10^6 cells and resuspend in complete MB#1 at a concentration of 1×10^6 cells/100 μL. Aliquot 100 μL of cells to a 1.5 mL Eppendorf tube. Incubate for 20 min on ice. Centrifuge for 5 min at $10,000 \times g$ at 4 °C. Discard the supernatant using a pipette tip, leaving 10–20 μL behind.

2. Wash the nuclear pellet in complete MB#1 at a concentration of 1×10^6 cells/100 μL. Resuspend the pellet by pipetting up and down. Centrifuge for 5 min at $10,000 \times g$ at 4 °C. Discard the supernatant using a pipette tip. Try to remove as much liquid as possible without disturbing the pellet.

3. Resuspend the nuclear pellet in complete MB#1 at a concentration of 1×10^6 cells/100 μL. Resuspend the pellet by pipetting up and down.

4. Add increasing amounts of MNase as follows (*see* **Note 15**).

   (a) 5 units: 5 μL of the 1 unit/μL stock.

   (b) 10 units: 0.5 μL of the 20 units/μL stock.

   (c) 20 units: 1 μL of the 20 units/μL stock.

   (d) 40 units: 2 μL of the 20 units/μL stock.

   (e) 60 units: 3 μL of the 20 units/μL stock.

5. Briefly vortex and spin the tubes. Incubate for 10 min at 37 °C while shaking in a thermomixer at 850–1000 rpm.

6. Transfer the tubes back to ice and stop the reaction by adding 0.8 μL of 500 mM EGTA. Vortex and briefly spin the tubes. Incubate for 10 min at 65 °C. Centrifuge for 5 min at $12,000 \times g$ at 4 °C. Discard the supernatant.

7. Resuspend each pellet in 200 μL of reverse cross-linking solution by pipetting up and down. Incubate overnight at 65 °C.

8. Add 200 μL of phenol–chloroform–isoamyl alcohol and vortex for 20 s. Centrifuge for 15 min at $19,800 \times g$ at room temperature.

9. Transfer ~200 μL of the aqueous phase to a new 1.5 mL tube and proceed with DNA purification by either ethanol precipitation (**step 10**) or Zymoclean DNA Clean & Concentrator Kit (**step 11**) (*see* **Note 16**).

10. For DNA purification by ethanol precipitation, add 500 μL of room-temperature 100% ethanol and 20 μL of 3 M sodium acetate pH 5.5. Invert the tube to mix and precipitate nucleic acids at −80 °C for at least 1 h. Centrifuge for 15 min at $19,800 \times g$ at 4 °C. Discard the ethanol using a pipette and wash the pellet with 1 mL of 80% room-temperature ethanol. Centrifuge for 5 min at $19,800 \times g$ at 4 °C. Remove as much ethanol as possible using a pipette. Centrifuge the tube once again briefly and remove as much ethanol residue as possible. Air dry the pellet for 5 min at 37 °C with an open lid. Resuspend the pellet in 15 μL of Zymo Elution Buffer or 10 mM Tris–HCl pH 7.5. Proceed to **step 12**.

11. For DNA purification using the Zymoclean DNA Clean & Concentrator Kit, add 1 mL of DNA Binding Buffer and load onto the column. Wash twice with 400 μL of wash buffer. Elute with 15 μL of Zymo Elution Buffer.

12. Quantify the DNA by NanoDrop (*see* **Note 17**).

13. Check fragment sizes on a 2% regular agarose gel. The desired MNase   concentration   produces   80–90%   nucleosome

monomer: 20–10% dimer ratio. In mammals, the monomer size is ~100 to 175 bp and the dimer size is ~200 to 400 bp (Fig. 1a) (*see* **Notes 18** and **19**).

3.3 Digestion of Cross-Linked Chromatin with Micrococcal Nuclease

1. Resuspend the cell pellet in complete MB#1 at a concentration of 1×10^6 cells/100 µL. Incubate for 20 min on ice. Proceed with centrifugation, washing, and resuspension in complete MB#1 as in "Micrococcal nuclease titration" **steps #1–4**.

2. Add the appropriate amount of MNase, as determined by the MNase titration experiment on the same batch (*see* **Note 19**). Briefly vortex the tubes and incubate for 10 min at 37 °C while shaking in a thermomixer at 850–1000 rpm.

3. Transfer the tubes to ice. Add 0.8 µL of 500 mM EGTA to stop the reaction. Vortex briefly to mix and incubate for 10 min at 65 °C. Centrifuge for 5 min at 10,000 × *g* at 4 °C. Discard the supernatant using a pipette tip.

4. Wash the nuclear pellet in 1 mL of cold MB#2 twice. Centrifuge for 5 min at 10,000 × *g* at 4 °C. Discard the supernatant using a pipette tip. After the second wash, try to remove as much liquid as possible without disturbing the pellet.

3.4 DNA Fragment End Repair

1. Resuspend the nuclear pellet with 50 µL of end chewing master mix. Add 2.5 µL of 10 U/µL T4 PNK, mix well by finger flicking, briefly spin down, and incubate for 15 min at 37 °C while shaking in a thermomixer at 1000 rpm for an interval of 15 s every 3 min (*see* **Note 20**).

2. Add 5 µL of 5 U/µL Klenow fragment to the reaction, mix well by finger flicking, and briefly spin down. Incubate for 15 min at 37 °C while shaking in a thermomixer at 1000 rpm for an interval of 15 s every 3 min (*see* **Note 21**).

3. Add 25 µL of end labeling master mix, mix well by finger flicking, and briefly spin down. Incubate for 45 min at 25 °C while shaking in a thermomixer at 1000 rpm for an interval of 15 s every 3 min.

4. Add 9 µL of 500 mM EDTA, mix well by finger flicking, briefly spin down, and incubate at 65 °C for 20 min without shaking.

5. Centrifuge for 5 min at 10,000 × *g* at 4 °C. Discard the supernatant using a pipette tip. Rinse once with 1 mL of cold MB#3 by pipetting up and down. Centrifuge for 5 min at ~10,000 × *g* at 4 °C. Discard the supernatant using a pipette tip.

3.5 Proximity Ligation, Removal of Unligated Ends, and Reverse Cross-Linking

1. Resuspend the nuclear pellet with 250 µL of ligation master mix, mix well by finger flicking, and briefly spin down. Incubate for at least 2.5 h at 25 °C with slow rotation on an orbital shaker (*see* **Note 22**). Centrifuge for 5 min at 16,000 × *g* at 4 °C. Discard the supernatant using a pipette tip.

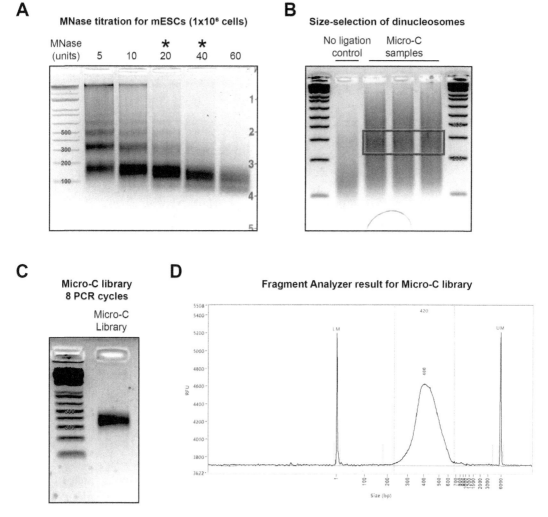

Fig. 1 (**a**) Example of an MNase titration for mouse embryonic stem cells. A million cells were digested by 5, 10, 20, 40, and 60 units of MNase separately for 10 min at 37 °C. After reverse cross-linking and DNA purification, the digestion pattern of DNA was analyzed by running the samples on a 2.5% agarose gel in TBE buffer. The asterisk highlights the optimal conditions for the Micro-C experiment (e.g., 20–40 units). The gel contains 500 ng of the 1-kb plus DNA marker. (**b**) Example of Micro-C samples. Micro-C samples (ligated) and a no-ligation control were analyzed by a 3% NuSieve GTG agarose gel in TBE buffer. The box highlights the size selection of dinucleosomes sized between 200–400 bp. The gel contains 500 ng of the 1-kb plus DNA marker. (**c**) Example of a Micro-C sequencing library. The Micro-C library was amplified with 8 PCR cycles and was subjected to DNA cleanup using the Zymoclean DNA Clean & Concentrator Kit. After pooling the samples with unique barcodes, the library was purified again using AMPure beads at a 0.9× ratio. The sample is ready for next-generation DNA sequencing. Here, to display the size distribution of the Micro-C library, we analyzed the Micro-C library with a 2% agarose gel in TBE buffer. The DNA band should run between 300 and 500 bp (i.e., 300 bp of dinucleosome + 110 bp of the Illumina adapters). There should be no detectable primer/adapter that runs around or lower than 100 bp. The gel contains 100 ng of the 1-kb plus DNA marker. (**d**) Example of a Fragment analyzer result. The Micro-C library shows a clear peak at ~400 bp

2. Resuspend the nuclear pellet with 100 μL of exonuclease III master mix, mix well by finger flicking, and briefly spin down. Incubate for 15 min at 37 °C while shaking at 1000 rpm for an interval of 15 s every 3 min (*see* **Note 23**).

3. Add 150 μL of reverse cross-linking solution and incubate at 65 °C overnight.

3.6 Dinucleosomal DNA Purification

1. Add 250 μL of phenol–chloroform–isoamyl alcohol to the sample. Vortex for 20 s. Centrifuge for 15 min at 19,800 × *g* at room temperature. Transfer the upper layer to a new tube and proceed with DNA purification by either ethanol precipitation (**step 2**) or using the Zymoclean DNA Clean & Concentrator Kit (**step 4**) (*see* **Note 16**).

2. For DNA purification by ethanol precipitation, add 625 μL of room-temperature 100% ethanol and 25 μL of 3 M sodium acetate pH 5.5. Invert the tube to mix and precipitate nucleic acids at −80 °C for at least 1 h. Centrifuge for 15 min at 19,800 × *g* at 4 °C. Discard the ethanol using a pipette and wash the pellet with 1 mL of 80% room-temperature ethanol. Centrifuge for 5 min at 19,800 × *g* at 4 °C. Remove as much ethanol as possible using a pipette. Centrifuge the tube once again briefly and remove as much ethanol residue as possible. Air dry the pellet for 5 min at 37 °C with an open lid. Add 50 μL of 1× TE and incubate at 37 °C for 30 min to dissolve the pellet.

3. Purify DNA again using the Zymoclean DNA Clean & Concentrator Kit (*see* **Note 24**). Add 300 μL of DNA Binding Buffer to 50 μL of DNA and load onto the column. Wash twice with 400 μL of wash buffer. Elute twice, each time with 15 μL of Zymo Elution Buffer for a total volume of 30 μL. Proceed to **step 5**.

4. For DNA purification by Zymoclean DNA Clean & Concentrator Kit alone, add 1.25 mL of DNA Binding Buffer and load onto the column. Wash twice with 400 μL of wash buffer. Elute twice, each time with 15 μL of Zymo Elution Buffer for a total volume of 30 μL.

5. Prepare a 3% TBE NuSieve GTG agarose gel (*see* **Note 7**). Separate nucleosome monomers from dimers by loading two large wells with 20 μL in each well (*see* **Notes 18** and **19**). Cut the bands sized between 250 and 400 bp (Fig. 1b). Avoid cutting below 200 bp to eliminate nucleosome monomers.

6. Use the Zymoclean Gel DNA Recovery Kit to extract DNA from the cut bands. Add 3× volume/weight of Agarose Dissolving Buffer to the cut gel and incubate at 50 °C for 10 min until the gel is completely dissolved. Load onto the column and

wash twice with 400 μL of wash buffer. Elute twice, each time with 9 μL of Elution Buffer for a total volume of 18 μL (*see* **Note 25**).

7. Quantify DNA using the Qubit dsDNA HS assay. There should be at least 50–200 ng of DNA for a high-coverage result (*see* **Note 26**).

3.7 Library Preparation

1. Add 8 μL of end polishing master mix to 17 μL of input DNA. Incubate for 45 min at 25 °C without shaking. Inactivate the enzyme mix by incubating for 10 min at 70 °C.

2. Wash 5 μL of streptavidin beads with 1 mL of 1× TBW (*see* **Note 27**). Mix on a nutator for 2 min, then place beads on a magnetic stand for 30 s. Remove the supernatant with a pipette tip. Resuspend beads in 150 μL of 2× BW. Add 125 μL of water to the DNA sample to bring the volume to 150 μL. Transfer an equal volume (150 μL) of the prewashed streptavidin beads to the sample. Mix the DNA and beads on a nutator for 20 min at room temperature. Centrifuge briefly in a microfuge and add 950 μL of 1× TBW. Invert the tube multiple times to resuspend the beads and incubate for 5 min at 55 °C while shaking at 1200 rpm. Centrifuge briefly in a microfuge and place the tube on the magnetic stand for 30 s. Remove the supernatant and repeat the 1× TBW wash. Centrifuge briefly in a microfuge and place the tube on the magnetic stand for 30 s. Remove the supernatant. Rinse the beads with 500 μL of 10 mM Tris–HCl pH 7.5 without disturbing the beads (*see* **Note 28**). Discard the supernatant and transfer the tubes to ice.

3. Add 30 μL of End Prep master mix and pipette up and down or vortex to resuspend the beads without creating bubbles. Incubate for 30 min at 20 °C while shaking at 1000 rpm for an interval of 15 s every 3 min. Incubate for 30 min at 65 °C to inactivate the enzymes. Transfer samples to ice.

4. Add 3.5 μL of NEB Illumina adapter, 15 μL of Ligation Master Mix, and 0.5 μL of Ligation enhancer sequentially to 30 μL of the sample from the previous step for a final volume of 49 μL. Vortex briefly and incubate for 30 min at 20 °C while shaking at 1000 rpm for an interval of 15 s every 3 min. Add 1.5 μL of the USER enzyme. Incubate for 15 min at 37 °C while shaking at 1000 rpm for an interval of 15 s every 3 min.

5. Add 950 μL of 1× TBW. Invert the tube multiple times to resuspend the beads and incubate for 3 min at 55 °C while shaking at 1000 rpm. Centrifuge briefly in a microfuge and place the tube on the magnetic stand for 30 s. Remove the supernatant and repeat the 1× TBW wash. Centrifuge briefly in a microfuge and place the tube on the magnetic stand for 30 s.

Remove the supernatant. Rinse the beads with 500 μL of 10 mM Tris–HCl pH 7.5 without disturbing the beads. Discard the supernatant. Resuspend the beads in 20 μL of Zymo Elution Buffer (*see* **Note 29**).

6. Take a 1 μL aliquot of beads and transfer to a 200 μL PCR tube. Sequentially add the following reagents on ice: 3.5 μL of nuclease-free water, 5 μL of 2× Q5 PCR Enzyme Mix, 0.25 μL of 10 μM PE1.0 primer, 0.25 μL of PE2.0 primer.

7. Run the following PCR setup:

   Denaturation: 98 °C, 30 s.

   12 cycles: 98 °C, 10 s; 65 °C, 1 min 15 s.

   Final extension: 65 °C, 5 min.

   Hold at 4 °C.

8. Check the library size and estimate library quantity by agarose gel (*see* **Note 30**). Image the gel and calculate the amount of DNA in the sample by using the ladder intensity for comparison. Adjust the number of PCR cycles to yield roughly 50–100 ng of product for sequencing.

9. Assemble full 50 μL PCR reactions by sequentially adding to 200 μL PCR tubes on ice: 19 μL of sample beads, 3.5 μL of nuclease-free water, 25 μL of 2× Q5 PCR Enzyme Mix, 1.25 μL of 10 μM universal primer, 1.25 μL of indexed primer (*see* **Note 31**).

10. Run the following PCR setup, with the minimal number of cycles (**N**) to yield 50–100 ng of PCR product:

    Denaturation: 98 °C, 30 s.

    N cycles: 98 °C, 10 s; 65 °C, 1 min 15 s.

    Final extension: 65 °C, 5 min.

    Hold at 4 °C.

11. Purify DNA with the Zymoclean DNA Clean & Concentrator Kit: Add 250 μL of DNA Binding Buffer to 50 μL of PCR sample. Transfer the tube to a magnetic stand and let the beads attach to the side of the tube. Load the supernatant onto the column. Wash twice with 400 μL of wash buffer. Elute twice, each time with 9 μL of Elution Buffer for a total volume of 18 μL.

12. Quantify the library using the Qubit dsDNA HS Assay Kit (*see* **Note 32**).

13. Pool the indexed samples; indexed samples are typically pooled at a 1:1 molar ratio. Consider using ~1 pmol of total DNA, splitting this amount between indexed samples. For example, you can pool 100 fmol DNA each from 10 samples, or 50 fmol DNA each from 20 samples. Bring the final volume to 50 μL with Zymo Elution Buffer.

14. Resuspend AMPure XP beads thoroughly before pipetting. Add 45 μL of beads (0.9×) to 50 μL of pooled samples. Pipette up and down until fully resuspended. Incubate for 10 min to allow DNA to bind to the beads. Place the tube on the magnetic stand for 5 min, until the solution is clear. Discard the supernatant without disturbing the beads.

15. Add 200 μL of freshly prepared 80% ethanol without disturbing the beads. Leave on the magnetic stand for 30 s. Discard the ethanol without disturbing the beads. Repeat the ethanol wash step. Remove residual ethanol using a pipette tip.

16. Air dry the beads for 5 min, which will avoid overdrying them. Remove the tube from the magnetic stand and add 25 μL of Zymo Elution Buffer for a final DNA concentration of 10–20 nM (i.e., assuming ~25–50% DNA loss during the AMPure cleanup). Thoroughly resuspend the beads by pipetting up and down ~10 times. Incubate for 2 min to allow the DNA to elute off of the beads. Place the tube on the magnetic stand until the solution is clear. Transfer the supernatant to a new 1.5-mL tube.

17. Check samples on an agarose gel or by Fragment Analyzer (Fig. 1c, d) and proceed with Illumina sequencing with the paired-end 50- or 100-bp kit.

3.8 Data Analysis

We use the Linux/MacOS command-line interface (CLI) to process Micro-C data. We recommend using Conda to create a virtual environment for Micro-C analysis.

3.8.1 Preprocessing Micro-C Contact Pairs

1. Install HiC-Pro v.2.11.4 from https://github.com/nservant/ HiC-Pro [16].

2. Set up the HiC-Pro configuration file. For example, we used the following settings for Micro-C data in mouse embryonic stem cells.

```
REFERENCE_GENOME = mm10
        BOWTIE2_GLOBAL_OPTIONS = --very-sensitive-local --
reorder --trim5 5 --trim3 5
      MIN_CIS_DIST = 200
      GET_ALL_INTERACTION_CLASSES = 1
      GET_PROCESS_SAM = 0
      RM_SINGLETON = 1
      RM_MULTI = 1
      RM_DUP = 1
      BIN_SIZE = 500 1000 2500 5000 10000
```

3. The output folder contains Micro-C mapping, pairing, and QC statistics, and all unique contacts at base-pair resolution and binned/normalized matrix files.

3.8.2 Convert Tab-Delimited Files to HDF5 (Cool) Format and Normalization

1. Install cooler (https://github.com/mirnylab/cooler) and cooltools (https://github.com/mirnylab/cooltools) packages [17].

2. Build a cool file from the allValidPairs file from the HiC-Pro output:

```
cooler cload pairs -c1 2 -p1 3 -c2 5 -p2 \
    mm10-bin-200bp-bed-file \
    allValidPairs-file \
    output-cool-file
```

3. Build cool files in a different resolution. For example, we are going to build a 1-kb data from 200-bp data:

```
cooler coarsen -k 5 -p 16 \
    input-200bp-cool-file \
    output-1kb-cool-file
```

4. Normalize the data using the matrix balancing method:

```
cooler balance -p 16 input-cool-file
```

Note that an optimal matrix balancing result may require fine-tuning with the 'MAD-MAX' function. An optimal normalization result is expected to have minimal background noise that could be artificially introduced during the process of matrix balancing.

5. Compute the expected matrix:

```
cooltools compute-expected -p 16 \
    input-cool-files > output-expected-file
```

6. Build multiple-resolution cool files for the HiGlass browser [18]:

```
cooler zoomify -p 16 -balance \
    input-cool-files \
    -o output-mcool-file
```

7. Load multiple-resolution cool files to the HiGlass server by following the tutorial available at https://docs.higlass.io/tutorial.html. Here is an example of Micro-C results (Fig. 2). Publicly available Micro-C data sets are summarized in Table 2.

Example of mESC Micro-C maps on the HiGlass browser

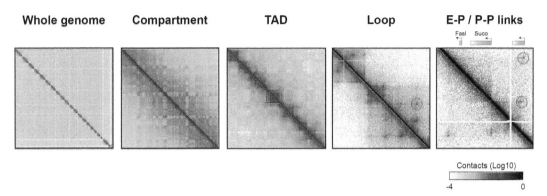

Fig. 2 Example of Micro-C maps on the HiGlass browser. From left to right, the Micro-C maps were zoomed-in from the whole genome scale to a 200-kb region in chr1 with increasing resolution. 3D genome features such as compartments, TADs, loops, and E-P/P-P links are clearly discernible on the maps

4 Notes

1. The quality of low-retention tubes and pipette tips varies with different manufacturers. We tested various brands and found that the 15 mL and 50 mL conical tubes by Eppendorf and the low-retention tips by Rainin perform robustly and consistently to reduce sample loss during Micro-C experiments. Using high-quality supplies is especially important if the amount of input is low, for example, <100,000 cells.

2. We tested a variety of cross-linking conditions in budding yeast [9] and mouse ES cells (Fig. 3), and identified a combination of protein-protein crosslinker DSG (disuccinimidyl glutarate, 7.7 Å) or EGS (ethylene glycol bis(succinimidyl succinate), 16.1 Å) with DNA-protein crosslinker FA (formaldehyde, weaker protein-protein crosslinker) that produces the highest signal-to-noise ratio in Micro-C maps. The irreversible protein-protein crosslinker (DSG or EGS) also prevents loss of cross-linking during the harsher incubation steps (e.g., 65 °C). The usage of DSG or EGS and the order of treating cells with FA and either long crosslinker does not show a significant effect on the results at the scale of chromosome and compartment (Fig. 3). Thus, we choose the combination of FA and DSG as the standard cross-linking procedure for Micro-C.

3. Freshly prepared FA and DSG are critical to achieve the best cross-linking outcomes. We recommend using FA in the ampule packaging and preventing exposure of DSG powder to moisture.

Table 2
Publicly available Micro-C data sets in mammalian cells (last updated 2020/12/20)

Cell type	Genotype/ Treatment	Estimated unique reads	Data links	Lab	References
Mouse embryonic stem cell (JM8.N4)	Wild-type	~2.64 B	GSE1302754DNES14CNC1I	Tjian and Darzacq lab	Hsieh et al. (2020) Mol Cell [14]
Mouse embryonic stem cell (JM8.N4)	Triptolide	~500 M	GSE1302754DNESSY8C22T	Tjian and Darzacq lab	Hsieh et al. (2020) Mol Cell [14]
Mouse embryonic stem cell (JM8.N4)	Flavopiridol	~500 M	GSE1302754DNES7X5GQUR	Tjian and Darzacq lab	Hsieh et al. (2020) Mol Cell [14]
Mouse embryonic stem cell (JM8.N4)	Wild-type	~250 M	GSE126112	Tjian and Darzacq lab	Xie et al. (2020) Nat Methods [19]
Mouse embryonic stem cell (JM8.N4)	CTCF depletion	~225 M	GSE126112	Tjian and Darzacq lab	Xie et al. (2020) Nat Methods [19]
Mouse embryonic stem cell (JM8.N4)	Wild-type	~668 M	GSE123636	Tjian and Darzacq lab	Hansen et al. (2019) Mol Cell [13]
Mouse embryonic stem cell (JM8.N4)	CTCF with RBRi-del	~697 M	GSE123636	Tjian and Darzacq lab	Hansen et al. (2019) Mol Cell [13]
Human embryonic stem cell (H1)	Wild-type	~3.23 B	4DNES21D8SP8	Oliver Rando lab	Krietenstein et al. (2020) Mol Cell [15]
Human foreskin fibroblast (HFFc6)	Wild-type	~4.57 B	4DNESWST3UBH	Oliver Rando lab	Krietenstein et al. (2020) Mol Cell [15]
WTC derived iPS cells	Wild-type	–	4DNESODGV2V24DNESAGG7EUC	Job Deeker lab	–

Various cross-linking conditions show similar Micro-C signals

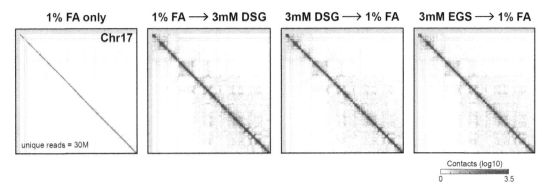

Fig. 3 Example of the effects of different cross-linking protocols. From left to right, cells were crosslinked by (1) 1% formaldehyde (FA) for 10 min at RT; (2) 1% FA for 10 min at RT and then subjected to 3 mM DSG for 40 min at RT after washes; (3) 3 mM DSG for 30 min at RT and then subjected to 1% FA directly in the sample for an additional 10 min at RT; (4) 3 mM EGS for 30 min at RT and then subjected to 1% FA directly in the sample for an additional 10 min at RT. All dual cross-linking protocols produce comparable Micro-C maps at the chromosome and compartment levels regardless of the order of FA and long crosslinkers, while the FA only condition fails to recapitulate the known 3D genome features

4. Prepare a stock without NP-40 or protease inhibitors and filter through a 0.22 μm filter. Add NP-40 and protease inhibitors directly prior to use.

5. Upon first use of the MNase solution, store the aliquot at −20 °C and use another two times at most (maximum three uses per −20 °C aliquot).

6. EGTA is a stronger Ca^{2+} chelator than EDTA.

7. NuSieve GTG agarose is ideal for separating mono- and dinucleosomes, and it yields a higher DNA recovery rate with most gel extraction kits. Typically, a 2.5–3% TBE gel can achieve a clean separation between mono- and dinucleosomes. When making the NuSieve agarose gel, shake at 100 rpm for 10–20 min after adding TBE to the powder to help with dissolving, then microwave as usual.

8. This Micro-C protocol is also compatible with cells sorted from a cell sorter. For this purpose, we prepare ~1–1.5 × 10^7 cells expressing a fluorescent protein (e.g., GFP) growing on a 10 cm² culture dish. After cell suspension, PBS wash, cross-linking with 1% formaldehyde, and quenching (*see* Method Subheading 3.1, **steps 2–5**), we resuspend cells with cell medium or PBS and pass the cells through a 35-μm strainer into a sterile 5 mL round-bottom polystyrene FACS tube. We routinely collect one million cells. However, if the targeted population is not abundant, using as few as 100 thousand

cells also produces high-quality Micro-C results. After sorting, proceed to Subheading 3.1, **step 6**. We note that this protocol has not been tested with ex vivo tissues.

9. If your applications or cell lines are sensitive to trypsin treatment, cells can alternatively be crosslinked directly on the dish as follows.

    (a) Wash cells twice with PBS or base culture media (-FBS) at room temperature.

    (b) In the fume hood, add freshly prepared 1% FA crosslinking media (-FBS) and incubate for 10 min at RT while shaking at 50 rpm.

    (c) Add 1 M Tris–HCl pH 7.5 to a final concentration of 375 mM to quench the reaction.

    (d) Wash cells twice with PBS at room temperature.

    (e) Add freshly prepared 3 mM DSG solution and incubate for 40 min at room temperature while shaking at 50 rpm.

    (f) Add 1 M Tris–HCl pH 7.5 to a final concentration of 375 mM to quench the reaction.

    (g) Wash cells twice with PBS at room temperature.

    (h) Scrape off the crosslinked cells, transfer to a low-binding tube, and snap freeze in liquid nitrogen.

10. FBS substantially reduces cross-linking efficiency. Before performing formaldehyde and DSG cross-linking, any FBS residue should be removed by washing cells with PBS.

11. Avoid using glass or plastic surgical pipettes to resuspend cell pellets. Cells easily stick to the pipettes, which results in a drastic loss of cells.

12. We routinely perform Micro-C with 1×10^6 cells. However, in some circumstances or cell types, one may find significant cell loss after DSG cross-linking (i.e., cells stick on the tube or tip). In that case, we recommend starting with 5×10^6 cells. We note that in this protocol, all the reactions are specified for using 1×10^6 cells. If using 5×10^6 cells, we suggest scaling up the volume by 2 for all reactions starting from Method Subheading 3.2.

13. Pellets can be stored at -80 °C for several months.

14. Optimally, every batch of cross-linked cells needs to undergo MNase titration. If not a full titration, perform a "best-guess" digestion and adjust experimental conditions accordingly. For example, when we carry out a new Micro-C experiment with a relatively low amount of input cells that are not amenable to a full titration, we usually run a "best-guess" digestion, that is, using 20 units MNase for 1 M cells. We then adjust the

digestion conditions for the actual Micro-C experiment based on the result of the "best-guess" digestion. Typically, if tri- or tetranucleosomes are clearly visible (an indicator of underdigestion), we will increase the amount of MNase by 1.5×; in contrast, if nucleosome dimers are faint (or invisible) and the majority of DNA is smaller than 100 bp (an indicator of overdigestion), we will reduce the amount of MNase by half. The MNase titration experiment requires good practice to gain an idea of the outcomes of chromatin digestion upon changes in parameters such as the amount of MNase, incubation time, or temperature. If you have a limited availability of your cells of interest, we recommend practicing MNase titration using common culture cells with similar properties.

15. For mouse embryonic stem cells, the optimal amount of MNase is typically around 20 U for 1×10^6 cells.

16. DNA purification by either ethanol precipitation or the Zymoclean DNA Clean & Concentration kit shows indistinguishable results regarding DNA yields and the pattern of the nucleosomal ladder. Purification with the Zymoclean kit saves ~2 h in the protocol.

17. The yield varies by cell type. Typically, one million mESCs with an optimal MNase digestion yields ~2.5 μg of total DNA. A DNA yield lower than 1 μg indicates overdigestion of chromatin. The majority of DNA fragments will run below 100 bp on an agarose gel. We strongly suggest not to use overdigested samples for Micro-C to avoid poor ligation efficiency.

18. For all gels in this protocol, use Orange G gel loading dye, which migrates at <75 bp, and so minimally interferes with the visualization of nucleosomal fragments.

19. A key factor that affects the quality of the Micro-C library is minimizing contamination from unligated dinucleosomes. Since dimers are relatively more abundant than ligated dinucleosomes, they could still remain at a significant amount after streptavidin purification. We strongly recommend using the digestion conditions that yield 90% monomers and 10% dimers. This reduces the ratio of dimer contamination in the sequencing library while retaining the length of nucleosome ends that are sufficient for ligation.

20. T4 PNK catalyzes the transfer of a phosphate group to the 5′ ends of DNA and removes phosphate groups from the 3′ ends of DNA. This reaction will convert the multiple types of MNase-digested ends to ligatable ends.

21. In the absence of dNTPs, the Klenow Fragment only possesses a 3′–5′ exonuclease activity, creating single-stranded DNA overhangs that can be labeled with biotin in the next step.

22. About 2 to 4 h of incubation for the ligase reaction is sufficient. In our hands, an incubation longer than 4 h is not necessary, as it does not lead to any noticeable improvement in the Micro-C library quality.

23. The exonuclease reaction does not chew off the entire nucleosomal DNA since the crosslinked nucleosome can block the digestion. Thus, extending the exonuclease reaction time from 15 min to 30 min sometimes reduces the ratio of un-ligated dimers in the library.

24. Purifying the ethanol-precipitated DNA with the Zymoclean DNA Clean & Concentrator Kit is critical to remove ethanol and salts that might affect DNA mobility on a gel.

25. The purification of dinucleosomes is the key step to eliminate unwanted contamination from mononucleosomes. The result is much cleaner by gel extraction than by two-sided AmpureXP purification. Using bead-based DNA purification is preferable for lower amounts of DNA.

26. At this point, you may freeze your sample at -20 °C and proceed the following day.

27. The streptavidin beads have a very high capacity, so 5 μL will be more than enough even when starting from 5×10^6 cells.

28. Remove the 10 mM Tris–HCl only once the End Repair mix for the next step is ready. Do not let the beads dry out.

29. At this point, you may freeze your sample at -20 °C and proceed the following day.

30. There is no need to remove streptavidin beads before this step, as they will stay in the well without disturbing DNA migration through the gel.

31. If processing more than one sample, you can choose different indexed primers to barcode each sample and pool samples before sequencing.

32. A high-quality Micro-C library should have ~50–100 ng of total DNA with <12 PCR cycles.

Acknowledgments

TSH developed the Micro-C protocol. ES, CC, TSH tested and optimized the protocols. ES, CC, and TSH drafted and edited the manuscript. We thank all members in the Tjian and Darzacq group for valuable discussions and suggestions.

References

1. Goel VY, Hansen AS (2020) The macro and micro of chromosome conformation capture. Wiley Interdiscip Rev Dev Biol 10:e395

2. Beagan JA, Phillips-Cremins JE (2020) On the existence and functionality of topologically associating domains. Nat Genet 52:8–16

3. Rowley MJ, Corces VG (2018) Organizational principles of 3D genome architecture. Nat Rev Genet 19:789–800

4. Bonev B, Cohen NM, Szabo Q et al (2017) Multiscale 3D genome rewiring during mouse neural development. Cell 171:557–572.e24

5. Hu J, Zhang Y, Zhao L et al (2015) Chromosomal loop domains direct the recombination of antigen receptor genes. Cell 163:947–959

6. Zheng H, Xie W (2019) The role of 3D genome organization in development and cell differentiation. Nat Rev Mol Cell Biol 20: 535–550

7. Lieberman-Aiden E, van Berkum NL, Williams L et al (2009) Comprehensive mapping of long-range interactions reveals folding principles of the human genome. Science 326: 289–293

8. Hsieh T-HS, Weiner A, Lajoie B et al (2015) Mapping nucleosome resolution chromosome folding in yeast by micro-C. Cell 162:108–119

9. Hsieh T-HS, Fudenberg G, Goloborodko A et al (2016) Micro-C XL: assaying chromosome conformation from the nucleosome to the entire genome. Nat Methods 13: 1009–1011

10. Hamdani O, Dhillon N, Hsieh T-HS et al (2019) Transfer RNA genes affect chromosome structure and function via local effects. Mol Cell Biol 39(8):e00432–e00418

11. Swygert SG, Kim S, Wu X et al (2019) Condensin-dependent chromatin compaction represses transcription globally during quiescence. Mol Cell 73:533–546.e4

12. Costantino L, Hsieh T-HS, Lamothe R et al (2020) Cohesin residency determines chromatin loop patterns. Elife 9:e59889

13. Hansen AS, Hsieh T-HS, Cattoglio C et al (2019) Distinct classes of chromatin loops revealed by deletion of an RNA-binding region in CTCF. Mol Cell 76:395–411.e13

14. Hsieh T-HS, Cattoglio C, Slobodyanyuk E et al (2020) Resolving the 3D landscape of transcription-linked mammalian chromatin folding. Mol Cell 78:539–553.e8

15. Krietenstein N, Abraham S, Venev SV et al (2020) Ultrastructural details of mammalian chromosome architecture. Mol Cell 78: 554–565.e7

16. Servant N, Varoquaux N, Lajoie BR et al (2015) HiC-Pro: an optimized and flexible pipeline for Hi-C data processing. Genome Biol 16:259

17. Abdennur N, Mirny LA (2019) Cooler: scalable storage for Hi-C data and other genomically labeled arrays. Bioinformatics 36(1): 311–316

18. Kerpedjiev P, Abdennur N, Lekschas F et al (2018) HiGlass: web-based visual exploration and analysis of genome interaction maps. Genome Biol 19:125

19. Xie L, Dong P, Chen X et al (2020) 3D ATAC-PALM: super-resolution imaging of the accessible genome. Nat Methods 17:430–436

Part II

Targeted Hi-C Approaches

Chapter 5

Targeted Chromosome Conformation Capture (HiCap)

Artemy Zhigulev and Pelin Sahlén

Abstract

Targeted chromosome conformation capture (HiCap) is an experimental method for detecting spatial interactions of genomic features such as promoters and/or enhancers. The protocol first describes the design of sequence capture probes. After that, it provides details on the chromosome conformation capture adapted for next-generation sequencing (Hi-C). Finally, the methodology for coupling Hi-C with sequence capture technology is described.

Key words Chromosome conformation capture, HiCap, Capture Hi-C, Promoter, Enhancer, Targeted sequencing

1 Introduction

Chromosome conformation capture (3C) [1] adapted for next-generation sequencing (Hi-C) [2] is the chief method for mapping the folded structure of chromatin in the nucleus. Capture Hi-C [3–6] combines this technology with sequence capture to resolve chromatin structures at higher resolution at selected regions. All 3C-based methods such as 4C [7, 8] (Chapter 2), 5C [8], Hi-C [2] (Chapter 3), ChiA-PET [9], CHi-C [6], and HiCap [5] rely on efficient fragmentation of the chromatin to generate free DNA ends that will then be ligated to each other. The ligation of these free ends captures the proximity information within the nucleus, that is, the ends that are close to each other in three-dimensional space are more likely to ligate to each other than those that are far away. Therefore, the average size of the fragmented chromatin dictates the resolution of the method since each ligation junction informs the spatial proximity of the two restriction fragments.

HiCap is often used to derive a genome-wide interaction profile of a set of regions of interest. Hi-C is a powerful method; however, its ability is limited for informing local chromatin loops involving promoter–enhancer contacts due to the quadratic relationship

Tom Sexton (ed.), *Spatial Genome Organization: Methods and Protocols*,
Methods in Molecular Biology, vol. 2532, https://doi.org/10.1007/978-1-0716-2497-5_5,

between resolution and sequencing depth (n ~ n^2) [2]. HiCap approximates a linear relationship between sequencing depth and the resolution by targeting one end of the interacting partner [5].

The sequence capture probes are 90–120-base oligonucleotides carrying a biotin moiety to enable their capture using streptavidin coupled paramagnetic beads. They are available to purchase from various vendors and most of the vendors also offer a service for the design of the sequence capture probes for the sequences of interest.

Promoters and enhancers are often the targeted regions to find their regulatory counterparts [5, 10]. However, other features such as disease or trait-associated noncoding variants can also be targeted to locate the gene(s) they regulate [11–14].

2 Materials

1. 1× phosphate-buffered saline (PBS).
2. 1% formaldehyde in 1× PBS, made fresh on day of experiment from methanol-free 16% ampoules.
3. 2 M glycine.
4. Cell lysis buffer: 10 mM Tris–HCl pH 8, 10 mM NaCl, 0.2% (v/v) Triton X-100, 1× complete EDTA-free protease inhibitors, made fresh on day of experiment.
5. Fast Digest MboI with supplied 10× Fast Digest buffer (Thermo Scientific) (*see* **Note 1**).
6. 20% (w/v) SDS.
7. 20% (v/v) Triton X-100.
8. 10 mM dCTP, made fresh from 100 mM stock.
9. 10 mM dGTP, made fresh from 100 mM stock.
10. 10 mM dTTP, made fresh from 100 mM stock.
11. 0.4 mM biotin-14-dATP.
12. 10 U/μL DNA polymerase I large (Klenow) fragment.
13. 0.5 M EDTA.
14. Double distilled water (ddH₂O).
15. 6 Weiss units/μL T4 DNA ligase with supplied 10× buffer.
16. 100 mM ATP (*see* **Note 2**).
17. 20 mg/mL proteinase K.
18. Phenol–chloroform–isoamyl alcohol (25:24:1), warmed to room temperature for 1 h protected from light before use.
19. 3 M sodium acetate pH 5.2.
20. 10 mg/mL glycogen.

21. 100%, 80%, and 70% ethanol.

22. 10 mg/mL RNase A.

23. Qubit fluorimeter and dsDNA HS and BR quantification assays (Thermo Fisher), or equivalent.

24. Tape station or Bioanalyzer.

25. 5 U/μL T4 DNA polymerase with supplied 5× buffer.

26. 10 mg/mL BSA.

27. 10 mM dATP, from 100 mM stock.

28. Genomic DNA purification columns and reagents (e.g., Macherey-Nagel).

29. Sonicator (e.g., Covaris S220).

30. AMPure XP beads.

31. Magnetic stands compatible with 1.5 mL or 2 mL Eppendorf tubes and PCR tubes or plates.

32. Streptavidin-coupled paramagnetic beads such as Dynabeads™ MyOne™ Streptavidin T1 (Thermo Fisher).

33. 2× No-Tween Wash Buffer (NTB): 10 mM Tris–HCl pH 8, 2 M NaCl, 1 mM EDTA. Also prepare 1× solution.

34. 1× Tween Wash Buffer (TWB): 0.0005% (v/v) Tween 20 in 1× NTB.

35. SureSelectXT HS2 Target Enrichment system (Agilent Technologies) or equivalent (*see* **Note 3**). Kit contains: Ligation Buffer, T4 DNA Ligase, End Repair/dA Tailing Buffer, End Pair/dA Tailing Enzyme Mix, SureSelect XT HS2 Adapter Oligo mix, 5× Herculase II Buffer, Herculase II Fusion DNA Polymerase, SureSelect XT HS2 Index Primer Pair, SureSelect Fast Hybridization Buffer, SureSelect XT HS2 Blocker Mix, SureSelect RNase block, SureSelect Binding Buffer, Wash Buffer 1 and 2, SureSelect Post-Capture Primer Mix.

36. Biotinylated sequence capture oligonucleotide library (*see* **Note 4**) (*see* Subheading 3.1).

37. Low TE buffer: 10 mM Tris–HCl pH 8, 0.1 mM EDTA.

38. Next-generation sequencing machine (e.g., Illumina HiSeq).

3 Methods

3.1 Probe Design

HiCapTools software suite [15] can be used for sequence probe design. The design also produces coordinates of a set of sequence capture probes that target genomic regions with no known promoter or enhancer annotation, which constitute negative controls. Around 5–10% of the features are dedicated to negative controls. They are used to derive an empirical interaction background

Table 1
The comparison of quality of calls between HiCapTools and CHiCAGO. Enrichment is calculated by overlapping high-quality enhancer datasets with the promoter-interacting sequences. (DCR: Data, distance and fragment length, Corrected Random set)

Method	HiCap			CHiCAGO		
Distance	<50 kb	50–500 kb	>500 kb	<50 kb	50–500 kb	>500 kb
Cut-off	$p < 0.001$	$p < 0.01$	$p < 0.01$	Score > =5		
Dataset size	10,129	27,167	2668	11,815	23,762	3335
Enrichment vs. DCR	2.10	2.88	3.32	2.63	2.88	2.02

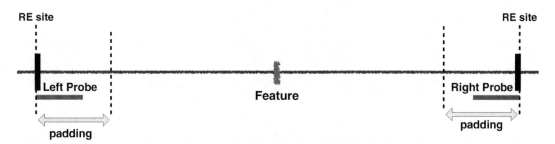

Fig. 1 The ideal placement of probes around a feature. The probes are designed at the end of the restriction enzyme (RE) site to increase the probability of capturing junctions. The closest restriction fragments closest to the feature are first candidates for probe design. If probe cannot be designed anywhere around the fragment end, the next restriction fragment site is used as long as the distant between the fragment site and feature is less than 2500 bases. This parameter can be changed in the configuration file. The padding distance is the number of base pairs around a probe that will be taken into account while deciding overlap with a read and can be adjusted in the configuration file

frequency for each sample which is then used to call significant interactions. This approach eliminates the need for multiple testing correction and simplifies the comparison of interaction strengths across different samples [15]. This method performs similarly for calling short and mid-range interactions compared with another method that uses theoretical interaction background frequency (CHiCAGO) [16]. However, HiCapTools are better at calling long-range interactions (>500 kb), possibly due to a better approximation of background frequencies at longer distances (Table 1 [15]). The repository of HiCapTools can be found here: https://github.com/sahlenlab/HiCapTools

We design two probes for each promoter (Fig. 1). However, some features might have only one or no probes depending on GC and repeat content of the sequence surrounding the feature. The probe design requires several files as an input to HiCapTools PD1 module that are described below.

3.1.1 Preparing Target Files	For transcripts that will be targeted, prepare the transcript files, using the UCSC Genome Browser Table Browser (http://genome.ucsc.edu/cgi-bin/hgTables), as follows:

1. Select the correct *genome* and *assembly.*

2. Select the "Genes and Gene Prediction" *group.*

3. Select the desired *track* (e.g., "RefSeq Genes").

4. Alternatively, one can select the *region* as "genome" or a list of "positions" that can be inserted/uploaded using the "define regions" tab, or enter a list of gene or transcript identifiers using the "paste list" or "upload list" options to include only transcripts or genes of interest.

5. Select *output format* as "all fields from selected table".

6. Press the "get output" tab to obtain the transcript list. Sort transcript list using the "sortTranscriptFile.sh" script in the HiCapTools repository.

For variants to be targeted, prepare a BED file that specifies the genomic coordinates of variants. The fourth column should be populated by the identification of the variant.

3.1.2 Completing Probe Design	1. Download the fasta file of the genome and index file from http://hgdownload.cse.ucsc.edu/downloads.html.

2. Prepare a digested genome file using the HiCUP pipeline. Use the hicup_digester module to generate the digest file using the correct restriction enzyme. HiCapTools can directly use the digested genome file.

3. Populate probe design parameters of the Probe Design (PD1) module of HiCapTools in the configuration file. More details on the probe design parameters can be found in the HiCapTools user manual.

3.2 Cell Fixation

The appropriate number of cultured cells (*see* **Note 5**) are harvested during their optimal growth phase. Handling ex vivo cells or tissues are beyond the scope of the protocol.

3.2.1 Adherent Cells (Grown on 100 Mm Cell Culture Dishes)	1. Remove cell culture medium, rinse with 10 mL 1× PBS, then remove the PBS.

2. Add 5 mL 1% formaldehyde in 1× PBS and incubate for 10 min at room temperature on a rocker (*see* **Note 6**).

3. Add 312.5 μL 2 M glycine (final glycine concentration should be 0.125 M) and incubate for 5 min at room temperature.

4. Pour off solution and scape cells from the surface of the plate in 2 mL ice-cold 1× PBS. Transfer cells to a 15 mL Falcon tube. Proceed to Subheading 3.3.

3.2.2 Suspension Cells

1. Transfer cells to a 15 mL Falcon, centrifuge for 5 min at 300 × *g*, room temperature and remove supernatant.

2. Resuspend cells in 5 mL 1× PBS, centrifuge for 5 min at 300 × *g*, room temperature and remove supernatant.

3. Resuspend cells in 5 mL 1% formaldehyde in 1× PBS and incubate for 10 min at room temperature on a rotator.

4. Add 312.5 μL 2 M glycine (final glycine concentration should be 0.125 M) and incubate for 5 min at room temperature on a rotator. Proceed to Subheading 3.3.

3.3 Hi-C

1. Centrifuge the cells for 5 min at 300 × *g*, 4 °C and remove supernatant.

2. Resuspend cells in 1 mL cell lysis buffer by gently pipetting up and down. Add an additional 3.5 mL cell lysis buffer and incubate on ice for 10 min, resuspending the solution by pipetting up and down after 5 min.

3. Centrifuge for 5 min at 600 × *g* at 4 °C and remove supernatant. At this stage, the nuclei pellet can be snap-frozen and stored at −70 °C for later use.

4. Resuspend cells in 550 μL 1× Fast Digest buffer and transfer to a 1.5 mL microtube. If the pellet contains more than five million cells, split the sample to maximum five million cell/ 550 μL aliquots (*see* **Note 5**).

5. Transfer a 50 μL aliquot ("Undigested control") to a new microtube and store at 4 °C until **step 15**.

6. Add 7.6 μL 20% SDS (final concentration 0.3%) and incubate for 1 h at 37 °C on a thermomixer at 950 rpm.

7. Add 56.4 μL 20% Triton X-100 (final concentration 2%) and incubate for 1 h at 37 °C on a thermomixer at 950 rpm.

8. Add 7 μL of 10× Fast Digest MboI buffer to compensate for the added volume. Add 6 μL Fast Digest MboI and incubate for 2 h at 37 °C on a thermomixer at 450 rpm (*see* **Note 1**).

9. Inactivate MboI by incubating for 10 min at 75 °C.

10. Transfer a 50 μL aliquot ("Digested control") to a new microtube and store at 4 °C until **step 14**.

11. Transfer a 75 μL aliquot ("3C control") to a new microtube and store at 4 °C until **step 14**.

12. Add 1.2 μL 10 mM dCTP, 1.2 μL 10 mM dGTP, 1.2 μL 10 mM dTTP, 30 μL 0.4 mM biotin-14-dATP, and 1.2 μL 10 U/μL DNA polymerase I large (Klenow) fragment and incubate for 4 h at 23 °C on a thermomixer at 450 rpm (*see* **Notes 7–9**).

13. Add 9.9 μL 0.5 M EDTA (10 mM final concentration) and incubate for exactly 10 min at 75 °C (*see* **Note 10**).

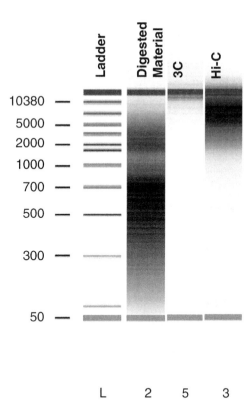

Fig. 2 The size distribution of digested chromatin, 3C and Hi-C experiments. The digested chromatin produces a smear centered around 650 bases. 3C produces very high molecular weight products that overlaps and exceeds the longest MW marker. Hi-C produces a smear centered around 7–8 kb long

14. Add 150 μL 10× T4 DNA ligase buffer, 15 μL 100 mM ATP, 8.5 μL 6 Weiss units/μL T4 DNA ligase and 829.8 μL ddH₂O (final volume is 1.5 mL) to each Hi-C sample. Add 50 μL 10× T4 DNA ligase buffer, 5 μL 100 mM ATP, 1 μL 6 Weiss units/μL T4 DNA ligase, and 369 μL ddH₂O (final volume is 0.5 mL) to each 3C control. Do the same for the 50 μL of digested control but omitting the ligase. Incubate for 4.5 h at 16 °C, then 30 min at room temperature (*see* **Note 11**).

15. Return the "Undigested" control tube to the protocol (**step 5**). Add 450 μL 1× Fast Digest buffer.

16. Add 9 μL 20 mg/mL proteinase K to the Hi-C sample and 3 μL 20 mg/mL proteinase K to each control sample. Seal the tubes with parafilm and incubate for 6 h at 65 °C.

17. Divide Hi-C reaction into three 1.5 mL Eppendorf tubes, each containing 500 μL. Add 500 μL phenol–chloroform–isoamyl

alcohol to each Hi-C sample and all the controls, mix well by vortexing and centrifuge for 5 min at $16,000 \times g$ at room temperature. Carefully transfer the upper aqueous layer to new 1.5 mL microtubes.

18. Repeat **step 17** once more but this time, transfer the upper aqueous layer to 2 mL microtubes.

19. Add 50 μL 3 M sodium acetate, 1 μL 10 mg/mL glycogen and 1250 μL 100% ethanol to each Hi-C sample. Add 10 μL 3 M sodium acetate, 0.2 μL 10 mg/mL glycogen, and 250 μL 100% ethanol to each control sample. Incubate all tubes at -20 °C for 30 min or overnight.

20. Centrifuge all samples for 20 min at $16,000 \times g$ at 4 °C and carefully discard supernatants without disturbing the pellets.

21. Add 200 μL 70% ethanol to the pellets, centrifuge for 5 min at $16,000 \times g$ at room temperature and carefully discard supernatant. Air-dry the pellet for 10 min at room temperature.

22. Dissolve one aliquot of Hi-C DNA in 100 μL ddH$_2$O, then transfer the sample to the other two Hi-C aliquots to combine all three Hi-C tubes. Dissolve each control sample in 30 μL ddH$_2$O.

23. Add 1 μL 10 mg/mL RNase A to Hi-C tube and 0.2 μL 10 mg/mL RNase A to control tubes. Incubate the tubes for 1 h at 37 °C.

24. Determine the concentration of the samples by Qubit fluorimeter/dsDNA BR assay.

25. Run Hi-C material and control samples on an agarose gel, Bioanalyzer or tape station to assess the Hi-C and control material (*see* **Note 12**; Fig. 2).

3.4 Removal of Biotin Moieties from Ends

This step is necessary to remove the biotin moieties at the end of fragments to reduce the number of fragments containing biotin moiety but no ligation junction.

1. Transfer 5 μg Hi-C DNA to a 1.5 mL microtube and complete the volume to 74 μL with ddH$_2$O. Add 0.5 μL 10 mg/mL BSA, 20 μL 5× T4 DNA Polymerase Buffer, 0.5 μL 10 mM dATP, 0.5 μL 10 mM dGTP, and 1 μL 5 U/μL T4 DNA polymerase and incubate for 15 min at 12 °C. The final DNA concentration should be 50 ng/μL (*see* **Note 9**).

2. Add 2 μL 0.5 M EDTA to quench the reaction.

3. Purify DNA with a genomic DNA column, following manufacturer's instructions and elute in 100 μL ddH$_2$O (*see* **Note 13**).

4. Quantify Hi-C DNA with Qubit fluorimeter and dsDNA BR assay.

3.5 Sample Shearing

Hi-C produces long fragments (>5 kb) due to the ligation step. These long fragments are sheared into shorter fragments to permit sequencing of the sample by short-read sequencing platforms.

1. Dilute 3 μg Hi-C in a volume of 130 μL in ddH$_2$O in a microtube (23.1 ng/μL final DNA concentration). Sonicate the material in a Covaris S220 sonicator with the following settings: Duty cycle 10%; Intensity 5; 200 Cycles per burst; 5 cycles of 40 s each, with frequency sweeping on; 4 °C (*see* **Note 14**).

2. Remove the AMPure XP beads from cold storage and equilibrate to room temperature at least 30 min before use. Do not freeze the beads at any time. Mix the bead suspension well so that the reagent appears homogeneous and consistent in colour.

3. Add 180 μL of homogeneous AMPure XP beads to the sheared DNA sample (approximately 130 μL). Pipet up and down 10 times to mix. Incubate samples for 5 min at room temperature.

4. Put the tube onto a magnetic stand. Wait for the solution to clear (approximately 3–5 min).

5. While keeping the tube in the magnetic stand, carefully remove and discard the cleared solution from the tube. Do not touch the beads while removing the solution.

6. Continue to keep the tube in the magnetic stand while adding 200 μL of 80% ethanol to the sample. Wait for 1 min to allow any disturbed beads to settle, then remove the ethanol solution.

7. Repeat **step 6** once more.

8. Remove the tube from the magnetic stand. Dry the sample by waiting for 5 min at room temperature (*see* **Note 15**).

9. Add 50 μL of ddH$_2$O to the beads and incubate for 2 min at room temperature.

10. Put the tube in the magnetic stand and leave for 2–3 min, until the solution is clear. Transfer the cleared supernatant (approximately 48 μL) to a fresh tube. Discard the beads.

11. Analyze the fragmented material using a TapeStation, BioAnalyzer, or agarose gel (*see* **Note 14**; Fig. 3).

3.6 Capture of Biotinylated Fragments

During Hi-C, biotin moieties are introduced during the end-filling of compatible ends created by the restriction enzyme. Therefore, most ligated fragments contain biotin moieties due to the ligation of the restriction fragments. In this step, these biotinylated fragments are selected/captured using streptavidin-coupled magnetic beads. This step significantly reduces the number of genomic

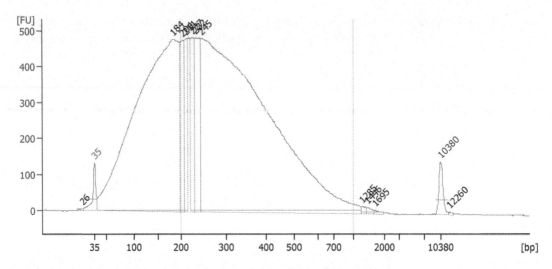

Fig. 3 The size distribution of fragmented Hi-C material using a sonicator

fragments (i.e., fragments with no digestion or ligation events) present in the captured material since genomic fragments constitute more than 90% of the material [4]. This, in turn, makes next-generation sequencing of the material more worthwhile, since most of the sequenced material contains a ligation event.

1. Mix well the streptavidin-coupled paramagnetic beads (Dyna-beads T1) and transfer 200 μL beads to a 1.5 mL microtube. Place on a magnetic stand, wait until the supernatant is clear (approximately 3–5 min) and remove the supernatant.

2. Resuspend the streptavidin beads in 400 μL TWB and incubate for 3 min at room temperature. Place beads on magnetic stand and wait until the supernatant is clear (approximately 3–5 min) and remove supernatant. Repeat this step once more.

3. Add 102 μL of ddH$_2$O to the Hi-C sample from Subheading 3.5 to complete the volume of each sample to 150 μL.

4. Resuspend beads in 150 μL 2× NTB. Add 150 μL Hi-C material from the previous step to the beads. Incubate bead–DNA mix for 15 min at room temperature on a rotator to allow DNA to bind to beads.

5. Place the beads on the magnetic stand and wait for 3 min at room temperature (or until the supernatant becomes clear) and remove the clear supernatant. The biotinylated Hi-C material is now bound to streptavidin beads.

6. Take the Hi-C-beads from the magnetic stand and resuspend them in 400 μL 1× NTB, incubate at room temperature for 3 min. Then place the tube on magnetic stand, and wait for 3 min at room temperature (or until the supernatant becomes clear) and remove the clear supernatant. Repeat this step.

7. Take the Hi-C beads from the magnetic stand and resuspend them in 200 μL 1× T4 DNA Ligase Buffer. Incubate at room temperature for 3 min. Place the beads in magnetic stand, wait for 3 min or until the supernatant is clear. Remove the supernatant.

8. Add 52 μL of ddH$_2$O to the Hi-C beads. The sequencing library is prepared while the Hi-C DNA is bound to beads.

3.7 Sequencing Library Preparation

From this point on, the content of SureSelect XT HS2 DNA Reagent Kit (Agilent Technologies) is used for sequencing library preparation. This step ligates indexed sequencing adapters to the sheared Hi-C fragments in the captured material (*see* **Notes 3** and **16**).

1. Vortex the thawed vial of Ligation Buffer for 15 s at high speed to ensure homogeneity, then slowly pipet 23 μL Ligation Buffer into a 1.5 mL Eppendorf tube, ensuring that the full volume is dispensed.

2. Slowly add 2 μL T4 DNA Ligase, rinsing the enzyme tip with the buffer solution after addition. Mix well by slowly pipetting up and down 15–20 times or seal the tube and vortex at high speed for 10–20 s. Spin briefly to collect the liquid. Keep the mix at room temperature for 30–45 min before use at **step 6** to activate the enzyme.

3. Vortex the thawed vial of End Repair/dA Tailing Buffer for 15 s at high speed to ensure homogeneity. Visually inspect the solution; if any solids are observed, continue vortexing until all solids are dissolved. Add 16 μL of End Repair/dA Tailing Buffer into a PCR tube (compatible with the thermocycler to be used) and add 4 μL of End Repair/dA Tailing Enzyme Mix.

4. Add 50 μL of homogenized Hi-C-bead suspension material from Subheading 3.6 to the End Repair/dA tailing reaction mix. Mix the solution by pipetting up and down 15–20 times (70 μL total volume).

5. Immediately place the tube in the thermocycler and start the following program with the heated lid on:

20 °C, 15 min;

72 °C, 15 min;

4 °C, hold.

6. Immediately transfer the PCR tube on ice and add 25 μL of Ligation master mix prepared in **step 2**. Mix by pipetting up and down at least 10 times (*see* **Note 17**).

7. Add 5 μL of SureSelect XT HS2 Adaptor Oligo Mix to the sample. Mix by pipetting up and down at least 15–20 times (*see* **Note 17**).

8. Briefly spin the sample and then immediately place the tube in the thermocycler and run the following program:

   20 °C, 30 min;

   4 °C, hold.

9. Place the bead-bound samples to a magnet and incubate until solution is clear. Transfer the supernatant and remove the tubes from the magnet.

10. Wash beads twice in 150 μL 1× NTB buffer: Incubate at room temperature for 3 min, return them to the magnet, remove the supernatant once clear, repeat this step.

11. Wash the beads with 150 μL of 1× Low TE buffer. Incubate at room temperature for 3 min, return them to the magnet, transfer the supernatant once clear.

12. Add 34 μL ddH$_2$O to each bead-bound sample.

3.8 Library Amplification

In this step, the material which now carries sequencing adapters, is PCR-amplified to obtain sufficient amount of material for sequence capture step. During the PCR-amplification, a sequencing index is also introduced to the sample (*see* **Note 16**).

1. Mix 10 μL of 5× Herculase II Buffer with dNTPs and 1 μL of Herculase II Fusion DNA Polymerase, and add the whole 11 μL to the bead-bound Hi-C sample from Subheading 3.7.

2. Add 5 μL of the chosen SureSelect XT HS2 Index Primer Pair to the reaction (*see* **Note 16**). Mix the sample well by pipetting up and down at least 15 times.

3. Transfer the sample to the thermocycler and run the following program (*see* **Note 18**):

   98 °C, 2 min;

   8× [98 °C, 30 s; 60 °C, 30 s; 72 °C, 1 min];

   72 °C, 5 min;

   4 °C, hold.

4. Reclaim the beads by placing them on a magnet stand. Transfer the supernatant containing the amplified DNA to a new tube. Add 34 μL ddH$_2$O to the beads and keep them at 4 °C (*see* **Note 19**).

5. Add 50 μL homogeneous AMPure XP beads to 50 μL amplified DNA sample (the supernatant from the previous step). Mix the sample well by pipetting up and down at least 15 times. Incubate samples for 5 min at room temperature.

6. Put the tube onto a magnetic stand. Wait for the solution to clear (approximately 3–5 min). Carefully remove and discard the cleared solution from the tube. Do not touch the beads while removing the solution.

7. Continue to keep the tube in the magnetic stand while adding 200 μL of freshly prepared 80% ethanol to the tube. Wait for 1 min to allow any disturbed beads to settle, then remove the ethanol. Repeat this step once more.

8. Seal the tube, then briefly spin the tube to collect the residual ethanol. Return the plate to the magnetic stand for 30 s. Remove the residual ethanol with a P20 pipette.

9. Dry the samples by placing the unsealed tube at room temperature for 1–2 min or until the residual ethanol completely evaporates (*see* **Note 15**).

10. Add 30 μL ddH$_2$O and seal the tube, then mix well on a vortex mixer and briefly spin the tube in a centrifuge to collect the liquid. Incubate for 2 min at room temperature.

11. Put the plate in the magnetic stand and leave for 2–3 min, until the solution is clear. Transfer the cleared supernatant (approximately 30 μL) to a fresh tube. The beads can be discarded.

12. Assess the yield and size distribution of the amplified library by Qubit fluorimeter/dsDNA BR assay and Tape station or BioAnalyzer (*see* **Note 6**).

3.9 Target Enrichment

In this step, the Hi-C library is hybridized to the capture probes targeting the sequences of interest. It is important to perform the following probe hybridization and wash steps in 0.2 mL PCR tubes or plate in a thermocycler. This is to avoid temperature fluctuations during washing steps. The hybridization step requires 500–1000 ng of prepared DNA in a volume of 12 μL (*see* **Note 18**). This might necessitate the use of a vacuum concentrator to achieve the above concentration. Use the maximum amount of prepared DNA available in this range.

1. Thaw the SureSelect Fast Hybridization buffer and keep at room temperature. Thaw the SureSelect XT HS2 Blocker Mix, SureSelect RNase Block and biotinylated probe oligonucleotides on ice.

2. Place 500–1000 ng of each prepared library into a PCR tube and then bring the final volume to 12 μL using ddH$_2$O.

3. To the Hi-C library tube, add 5 μL SureSelect XT HS2 Blocker Mix (*see* **Note 3**). Seal the tube, then vortex at high speed for 5 s. Spin briefly to collect the liquid and release any bubbles.

4. Transfer the sealed tube to the thermal cycler and run the following program with the heated lid set to 105 °C.

   95 °C, 5 min.

   65 °C, 10 min.

   65 °C, hold, while preparing the hybridization reagents in **steps 5** and **6**.

5. Prepare a 25% solution of SureSelect RNAse Block by adding 0.5 μL RNase Block to 1.5 μL ddH$_2$O. Mix well and keep on ice.

6. Add 6 μL SureSelect Fast Hybridization Buffer, and 5 μL biotinylated probe oligonucleotide library to the 25% SureSelect RNase Block solution (*see* **Note 4**). Mix well by vortexing at high speed for 5 s then spin down briefly and keep at room temperature.

7. Immediately transfer 13 μL of the room temperature solution from **step 6** to the Hi-C library sample, still within the thermal cycler at 65 °C. Mix well by pipetting up and down slowly 8 to 10 times. The hybridization reaction tube now contains approximately 30 μL.

8. Seal the sample tube with fresh cap, and make sure it is completely sealed. Vortex briefly, spin down to remove any bubbles and return the tube to the thermal cycler immediately. Run the following thermal program with the heated lid set at 105 °C to hybridize the probes (*see* **Note 20**).

   60×: [65 °C, 1 min; 37 °C, 3 s].

   65 °C, hold, while preparing beads in **steps 9–12**.

9. Vigorously resuspend the vial of streptavidin beads (Dynabeads T1) on a vortex mixer and add 50 μL of the resuspended streptavidin beads to a fresh PCR tube.

10. Add 200 μL of SureSelect Binding Buffer to the beads, mix by pipetting up and down 20 times or vortexing at high speed for 5–10 s, then briefly spin down.

11. Place beads on a magnetic stand for 5 min, or until solution is clear, then remove and discard supernatant.

12. Repeat **steps 10** and **11** two more times for a total of three washes. Resuspend beads in 200 μL of SureSelect Binding Buffer.

13. Cool the hybridization mixture to room temperature and transfer entire volume (~30 μL) of each hybridization mixture to the tube containing 200 μL of washed streptavidin beads. Pipet up and down 5–8 times to mix, then seal the tube.

14. Incubate the tube for 30 min at room temperature while mixing at 1400–1900 rpm.

15. During the 30-min incubation, preheat six 200 μL aliquots of Wash Buffer 2 in capped wells of a PCR plate within a thermal cycler, held at 70 °C.

16. After the 30-min capture incubation, spin the sample briefly to collect the liquid. Put the sample tube in a magnetic stand to collect the beads. Wait until solution is clear, then remove and discard the supernatant.

17. Resuspend the bead-bound hybridization mix in 200 μL of SureSelect Wash Buffer 1. Mix by pipetting up and down 15–20 times, until the beads are fully resuspended.

18. Put the tube in the magnetic stand. Wait for the solution to clear (approximately 3–5 min), then remove and discard the supernatant.

19. Remove the tube from the magnetic stand and transfer to a rack at room temperature. Resuspend the beads in 200 μL preheated Wash Buffer 2. Pipet up and down 15–20 times, until beads are fully resuspended.

20. Seal the tube with fresh cap and then vortex at high speed for 8 s. Spin the tube briefly to collect the liquid without pelleting the beads.

21. Incubate the samples for 5 min at 70 °C on the thermal cycler with the lid heated to 105 °C.

22. Put the tube in the magnetic stand at room temperature. Wait 1 min for the solution to clear, then remove and discard the supernatant.

23. Repeat **steps 19–22** with heated Wash Buffer 2 five more times for a total of six washes. Resuspend beads in 25 μL ddH$_2$O by pipetting up and down eight times. Keep samples on ice until used.

24. Thaw the stocks of 5× Herculase II Buffer with dNTPs and SureSelect Post-Capture Primer Mix, mix by vortexing, and keep on ice.

25. Mix 10 μL 5× Herculase II Buffer with dNTPs, 1 μL Herculase II Fusion DNA polymerase, 1 μL SureSelect Post-Capture Primer Mix, and 13 μL ddH$_2$O on ice. Mix well on a vortex mixer, then add entire mixture to the tube containing bead-bound target-enriched Hi-C DNA (from **step 23**). Mix by pipetting up and down until bead suspension is homogeneous. Do not spin samples at this point.

26. Place the tube in the thermal cycler and run the following program, ensuring that the thermal cycler has reached temperature before adding the samples (*see* **Note 21**).

    98 °C, 2 min.

    11–12×: [98 °C, 30 s; 60 °C, 30 s; 72 °C, 1 min].

    72 °C, 5 min.

    4 °C, hold.

27. Spin the sample briefly and place the tube on the magnetic stand at room temperature. Wait for 2 min for the solution to clear, then transfer the supernatant containing amplified material to a fresh tube (approximately 50 μL) (*see* **Notes 19 and 21**).

28. Mix the room temperature AMPure XP bead suspension well so that the suspension appears homogeneous and consistent in color. Add 50 μL of the homogeneous AMPure XP bead suspension to the amplified DNA sample from step 27 (approximately 50 μL). Mix well by pipetting up and down 15–20 times or cap the tube and vortex at high speed for 5–10 s.

29. Incubate sample for 5 min at room temperature.

30. Put the sample tube on the magnetic stand at room temperature. Wait for the solution to clear (3–5 min). Keeping the tube in the magnetic stand, carefully remove and discard the cleared solution. Do not disturb the beads.

31. Continue to keep the tube in the magnetic stand while adding 200 μL of freshly prepared 80% ethanol. Wait for 1 min to allow any disturbed beads to settle, then remove the ethanol. Repeat this step once more for a total of two washes.

32. Seal the tube, then briefly spin to collect the residual ethanol. Return the tube to the magnetic stand for 30 s. Remove the residual ethanol with a P20 pipette.

33. Dry the samples by placing the tube at room temperature, until the residual ethanol has just evaporated (usually 1–2 min) (*see* **Note 15**).

34. Add 25 μL Low TE buffer to the sample, seal the tube, mix well on a vortex mixer and briefly spin to collect the liquid without pelleting the beads.

35. Incubate for 2 min at room temperature. Put the tube in the magnetic stand and leave for 2 min or until solution is clear. Transfer the cleared supernatant (approximately 25 μL) to a fresh tube.

36. Analyze the library using Agilent 2100 Bioanalyzer system using High Sensitivity DNA Kit. Expected DNA fragment size is ~200–400 bp.

37. Perform paired-end high-throughput sequencing of the captured library.

38. The consequent data analyses can be then performed by HiCapTools ProximityDetector module (PD2) [15]. The files necessary to input for interaction calling are produced by the PD1 module (*see* Subheading 3.1).

4 Notes

1. MboI is not sensitive to CpG methylation or dcm methylation, but is dam methylation sensitive. 1 μL of FastDigest MboI can digest up to 5 μg of genomic DNA in 60 min. In this protocol, sufficient amount of enzyme is used to digest the genomic

DNA of five million diploid human cells. It might be possible to reduce the amount of enzyme used if lesser number of cells are used. However, we routinely use the same amount of enzyme (6 μL) for up to five million cells.

2. Repeated freeze-thaw cycles of ligase buffer can affect the stability of the ATP present in the buffer. Therefore, the ligation reaction is supplemented with additional ATP to ensure the presence of sufficient ATP present during ligation reaction.

3. We use the SureSelect XT HS2 kit that provides custom enrichment probes and related hybridization and wash kits available from Agilent Technologies. The sequencing adaptors with indexes are also provided by the SureSelect XT HS2 library preparation kit. We also use the blocking oligonucleotides that are part of the same kit. Blocking oligonucleotides bind to sequencing adapter sequences to minimize off-target hybridization. The library preparation protocol slightly deviates from the manufacturer's protocol to enable the processing of biotinylated fragments. The hybridization and wash protocols of the kit are used without modification in the manufacturer's protocol. It is possible to use probes from other vendors; however, then the hybridization and wash kits should be purchased also accordingly. The same library kit (Agilent Technologies) can be used as long as the blocking oligonucleotides are compatible; please refer to the manufacturer's instructions.

4. Capture probes can be ordered directly from Agilent Technologies or other vendors that provide biotinylated capture probe pools. Most companies also provide a solution for designing probes to regions of interest provided by the user. However, it is also possible to design the probes using a software tool such as HiCapTools [15] and order predesigned probes (Subheading 3.1).

5. In our hands, 2–3 million cells for a capture set composed of 50,000 probes (3–6 Mb) provide complex libraries provided that on-target capture rate is greater than 70% and the captured library is sequenced at a depth of around 300 million read pairs. This protocol is optimized to use maximum five million cells, therefore it is not recommended to process more than five million cells using the protocol.

6. It is recommended at this step to ensure that the cells are not under or over cross-linked. Undercrosslinking will result in higher background noise (i.e., overrepresentation of random collision events instead of protein-mediated proximities). Overcrosslinking will result in inefficient digestion and ligation, loss in complexity and resolution of the material. To avoid overcrosslinking, it is recommended to try several cross-linking conditions to optimize the protocol, by trying different

crosslinking time durations, for example, 2 min, 5 min, 10 min or 15 min. Then, in the protocol, the cells crosslinked at different conditions can be processed by omitting the steps in the Hi-C protocol using the volumes for the digested control samples (Subheading 3.3, **steps 10–15**). The optimal digestion pattern depends on the cut frequency of the restriction enzyme used; however, it is normal to obtain slightly longer fragments due to crosslinking. The rest of the samples can be kept at 4 °C until the next day. Once the expected digestion pattern is observed, the optimally crosslinked sample can be processed through the Hi-C protocol.

The undercrosslinking (not crosslinking the sample efficiently) results in high rates of interchromosomal proximities (>40%); it is not recommended to use such a sample for data production (unless the sample comes from a cell line/tissue with highly rearranged genome such as cancer cells or tumor tissue). To avoid processing such a sample to the end, it is recommended to shallow sequence the Hi-C library produced at Subheading 3.8 before continuing sequence-capture steps.

7. It is important not to use excess amount of Klenow enzyme since this can inhibit T4 DNA ligase via unspecific protein interaction [17].

8. We always include a 3C control sample to process in parallel with the Hi-C sample. This is to check the ligation efficiency of the blunt-end Hi-C sample in comparison to sticky end 3C sample (Fig. 2).

9. Note that the deoxyribonucleotides are often provided in 100 mM stock solution, therefore they need to be diluted 1:10 to obtain 10 mM working solutions. The dilutions should be prepared fresh at the time of the experiment.

10. The incubation time should be kept as indicated as longer incubation will result in reversal of crosslinks in the material.

11. The 16 °C ligation incubation can be extended to overnight. In that case we observe better ligation products.

12. Hi-C material often produces a smear with an average size above 7 kb for material prepared with a four-cutter restriction enzyme such as Mbo I. The smear produced by 3C material generally possesses a much higher average size (more than 10–12 kb) since 3C material is produced via sticky-end ligation whereas Hi-C material is produced by blunt-end ligation due to the end-filling step.

13. Phenol–chloroform–isoamyl extraction and ethanol precipitation can alternatively be used in this step for DNA clean-up.

14. It is important to ensure that the DNA is fragmented successfully. Figure 3 displays the smear produced by a successful

sonication operation. In case of over- or undersonication, the sonication parameter can be adjusted. Using too much (>3 μg) or too little (<500 ng) DNA can also result in poor sonication efficiency. There are other sonicator systems available to use; however, it is important to adjust the sonication parameters accordingly.

15. It is very important not to overdry the beads, this will significantly decrease the DNA elution yield.

16. There are 16 different indexes available within the library kit used in this protocol. Indexing of a sample enables to sequence multiple samples within the same sequencing run. The sequencing output from each sample can be separated using bioinformatic analysis.

17. It is vital to add the Ligation master mix and the Adaptor Oligo Mix to the samples in separate addition steps, mixing after each addition.

18. In our hands, 8 cycles of precapture PCR produces enough DNA (500–1000 ng) for a sequence capture experiment. However, the number of cycles might need to be optimized if sufficient amplification is not achieved. We recommend to start with 6–8 cycles of PCR first. If not enough material is obtained, it is possible to repeat the amplification using the beads (*see* **Note 19**) for 2–5 additional amplification cycles. We do not recommend to use more than 15 cycles in total for library amplification since this will seriously hamper the complexity of the library.

19. Most of the biotinylated DNA will still be bound to the streptavidin beads. Therefore, the PCR reaction can be repeated if the material is not amplified sufficiently using the beads.

20. Set the reaction volume of 30 μL on the thermocycler settings. The hybridization may be run overnight with the following protocol modification in the thermocycler program: Replace the final 65 °C step with a 21 °C hold for up to 16 h.

21. The number of cycles used for the postcapture amplification of the material depends on the size of capture probe. We routinely use a capture probe that targets around 6 Mb of genome, sufficient to target around 25,000 genomic features and 9–12 cycles of PCRs produces sufficient material for sequencing. However, smaller or larger probe sets might require lesser or greater number of cycles; please refer to probe manufacturer's instructions for choosing the correct number of cycles. Very low amounts of amplified products are expected after postcapture amplification: around 1–2 ng. In our hands, higher amount of DNA obtained after postcapture amplification tend to indicate lower on-target rates (i.e., the fraction of sequenced material complementary to the probes).

References

1. Dekker J, Rippe K, Dekker M et al (2002) Capturing chromosome conformation. Science 295(5558):1306–1311

2. Lieberman-Aiden E, van Berkum NL, Williams L et al (2009) Comprehensive mapping of long-range interactions reveals folding principles of the human genome. Science 326(5950):289–293

3. Schmitt AD, Hu M, Ren B (2016) Genome-wide mapping and analysis of chromosome architecture. Nat Rev Mol Cell Biol 17(12): 743–755

4. Hughes JR, Roberts N, McGowan S et al (2014) Analysis of hundreds of cis-regulatory landscapes at high resolution in a single, high-throughput experiment. Nat Genet 46(2): 205–212

5. Sahlen P, Abdullayev I, Ramskold D et al (2015) Genome-wide mapping of promoter-anchored interactions with close to single-enhancer resolution. Genome Biol 16:156

6. Dryden NH, Broome LR, Dudbridge F et al (2014) Unbiased analysis of potential targets of breast cancer susceptibility loci by capture Hi-C. Genome Res 24(11):1854–1868

7. Zhao Z, Tavoosidana G, Sjolinder M et al (2006) Circular chromosome conformation capture (4C) uncovers extensive networks of epigenetically regulated intra- and interchromosomal interactions. Nat Genet 38(11): 1341–1347

8. Simonis M, Klous P, Splinter E et al (2006) Nuclear organization of active and inactive chromatin domains uncovered by chromosome conformation capture-on-chip (4C). Nat Genet 38(11):1348–1354

9. Fullwood MJ, Liu MH, Pan YF et al (2009) An oestrogen-receptor-alpha-bound human chromatin interactome. Nature 462(7269):58–64

10. Javierre BM, Burren OS, Wilder SP et al (2016) Lineage-specific genome architecture links enhancers and non-coding disease variants to target gene promoters. Cell 167(5): 1369–84 e19

11. Sahlen P, Spalinskas R, Asad S et al (2021) Chromatin interactions in differentiating keratinocytes reveal novel atopic dermatitis and psoriasis-associated genes. J Allergy Clin Immunol 147(5):1742–1752

12. Pradhananga S, Spalinskas R, Poujade FA et al (2020) Promoter anchored interaction landscape of THP-1 macrophages captures early immune response processes. Cell Immunol 355:104148

13. Akerborg O, Spalinskas R, Pradhananga S et al (2019) High-resolution regulatory maps connect vascular risk variants to disease-related pathways. Circ Genom Precis Med 12(3): e002353

14. Beesley J, Sivakumaran H, Moradi Marjaneh M et al (2020) Chromatin interactome mapping at 139 independent breast cancer risk signals. Genome Biol 21(1):8

15. Anil A, Spalinskas R, Akerborg O et al (2018) HiCapTools: a software suite for probe design and proximity detection for targeted chromosome conformation capture applications. Bioinformatics 34(4):675–677

16. Cairns J, Freire-Pritchett P, Wingett SW et al (2016) CHiCAGO: robust detection of DNA looping interactions in capture Hi-C data. Genome Biol 17(1):127

17. Yang Y (2011) Thermodynamics of DNA binding and break repair by the Pol I DNA polymerases from Escherichia coli and Thermus aquaticus

Chapter 6

Assessment of Multiway Interactions with Tri-C

A. Marieke Oudelaar, Damien J. Downes, and Jim R. Hughes

Abstract

Tri-C is a chromosome conformation capture (3C) approach that can efficiently identify multiway chromatin interactions with viewpoints of interest. As opposed to pair-wise interactions identified in methods such as Hi-C, 4C, and Capture-C, the detection of multiway interactions allows researchers to investigate how multiple *cis*-regulatory elements interact together in higher-order structures in single nuclei and address questions regarding structural cooperation between these elements. Here, we describe the procedure for designing and performing a Tri-C experiment.

Key words Genome organization, Gene regulation, *cis*-regulatory elements, Chromosome conformation capture, Multiway interactions

1 Introduction

Chromosome conformation capture (3C) techniques have transformed our understanding of how the genome is organized in the cell nucleus and how *cis*-regulatory elements, including promoters, enhancers and boundary elements, interact and communicate when switching genes on and off [1]. However, most conventional 3C methods, such as Hi-C [2] (Chapter 3), 4C [3, 4] (Chapter 2), and Capture-C [5, 6], are based on the analysis of pair-wise interactions in populations of cells, and therefore do not provide information about dynamic higher-order chromatin structures formed in individual cells. More recently, innovative new techniques such as chromosomal walks (C-walks) [7], genome architecture mapping (GAM) [8], split-pool recognition of interactions by tag extension (SPRITE) [9], and single-cell Hi-C [10–12] (Chapter 10), have started to provide insights into large-scale chromosomal structures in single cells. However, the resolution of these techniques is currently insufficient to be informative at the level of individual *cis*-regulatory elements. These methods therefore do not provide information about how these elements interact in individual cells during the regulation of gene expression.

Tom Sexton (ed.), *Spatial Genome Organization: Methods and Protocols*,
Methods in Molecular Biology, vol. 2532, https://doi.org/10.1007/978-1-0716-2497-5_6,

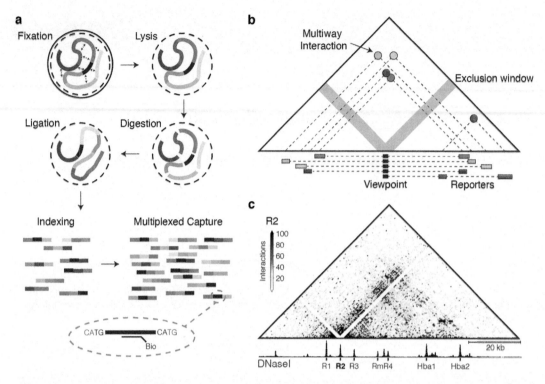

Fig. 1 Overview of Tri-C procedure. (**a**) Overview of the experimental procedure. (**b**) Overview of the analysis strategy to identify multiway interactions. (**c**) Example of a Tri-C matrix from the viewpoint of R2, the strongest enhancer of the murine α-globin genes, in erythroid cells

The Tri-C method overcomes these limitations by using a different strategy to analyze chromatin structures in single cells. After fixation, digestion and in situ proximity ligation, 3C libraries contain long DNA concatemers in which neighboring fragments represent chromatin regions that were in close proximity in individual nuclei. Single-allele chromatin conformations can therefore be derived from population-based assays in which several neighboring fragments in 3C concatemers are detected simultaneously as multiway chromatin interactions. Tri-C can identify such multiway interactions with viewpoints of interest with high sensitivity and at high resolution [13, 14]. Compared to other innovative methods to detect multiway chromatin interactions, such as three-way 4C [7] and multi-contact 4C (MC-4C) [15, 16], Tri-C offers advantages in throughput, sensitivity, and resolution [13].

The Tri-C procedure contains three stages: (I) 3C library preparation, (II) sonication and addition of sequencing adaptors, and (III) capture enrichment (*see* Fig. 1). To enable efficient detection of multiway chromatin interactions, 3C libraries for Tri-C are generated using a restriction enzyme selected to create relatively small DNA fragments (~200 bp) for the viewpoints of interest. We usually recommend *Nla*III, which has a recognition sequence of 4 bp and produces fragments with a median size of ~130 bp in the mouse and human genome [13]. Since the original publication of Tri-C [13], we have implemented a new protocol for 3C library preparation which generates high-quality libraries from intact nuclei [17, 18]. Sonication of 3C libraries to ~450 bp produces fragments of which ~50% contain multiple ligation junctions from which multiway chromatin interactions can be derived. After addition of sequencing adaptors, the interactions with selected viewpoints of interest are enriched using an optimized oligonucleotide-mediated capture procedure. The enriched multiway interactions can be identified using the Illumina sequencing platforms, allowing for the generation of deep 3C profiles at high resolution. Since Illumina sequencing provides accurate identification of the random sonication ends of the fragments, these can be used as unique molecular identifiers to filter PCR duplicates, thus allowing for quantitative analysis of the detected multiway interactions. Furthermore, Tri-C allows for multiplexing of both viewpoints and samples and therefore enables high-throughput analyses of multiple genomic regions and cell types of interest in a single experiment.

2   Materials

Prepare all solutions using PCR-grade water and analytical grade reagents. We recommend working with safe-lock microtubes throughout the protocol to prevent loss of material due to evaporation.

2.1   3C Library Preparation

1. Cell culture medium (*see* **Note 1**).

2. 37% (w/v) formaldehyde.

3. 1 M glycine.

4. 1× Phosphate-buffered saline (PBS).

5. Lysis buffer: 10 mM Tris–HCl pH 8, 10 mM NaCl, 0.2% Igepal CA-630, 1× EDTA-free cOmplete Protease Inhibitor Cocktail (Sigma-Aldrich; 1 tablet per 50 mL). Prepare fresh each time, prior to starting the experiment, and put on a rolling incubator in the cold room (4 °C) to dissolve and cool.

6. 10 U/μL *Nla*III with manufacturer-provided 10× restriction buffer.

7. Safe-lock microtubes.

8. Nuclease-free water.

9. 20% (w/v) SDS.

10. 20% (v/v) Triton X-100.

11. ≥600 U/mL proteinase K.

12. Ligation solution: 0.4 U/μL T4 DNA ligase in 2.1× manufacturer-supplied T4 DNA ligase buffer.

13. TE buffer: 10 mM Tris–HCl pH 8.0, 1 mM EDTA-NaOH pH 8.0.

14. ≥15 U/mL DNase-free RNase A.

15. Light PhaseLock gel tubes (1.5 mL).

16. Phenol–chloroform–isoamyl alcohol (25:24:1).

17. 3 M sodium acetate pH 5.5.

18. 20 mg/mL glycogen.

19. 100%, 80% and 70% ethanol.

20. Qubit spectrophotometer and dsDNA BR and HS assay reagents.

2.2 Sonication and Addition of Sequencing Adaptors

1. Sonicator and compatible tubes (e.g., Covaris S220).

2. AMPure XP SPRI Beads.

3. Magnetic stand (e.g., DynaMag).

4. Agilent 2200 TapeStation or Agilent BioAnalyzer and reagents.

5. NEBNext Ultra II DNA Library Prep Kit for Illumina (New England Biolabs), including End Prep, Adaptor Ligation and PCR amplification reagents.

6. NEBNext Multiplex Oligonucleotides for Illumina (Index Primers Set 1 and 2).

7. Herculase II Fusion DNA polymerase.

2.3 Capture Enrichment

1. Biotinylated capture oligonucleotides (*see* **Note 2**).

2. 1 mg/mL C_0t DNA of relevant species.

3. Vacuum concentrator.

4. HyperCap Target Enrichment Kit (Roche), including Universal Blocking Oligonucleotides, $2\times$ Hybridization buffer, Component H, Beads wash buffer, Strict wash buffer, Wash buffer I, Wash buffer II, Wash buffer III and KAPA HiFi PCR reagents.

5. Streptavidin Dynabeads M-270.

6. High-throughput sequencing machine, able to do long paired-end reads (e.g., Illumina HiSeq).

3 Methods

It is important to carefully select viewpoints of interest before starting the experiment. The viewpoints should be located on small restriction fragments generated by the restriction enzyme used for chromatin digestion. We recommend viewpoint fragments in the range of 150–250 bp and selecting the restriction enzyme used for digestion based on the fragment size distribution at the viewpoints of interest. We usually use *Nla*III or *Dpn*II for chromatin digestion. This protocol is written based on *Nla*III digestion and can be modified for *Dpn*II digestion using information provided in the Notes.

The capture oligonucleotides should be designed at the middle of the restriction fragments on which the viewpoints of interest are located. We recommend using capture oligonucleotides which are 120 nt in size. Repetitive sequences should be avoided. We usually use BLAT to make sure the selected oligonucleotide sequences are suitable.

Tri-C allows for multiplexing both viewpoints and samples. During the hybridization procedure, up to six samples (with unique indices) can be included in a single tube and up to three tubes can be handled in parallel comfortably, thus allowing for multiplexing up to 18 samples in a single hybridization experiment. The number of targeted viewpoints does not affect the experimental procedure and is only relevant for the amount of sequencing researchers are prepared to do. Depending on the quality of the library and the required data depth, 1–10 M reads are required per viewpoint per sample.

3.1 3C Library Preparation

3.1.1 Fixation

1. Collect cells from tissue or culture and make single-cell suspensions of $10–20 \times 10^6$ cells in 10 mL of medium (*see* **Note 1**).

2. Add 540 µL 37% formaldehyde and incubate for 10 min at room temperature on a rolling incubator.

3. Quench by adding 1.5 mL 1 M cold glycine. Centrifuge for 10 min at $500 \times g$, 4 °C.

4. Wash pellet by gently resuspending in 10 mL cold PBS. Centrifuge for 10 min at $500 \times g$, 4 °C, and gently remove supernatant without disturbing pellet.

5. Resuspend pellet in 5 mL cold lysis buffer. Incubate for 20 min on ice.

6. Centrifuge for 10 min at $500 \times g$, 4 °C, then gently remove supernatant without disturbing pellet.

7. Wash pellet by gently resuspending in 10 mL cold PBS. Centrifuge for 10 min at $500 \times g$, 4 °C, then gently remove supernatant without disturbing pellet.

8. Resuspend pellet in 1 mL PBS and transfer to a microtube. Snap freeze with ethanol and dry ice or liquid nitrogen (*see* **Note 3**).

3.1.2 Digestion (See Note 4)

1. Thaw fixed cells on ice.

2. Centrifuge the cells for 15 min at $500 \times g$, 4 °C, and gently remove supernatant without disturbing pellet.

3. Resuspend pellet in 650 µL $1\times$ *Nla*III restriction buffer.

4. Transfer three aliquots of 200 µL each to safe-lock microtubes, and add sequentially to each tube: 404 µL nuclease-free water, 60 µL $10\times$ *Nla*III restriction buffer, 10 µL 20% SDS.

5. Use the remaining 50 μL fixed nuclei as nondigested control (Control 1). Add 192.5 μL nuclease-free water, 28.5 μL 10× NlaIII restriction buffer, and 4 μL 20% SDS.

6. Shake all tubes horizontally at 37 °C, 500 rpm (intermittent: 30 s on/30 s off) for 1 h.

7. If nuclei have clumped, gently use a P200 pipette to disaggregate.

8. Add 66 μL 20% Triton X-100 to each digestion (25 μL to Control 1) to neutralize the SDS. Incubate at 37 °C, 500 rpm (intermittent: 30 s on/30 s off) for 1 h.

9. Add 20 μL 10 U/μL NlaIII to each digestion reaction. Shake until end of the day and then add a further 20 μL NlaIII to each digestion reaction.

10. Shake at 37 °C, 500 rpm (intermittent: 30 s on/30 s off) overnight.

11. Add 20 μL NlaIII to each digestion reaction and incubate at 37 °C, 500 rpm (intermittent: 30 s on/30 s off) for another 6 h.

3.1.3 Ligation and Decrosslinking

1. Take 100 μL from each digestion reaction and pool to one microtube to make the digested, nonligated control (Control 2).

2. Add 3 μL 600 U/mL proteinase K to controls 1 and 2 and incubate at 65 °C overnight (while proceeding with **steps 3** and **4**), then store the controls at −20 °C.

3. Incubate the rest of the digestion reactions at 65 °C for 20 min to heat-inactivate the restriction enzyme, then immediately cool on ice.

4. Add 642 μL ligation solution to each reaction and incubate at 16 °C, 500 rpm (intermittent: 30 s on/30 s off) for ~22 h.

5. Centrifuge the ligation reactions for 15 min at $500 \times g$, room temperature. Gently remove all of the supernatant without disturbing the nuclear pellet (*see* **Note 5**).

6. Resuspend each pellet in 300 μL TE buffer. Add 5 μL 600 U/ mL proteinase K to each ligation reaction and incubate at 65 °C overnight.

3.1.4 DNA Extraction

1. Cool ligation reactions to 37 °C and defrost Controls 1 and 2.

2. Add 5 μL 15 U/mL RNAse A to each ligation reaction and to Controls 1 and 2, and incubate at 37 °C, 500 rpm (intermittent: 30 s on/30 s off) for 30 min.

3. Prepare PhaseLock tubes by centrifuging at $5000 \times g$ for 2 min.

4. Add 310 µL phenol–chloroform–isoamyl alcohol to each ligation reaction and Control. Vortex thoroughly to mix.

5. Transfer the mixture to a prespun PhaseLock tube and centrifuge for 10 min at 12,600 × g, room temperature.

6. Transfer the upper layer to a microtube, avoiding the viscous interface.

7. Add 30 µL of 3 M sodium acetate and 1 µL of glycogen and mix by inversion.

8. Add 900 µL of 100% ethanol and mix thoroughly by inversion. Freeze at −20 °C for at least 2 h.

9. Centrifuge for 30 min at 21,000 × g, 4 °C.

10. Discard supernatant and wash pellets in 1 mL 70% cold ethanol. Centrifuge for 2 min at 21,000 × g, 4 °C.

11. Remove ethanol and repeat ethanol wash for a total of two washes.

12. Remove ethanol, briefly centrifuge, and remove residual ethanol.

13. Dry at room temperature and dissolve pellets in TE buffer (100 µL for ligation reactions; 30 µL for Controls) at 4 °C overnight.

14. Pool ligation reactions into a single microtube and proceed with the quality control of the 3C library or store at −20 °C.

3.1.5 Quality Control of 3C Library

1. Run ~10–15 µL of Controls 1 and 2 and ~5–10 µL of 3C library on a 1% agarose gel (*see* **Note 6** and Fig. 2). Control 1 should contain undegraded DNA at high molecular weight; Control 2 should contain digested DNA; the 3C library should contain a ligated product at relatively high molecular weight.

2. Determine the concentration of the 3C library with the Qubit dsDNA BR assay, following the manufacturer's instructions. 1×10^7 cells should produce approximately 15–30 µg of 3C library.

3.2 Sonication and Addition of Sequencing Adaptors

In this step, it is important to maximize input DNA and minimize losses to maintain optimal library complexity. We recommend using the NEBNext Ultra II DNA Library Prep Kit for Illumina. The protocol of the kit states that it is compatible with up to 1 µg starting material, but in our experience, it works well with up to 2 µg. We recommend sonicating 3 µg of 3C library and using all recovered material (~2 µg) as starting material for the NEBNext reactions. We recommend performing multiple parallel reactions to maximize library complexity. When there is 6 µg of 3C library available, we recommend sonicating all material in one tube and splitting it over 2 NEBNext reactions. These reactions can be amplified with the same index primer and pooled prior to hybridization.

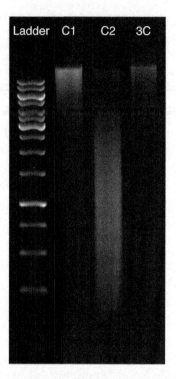

Fig. 2 Quality control of 3C libraries. DNA gel electrophoresis (1% agarose) of a representative nondigested chromatin control (C1), digested nonligated chromatin control (C2), and 3C library (3C). The ladder shows bands ranging from 10,000 bp at the top to 250 bp at the bottom, with the thick bands representing 6000, 3000, and 1000 bp

1. Sonicate 6 μg of 3C library to 400–500 bp fragments. We recommend using the Covaris S220 Focused-ultrasonicator with the following settings: duty cycle 10%; intensity 4; cycles per burst 200; time 55 s; set mode frequency sweeping. Prepare the 3C library for sonication by diluting 6 μg in a total of 120 μL TE buffer in a Covaris microtube, avoiding making any bubbles (*see* **Note** 7).

2. Transfer the sample to a low-bind microtube and add 0.7× volume (~85 μL) AMPure XP beads. Pipette up and down ten times and incubate at room temperature for 5 min.

3. Place the tube on a magnetic stand and discard the liquid when clear. Add 500 μL of fresh 80% ethanol without removing the tube from the magnetic stand. Avoid disturbing the beads by running the ethanol down the front of the tube. Incubate for 30 s at room temperature, then remove all the ethanol.

4. Repeat **step 3** without removing the beads from the magnetic stand.

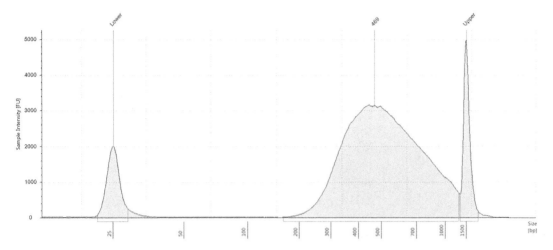

Fig. 3 Sonication. Assessment of processed 3C libraries after sonication using the Agilent Tapestation (D1000)

5. Briefly centrifuge the tube and replace it on the magnetic stand. Remove the residual ethanol with a P20 pipette, while taking care not to remove any beads.

6. Air dry the beads at room temperature on the magnetic stand until the beads are matt in appearance. Be careful not to over-dry the beads as this increases DNA loss. Cracks will appear when beads have dried too long.

7. Remove the tubes from the magnetic stand and resuspend the beads in 53 μL water. Mix by pipetting 10 times then incubate at room temperature for 2 min to elute the DNA off the beads.

8. Replace the tubes on the magnetic stand. Once the liquid is clear, recover 51 μL in a PCR tube.

9. Assess 1 μL of sonicated material using the TapeStation or BioAnalyzer (*see* **Note 8** and Fig. 3).

10. Split the sonicated material in two aliquots (25 μL each) in PCR tubes and perform two NEBNext End Prep reactions in parallel. To each tube, add 25 μL nuclease-free water, 7 μL NEBNext Ultra II End Prep reaction buffer and 3 μL NEBNext Ultra II End Prep enzyme mix. In a thermal cycler with heated lid, incubate for 30 min at 20 °C, then for 30 min at 65 °C.

11. Perform two NEBNext Adaptor Ligation reactions in parallel. To each tube, add 2.5 μL NEBNext Adaptor for Illumina, 30 μL NEBNext Ultra II Ligation Master Mix and 1 μL NEBNext Ultra II Ligation Enhancer. Incubate for 15 min at 20 °C.

12. Add 3 μL USER enzyme and incubate for 15 min at 37 °C with a heated lid.

13. Purify the DNA. We recommend using AMPure XP beads, following the procedure described in steps 2–8, using 1.8× volume of beads (~180 μL) and eluting in 50 μL. Size selection is not necessary, as adaptor dimers will not be captured.

14. Amplify half of the material by PCR with Herculase II DNA polymerase (*see* **Note 9**). Make up the reactions on PCR tubes over ice by sequentially adding: 25 μL adaptor-ligated 3C library, 3.5 μL PCR-grade water, 5 μL NEB Universal primer, 5 μL NEB Index primer (from NEBNext Multiplex Oligonucleotides for Illumina set), 10 μL 5× Herculase II reaction buffer, 0.5 μL 25 mM dNTP mix, 1 μL Herculase II Fusion DNA polymerase. Run the following program in a thermal cycler.

98 °C, 30 s.

6× [98 °C, 10 s; 65 °C, 30 s; 72 °C, 30 s].

72 °C, 5 min.

4 °C, hold.

15. Purify the DNA. We recommend using AMPure XP beads, following the procedure described in steps 2–8, using 1.8× volume of beads (~90 μL) and eluting in 52 μL.

16. Assess 1 μL of material using the TapeStation or BioAnalyzer (*see* Fig. 4).

17. Repeat **steps 14** and **15** with the remaining half of adaptor-ligated sample (*see* **Note 9**).

18. Pool parallel reactions that were amplified with the same index.

19. Quantify the library. We recommend using the Qubit dsDNA BR assay kit.

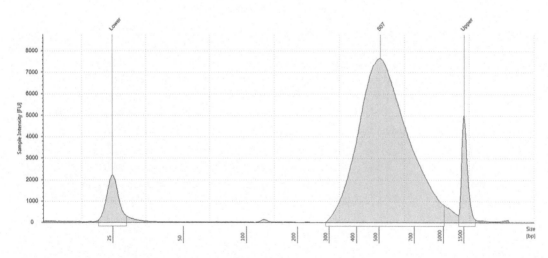

Fig. 4 PCR amplification and indexing. Assessment of processed 3C libraries after PCR amplification and indexing using the Agilent TapeStation (D1000)

3.3 Capture Enrichment

In this step, interactions with viewpoints of interest are enriched by hybridization with biotinylated oligonucleotides. For enrichment with single-stranded capture oligonucleotides, we recommend using KAPA HyperCap reagents (Roche). Enrichment can also be performed using double-stranded oligonucleotides [19, 20] or noncommercial hybridization reagents [20]. Two important considerations for capture enrichment are selection of appropriate targets without confounding bias (*see* **Note 10**) and the concentration of oligonucleotides (*see* **Note 11**) [17].

Tri-C allows for multiplexing any number of samples. The protocol below assumes an experimental setup in which three replicates of two different cell types or conditions (six samples in total) are analyzed. However, all steps can be scaled up or down to suit any number of multiplexed libraries. Up to six libraries can usually be combined in a single tube; if more than six libraries are used, divide the mixture of library/C_0t/blocking DNA equally over multiple tubes.

3.3.1 Preparation of Capture Oligonucleotides (See **Notes 11** and **12**)

1. Reconstitute individual or pools of oligonucleotides to a stock concentration of 1 μM. If using individual oligonucleotides, generate pools by mixing in exact 1:1 molar ratio.

2. Dilute the oligonucleotide pool so that each individual oligonucleotide is at a concentration of 2.9 nM and the total concentration of the pool is 2.9 nM multiplied by the number of unique oligonucleotides in the pool.

3.3.2 Hybridization Reaction

1. In a PCR tube, combine 1–2 μg of each uniquely indexed sample (from Subheading 3.2) in exact 1:1 mass ratio.

2. Add 30 μL of 1 μg/μL species-specific C_0t DNA (5 μg per library). Mix by pipetting up and down.

3. Vacuum centrifuge at 50 °C with tube lids open until the sample is completely dry (*see* **Note 13**).

4. Add 40.2 μL of Universal Enhancing Oligonucleotides (6.7 μL per library) to the desiccated DNA and mix by pipetting up and down.

5. Add 84 μL of 2× Hybridization buffer (14 μL per library) and 36 μL of Hybridization Component H (6 μL per library), mix by pipetting up and down, briefly centrifuge, and incubate at room temperature for 2 min (*see* **Note 12**).

6. Add 27 μL of pooled biotinylated capture oligonucleotides at working concentration (4.5 μL per library) to the hybridization mixture, mix by pipetting up and down and briefly centrifuge.

7. Transfer the tube to the thermocycler (with lid heated to 105 °C) and incubate at 95 °C for 5 min and then at 47 °C for 68–72 h.

3.3.3 Binding to
Streptavidin Beads

1. Place 300 µL of Dynabeads M-270 streptavidin beads (50 µL per library) in a low-bind microtube (*see* **Note 14**). Place beads on a magnetic stand and remove the liquid once clear.

2. Add 600 µL of 1× Bead Wash buffer (100 µL per library), vortex to resuspend the beads, and centrifuge briefly (*see* **Note 15**). Replace the tube on the magnetic stand and remove the liquid once clear (~30 s).

3. Wash once more with 600 µL 1× Bead Wash buffer by repeating **step 2**.

4. Remove the tube from the magnetic stand and resuspend the beads in 300 µL of 1× Bead Wash buffer (50 µL per library).

5. When ready to add the hybridization reaction, place the beads on the magnetic stand, and remove the Bead Wash buffer from the beads once clear. Transfer the hybridization reaction to the streptavidin beads and mix carefully by pipetting (*see* **Note 16**).

6. Place the tube on the thermomixer and incubate at 47 °C, 600 rpm for 45 min (*see* **Note 17**).

7. Add 300 µL of preheated (47 °C) 1× Wash buffer I (50 µL per library) to the beads and bound DNA and mix by pipetting. Place the tube in the magnetic stand and discard all the liquid when clear (~30 s).

8. Remove the tube from magnetic stand, add 600 µL of preheated (47 °C) Stringent Wash buffer (100 µL per library), mix by vortexing, and briefly centrifuge to remove any liquid from the lid.

9. Incubate the tube at 47 °C for 5 min. If there is condensation in the lid, briefly centrifuge the tube to remove any liquid from the lid.

10. Place the tube in the magnetic stand and discard all the liquid when clear (~30 s). Remove the tube from the magnetic stand, perform a second stringent wash with 600 µL of preheated (47 °C) Stringent Wash buffer (100 µL per library), mix by vortexing, and briefly centrifuge to remove any liquid from the lid. Work quickly to maintain the temperature at 47 °C.

11. Incubate the tube at 47 °C for 5 min. If there is condensation in the lid, briefly centrifuge the tube to remove any liquid from the lid.

12. Place the tube in the magnetic stand and discard all the liquid when clear (~30 s). Remove the tube from the magnetic stand and add 600 µL of 1× Wash Buffer I (100 µL per library; room temperature).

13. Mix by vortexing for 10 s, briefly centrifuge, and incubate at room temperature for 1 min.

14. Place the tube in the magnetic stand and discard all the liquid when clear (~30 s). Remove the tube from the magnetic stand and add 600 μL of 1× Wash Buffer II (100 μL per library; room temperature).

15. Mix by vortexing for 10 s, briefly centrifuge, and incubate at room temperature for 1 min.

16. Place the tube in the magnetic stand and discard all the liquid when clear (~30 s). Remove the tube from the magnetic stand and add 600 μL of 1× Wash Buffer III (100 μL per library; room temperature).

17. Mix by vortexing for 10 s, briefly centrifuge, and incubate at room temperature for 1 min.

18. Place the tube in the magnetic stand and discard all the liquid when clear (~30 s). Remove the tube from the magnetic stand and resuspend the beads in 240 μL PCR-grade water (40 μL per library) (*see* **Note 18**).

3.3.4 Amplification of Captured DNA

1. Amplify half of the material (*see* **Note 19**). Make up the master mix in a microtube over ice by sequentially adding: 120 μL bead-bound DNA, 150 μL KAPA HiFi Hot Start Ready mix, 30 μL POST-LM-PCR Oligos 1&2. Aliquot to 50 μL reaction volumes in PCR tubes. Run the following PCR setup.

   98 °C, 45 s

   10×: [98 °C, 15 s; 60 °C, 30 s; 72 °C, 30 s]

   72 °C, 1 min

   4 °C, hold.

2. Pool the six reactions into a microtube and place on the magnetic stand. Purify the DNA from the supernatant using 1.8× (540 μL) AMPure XP beads, as in Subheading 3.2 (Sonication and addition of sequencing adaptors), **steps 2–8**, eluting in 54 μL PCR-grade water (recovering 52 μL).

3. Confirm size and yield of amplified DNA with the TapeStation and high-sensitivity D1000 reagents or BioAnalyzer (*see* Fig. 5).

4. Repeat PCRs and DNA purification (**steps 1** and **2**) with the remaining material. Combine DNA from both amplifications.

5. Quantify the library with the Qubit dsDNA HS assay kit.

3.3.5 Double Capture

When using optimally titrated oligonucleotides, repeating the capture increases the on-target sequencing efficiency by 2–3 fold compared to a single capture step. The amount of DNA recovered after the first capture is generally <2 μg. The second capture is therefore usually performed as described for a single library using all of the recovered material.

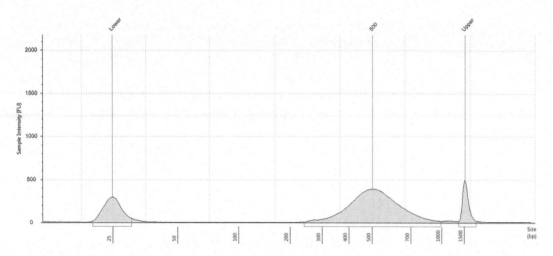

Fig. 5 Capture enrichment. Assessment of processed 3C libraries after capture enrichment using the Agilent Tapestation (D1000)

1. Perform *Hybridization Reaction* (*see* Subheading 3.3.2) as described using volumes for a single library and hybridizing at 47 °C for 18–22 h (**step 7**).

2. Perform *Binding to streptavidin beads* (*see* Subheading 3.3.3) as described using volumes for a single library.

3. Perform *Amplification of captured DNA* (*see* Subheading 3.3.4) as described, except that during the DNA purification with AMPure XP beads (**step 2**), resuspend the washed and dried beads in 25 μL of water and recover 23 μL.

3.3.6 Sequencing

1. Using the measured DNA concentration, make a 10 nM dilution of amplified captured DNA.

2. Perform accurate library quantification of the 10 nM dilution using quantitative PCR with size correction. We recommend using the KAPA Library Quantification Kit with 1:10,000 and 1:20,000 dilutions.

3. Dilute the DNA to an appropriate concentration for sequencing (usually 4 nM) and perform sequencing (300 cycles; 150 bp paired-end reads; *see* **Note 20**).

3.3.7 Analysis

1. The data can be analyzed using scripts available at https://github.com/oudelaar/TriC or the software package described in Telenius et al 2020 [21]. Briefly, reads are processed and in silico digested, after which fragments interacting with viewpoints of interested are identified and plotted (*see* Fig. 1).

4   Notes

1. It is not critical to use serum-free media. Adherent cells or tissues need to be trypsinized or dissected into a single-cell suspension in media usually supporting cell growth.

2. It is important to use high-purity biotinylated capture oligonucleotides. We recommend xGen Lockdown Pools from IDT or HPLC-purified oligonucleotides from Sigma.

3. Snap freezing aids digestion, so cells can be thawed again for digestion at this point or stored long-term at −80 °C.

4. To digest with *Dpn*II instead of *Nla*III, set the digestion reaction up in 434 μL nuclease-free water, 60 μL 10× *Dpn*II restriction buffer, and 10 μL 20% SDS. Add 10 μL 50 U/μL *Dpn*II to the digestion reactions at each of the set time points.

5. It is important to carefully remove the supernatant, as residual digestion and ligation buffers will contribute high levels of DTT to the DNA precipitation and impair accuracy of 3C library quantification. We recommend keeping the supernatant and extracting the DNA from the supernatant in the case of low yield. A good 3C library should have >90% DNA within the nuclear pellet.

6. It is also possible to assess the quality of the digestion using qPCR as previously described [22] (Chapter 1). Make sure to design primer pairs which are appropriate for the used restriction enzyme and species.

7. If using a different model of sonicator, use high molecular weight genomic DNA to optimize sonication for a modal distribution around 450 bp in size (*see* Fig. 3).

8. The material can be stored at −20 °C at this stage.

9. We recommend using the Herculase II Fusion Polymerase Kit instead of the NEBNext reagents as it gives a better yield in our hands. We recommend performing the PCR in duplicate on half the sample at a time. This provides backup material in case the PCR fails and allows for adjustment of the second PCR if amplification in the first reaction is not sufficient. Make sure to choose the indices carefully to allow for multiplexing using the Illumina pooling guidelines.

10. Because methods of targeted 3C library enrichment are not 100% efficient, a variable level of bias is introduced between enrichment of viewpoint fragments which are simultaneously targeted. When designing oligonucleotide pools for Tri-C, it is important not to include probes for two viewpoints between which you wish to measure interactions (e.g., a promoter and its cognate enhancer). For more information, we recommend reading the Supplementary Note associated with Downes et al. 2020 [17].

11. The specificity of the enrichment is dependent on the concentration of the capture oligonucleotides. Using the reagents described in this protocol, we have found that the optimal concentration of each individual oligonucleotide is ~2.9 nM [17]. With different reagents, the optimal concentration might differ, but we suspect it will be in a similar range. To determine the appropriate concentration of an oligonucleotide pool for enrichment, we multiply 2.9 nM by the number of unique oligonucleotides in a pool. We usually create a 1 μM pool by combining equal volumes of oligonucleotide stocks at 1 μM and dilute this pool to the calculated concentration.

12. The capture oligonucleotides are extremely efficient. Is therefore important to avoid ordering, storing and working with concentrated stocks of oligonucleotides that you do not wish to multiplex in a single experiment to prevent contamination. However, oligonucleotides that will be mixed in a single experiment can be ordered and stored together, either individually or as a pooled set. Avoid working with open tubes containing capture oligonucleotides and buffers or blocking reagents simultaneously on the bench to avoid contamination.

13. If a vacuum centrifuge is not available, it is possible to isolate prehybridization DNA using AMPure XP beads. Add the multiplexed library, C_0t DNA, and blocking oligonucleotide mix to 1.8 volumes of beads and perform a standard cleanup reaction (see Subheading 3.2, **steps 2–8**). Elute in 40.2 μL of Universal Enhancing Oligonucleotides (6.7 μL per library).

14. The streptavidin beads tend to stick to the walls of the microtubes. We have found that Safeseal Microcentrifuge Tubes (Sorenson BioScience) have a lower affinity for the beads and therefore reduce bead losses. When resuspending the beads in water before PCR amplification, make sure to manually remove the beads from the walls of the tube using a pipette tip.

15. Some of the wash buffers contain high levels of SDS, which can take some time to dissolve after being frozen. 10× stocks can be warmed on a thermomixer to dissolve the SDS.

16. To transfer total volume of the hybridization reaction (90 μL for initial capture and 15 μL for second capture), we set the pipette to a higher volume to ensure transfer of the entire solution. Work quickly to maintain the temperature and prevent the beads from drying out. When mixing by pipetting, take care not to lose too many beads in the tip due to their high affinity for plastic.

17. During the incubation of the capture reaction with the streptavidin beads, periodically check to make sure that the beads have not settled and carefully pipette to resuspend if they have.

18. DNA is not eluted but will be amplified off the beads; either store the DNA bound to the beads at -20 °C or proceed to amplification.

19. To amplify the captured DNA off the streptavidin beads, we perform two PCR reactions per library. This can be done simultaneously or separately on half of the hybridized beads at a time (as described). Separating the amplification over two reactions allows for checking for sufficient amplification cycles and provides backup material in case of errors. 10 cycles should be sufficient for six multiplexed samples; if working with fewer multiplexed libraries, consider increasing the number of cycles to 11–12.

20. Long reads are required to sequence multiple ligation junctions, which allow for the identification of multiway (≥ 3) chromatin interactions.

References

1. Oudelaar AM, Higgs DR (2021) The relationship between genome structure and function. Nat Rev Genet 22:154–168

2. Lieberman-Aiden E, van Berkum NL, Williams L et al (2009) Comprehensive mapping of long-range interactions reveals folding principles of the human genome. Science 326: 289–293

3. Zhao Z, Tavoosidana G, Sjölinder M et al (2006) Circular chromosome conformation capture (4C) uncovers extensive networks of epigenetically regulated intra- and interchromosomal interactions. Nat Genet 38: 1341–1347

4. Simonis M, Klous P, Splinter E et al (2006) Nuclear organization of active and inactive chromatin domains uncovered by chromosome conformation capture-on-chip (4C). Nat Genet 38:1348–1354

5. Davies JOJ, Telenius JM, McGowan SJ et al (2015) Multiplexed analysis of chromosome conformation at vastly improved sensitivity. Nat Methods 86:1202–1210

6. Hughes JR, Roberts N, McGowan S et al (2014) Analysis of hundreds of cis-regulatory landscapes at high resolution in a single, high-throughput experiment. Nat Genet 46: 205–212

7. Olivares-Chauvet P, Mukamel Z, Lifshitz A et al (2016) Capturing pairwise and multi-way chromosomal conformations using chromosomal walks. Nature 540:296–300

8. Beagrie RA, Scialdone A, Schueler M et al (2017) Complex multi-enhancer contacts captured by genome architecture mapping. Nature 543:519–524

9. Quinodoz SA, Ollikainen N, Tabak B et al (2018) Higher-order inter-chromosomal hubs shape 3D genome organization in the nucleus. Cell 174:744–57.e24

10. Stevens TJ, Lando D, Basu S et al (2017) 3D structures of individual mammalian genomes studied by single-cell Hi-C. Nature 544:59–64

11. Flyamer IM, Gassler J, Imakaev M et al (2017) Single-nucleus Hi-C reveals unique chromatin reorganization at oocyte-to-zygote transition. Nature 544:110–114

12. Nagano T, Lubling Y, Stevens TJ et al (2013) Single-cell Hi-C reveals cell-to-cell variability in chromosome structure. Nature 502:59–64

13. Oudelaar AM, Davies JOJ, Hanssen LLP et al (2018) Single-allele chromatin interactions identify regulatory hubs in dynamic compartmentalized domains. Nat Genet 50: 1744–1751

14. Oudelaar AM, Harrold CL, Hanssen LL et al (2019) A revised model for promoter competition based on multi-way chromatin interactions at the α-globin locus. Nat Commun 10: 1–8

15. Allahyar A, Vermeulen C, Bouwman BAM et al (2018) Enhancer hubs and loop collisions identified from single-allele topologies. Nat Genet 50:1151–1160

16. Vermeulen C, Allahyar A, Bouwman BAM et al (2020) Multi-contact 4C: long-molecule sequencing of complex proximity ligation products to uncover local cooperative and

competitive chromatin topologies. Nat Protoc 15:364–397

17. Downes DJ, Beagrie RA, Gosden ME et al (2021) High-resolution targeted 3C interrogation of cis-regulatory element organization at genome-wide scale. Nat Commun 12:531

18. Downes DJ, Smith AL, Karpinska MA et al (2022) Capture-C: a modular and flexible approach for high-resolution chromosome conformation capture. Nat Protoc 17:445–475

19. Oudelaar AM, Beagrie RA, Gosden M et al (2020) Dynamics of the 4D genome during in vivo lineage specification and differentiation. Nat Commun 11:2722

20. Golov AK, Ulianov SV, Luzhin AV et al (2020) C-TALE, a new cost-effective method for targeted enrichment of Hi-C/3C-seq libraries. Methods 170:48–60

21. Telenius JM, Downes DJ, Sergeant M et al (2020) CaptureCompendium: a comprehensive toolkit for 3C analysis. bioRxiv 2020: 2020.02.17.952572

22. Oudelaar AM, Downes DJ, Davies JOJ et al (2017) Low-input capture-C: a chromosome conformation capture assay to analyze chromatin architecture in small numbers of cells. Bio Protoc 7:e2645

Chapter 7

Assessing Specific Networks of Chromatin Interactions with HiChIP

Dafne Campigli Di Giammartino, Alexander Polyzos, and Effie Apostolou

Abstract

The introduction of chromosome conformation capture (3C)-based technologies coupled with next-generation sequencing have significantly advanced our understanding of how the genetic material is organized within the eukaryotic nucleus. Three-dimensional (3D) genomic organization occurs at hierarchical levels, ranging from chromosome territories and subnuclear compartments to smaller self-associated domains and fine-scale chromatin interactions. The latter can be further categorized into different subtypes, such as structural or regulatory, based either on their presumed functionality and/or the factors that mediate their formation. Various enrichment strategies coupled with 3C-based technologies have been developed to prospectively isolate and quantify chromatin interactions around regions occupied by specific proteins or marks of interest. These approaches not only enable high-resolution characterization of the selected chromatin contacts at a cost-effective manner, but also offer important biological insights into their organizational principles and regulatory function. In this chapter, we will focus on the recently developed HiChIP technology with an emphasis on the discovery of putative active enhancers and promoter interactions in cell types of interest. We will describe the specific steps for designing, performing and analyzing successful HiChIP experiments as well as important limitations and considerations.

Key words 3D chromatin architecture, Enhancer–promoter contact, Chromatin loop, Chromosome conformation capture, HiChIP, H3K27ac, Gene regulation

1 Introduction

3D chromatin architecture plays important roles in regulating cellular functions [1]. Technological advances during the past decade have helped us appreciate that regulatory elements that are distal in the linear genome map could be in physical proximity as a result of the 3D chromatin folding [2, 3]. In particular, the development of 3C-based techniques, such as Hi-C [4] and its derivatives, have provided useful tools to investigate the regulatory principles that govern 3D genome architecture, leading to the identification of

---

Dafne Campigli Di Giammartino and Alexander Polyzos contributed equally to this work.

Tom Sexton (ed.), *Spatial Genome Organization: Methods and Protocols*,
Methods in Molecular Biology, vol. 2532, https://doi.org/10.1007/978-1-0716-2497-5_7,

different hierarchical layers of 3D chromatin topology. Those comprise chromosome compartments, (sub)megabase-sized topologically associated domains [5, 6] (TADs) and long-range chromatin contacts, including interactions among active or repressive regulatory elements and target genes [2]. While Hi-C has been pivotal to unravel genome organization principles at the levels of compartments, TADs and structural chromatin loops [4], it has limited ability to detect fine-scale enhancer–promoter contacts.

Mapping the spatial organization of *cis*-regulatory elements is critical to understand how enhancers regulate their target genes over distance in order instruct cell-type specific transcriptional programs in normal physiological conditions or in disease. Recent technological advances that combine Hi-C with antibody-based or sequence-based enrichment steps, such as ChIA-PET [7], PLAC-seq [8], HiChIP [9], and Capture Hi-C [10] (Chapter 5), have drastically improved the resolution and the discovery rate of promoter–enhancer interactions. Several recent studies have used these technologies combined with antibodies against architectural proteins, transcription factors or specific histone modifications to generate high-resolution 3D interaction networks around loci with specific chromatin features of interest [3, 8, 10–14]. These studies have offered novel insights into the nature, cell-type specificity and overall dynamics of chromatin contacts and their association to gene regulation and cell fate control.

In this chapter, we will describe HiChIP [9], a recently published method that combines in situ Hi-C with chromatin immunoprecipitation (HiChIP) to specifically capture and quantify DNA interactions between regions containing a particular protein or histone mark. As outlined in Fig. 1 the HiChIP protocol involves cross-linking DNA–protein contacts, permeabilizing nuclei and digesting DNA with a restriction enzyme. The resulting chromatin fragments are end-repaired and tagged with biotin prior to proximity ligation, which is carried out in the intact nucleus to minimize false-positive interactions. This process generates chimeric DNA fragments that were originally in physical 3D proximity in the nucleus and are marked with biotin at the junction. The nuclei are then lysed and the DNA sonicated. At this point an antibody for either protein or histone mark of interest is used in order to immunoprecipitate the protein–DNA complexes. After reverse cross-linking and DNA purification, the immunoprecipitated DNA is pulled-down using streptavidin beads. Library preparation is carried out on-beads through DNA tagmentation using the Tn5 transposase (which cleaves the DNA tagging it with a universal overhang). The tagmented template is then amplified by PCR using appropriate primers, followed by paired-end next-generation sequencing. The resulting reads are then processed for alignment, mapping and filtering to determine all valid paired interactions prior to calling statistically significant contacts. In case of

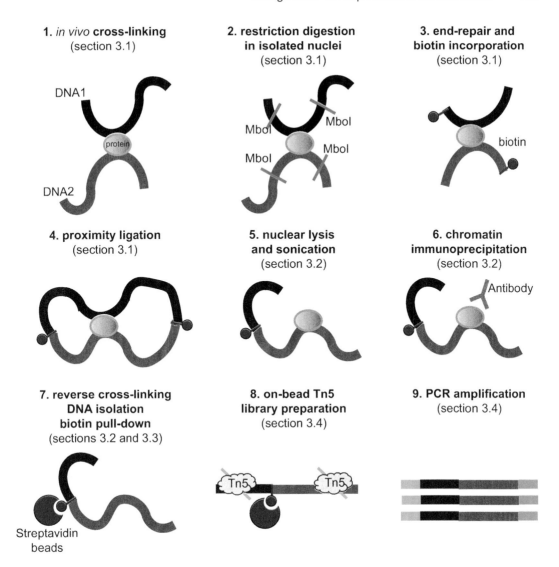

1. *in vivo* **cross-linking**
(section 3.1)

DNA1

protein

DNA2

**2. restriction digestion
in isolated nuclei**
(section 3.1)

MboI

MboI

MboI

MboI

**3. end-repair and
biotin incorporation**
(section 3.1)

biotin

4. proximity ligation
(section 3.1)

**5. nuclear lysis
and sonication**
(section 3.2)

**6. chromatin
immunoprecipitation**
(section 3.2)

Antibody

7. *reverse cross-linking*
**DNA isolation
biotin pull-down**
(sections 3.2 and 3.3)

Streptavidin
beads

**8. on-bead Tn5
library preparation**
(section 3.4)

Tn5   Tn5

9. PCR amplification
(section 3.4)

Fig. 1 HiChIP overview. Schematic representation of the main steps of a typical HiChIP procedure. Matching step numbers are reported accordingly to the protocol provided

H3K27ac (acetylated lysine-27 of histone H3) HiChIP, these contacts include putative enhancer–promoter, enhancer–enhancer, and promoter–promoter interactions.

We provide a detailed step-by-step experimental protocol and bioinformatics pipeline as we have optimized them specifically for H3K27ac HiChIP in mouse embryonic stem cells (ESCs) [14]. However, we will also mention in the Notes section how to optimize the protocol in order to capture chromatin contacts bound by transcription factors, such as KLF4, or in different cell types, such as mouse embryonic fibroblasts (MEFs).

2 Materials

2.1 Hi-C (Cross-Linking, Digestion, and Ligation)

1. Mouse embryonic stem cells (mESCs), cultured on irradiated feeder cells in KO-DMEM media, supplemented with 15% heat-inactivated fetal bovine serum, 1× GlutaMAX, 100 U/mL penicillin–streptomycin, 1× MEM nonessential amino acids, 0.1 mM β-mercaptoethanol, and 1000 U/mL leukemia inhibitory factor. See **Note 1** regarding culture conditions for mouse embryonic fibroblasts (MEFs).

2. 0.25% trypsin–EDTA.

3. 1× phosphate-buffered saline (PBS).

4. 1% formaldehyde in cell culture medium (as described above), made from 37% commercial stock (within 3 months of opening; store in dark conditions at room temperature), or fresh from 16% paraformaldehyde ampoules).

5. 2.5 M glycine.

6. LoBind microtubes.

7. Hi-C lysis buffer: 10 mM Tris–HCl pH 8, 10 mM NaCl, 0.2% (v/v) Igepal-CA630, 1× EDTA-free protease inhibitor cocktail (Roche).

8. 0.5% (w/v) SDS.

9. Molecular biology-grade water.

10. 10% (v/v) Triton X-100.

11. 1× NEBuffer 2 (New England Biolabs).

12. 25 U/μL MboI (see **Note 2**).

13. Fill-in master mix: 0.3 mM biotin-dATP, 0.3 mM dCTP, 0.3 mM dGTP, 0.3 mM dTTP, 1 U/μL DNA polymerase I large (Klenow) fragment, in nuclease-free water.

14. 10× ligation buffer: 50 mM Tris–HCl pH 7.5, 10 mM MgCl$_2$, 1 mM ATP, 10 mM DTT.

15. 400 U/μL T4 DNA ligase (NEB).

16. 10 mM ATP.

17. Ligation master mix: 1.6 mM ATP, 0.16 mg/mL BSA, 1.3% (v/v) Triton X-100, 4.2 U/μL T4 DNA ligase in 1.6× ligation buffer.

18. 10 mM Tris–HCl pH 7.5.

19. 20 mg/mL proteinase K.

20. 20 mg/mL RNase A.

2.2 Sonication and Chromatin Immunoprecipitation

1. Nuclear lysis buffer: 50 mM Tris–HCl pH 7.5, 10 mM EDTA, 0.5% (w/v) SDS, 1× protease inhibitors (Roche).

2. Bioruptor Pico sonicator (Diagenode) or equivalent.

3. ChIP dilution buffer: 16.7 mM Tris–HCl pH 7.5, 167 mM NaCl, 2 mM EDTA, 0.01% (w/v) SDS, 1.1% (v/v) Triton X-100.

4. Protein G magnetic Dynabeads (*see* **Note 3**).

5. Magnetic stand.

6. ChIP-grade H3K27ac antibody from Abcam (ab#4729) (*see* **Note 4**).

7. 0.1 mg/mL BSA in ChIP dilution buffer.

8. Low salt wash buffer: 20 mM Tris–HCl pH 7.5, 150 mM NaCl, 2 mM EDTA, 0.1% (w/v) SDS, 1% (v/v) Triton X-100.

9. High salt wash buffer: 20 mM Tris–HCl pH 7.5, 500 mM NaCl, 2 mM EDTA, 0.1% (w/v) SDS, 1% (v/v) Triton X-100.

10. LiCl wash buffer: 10 mM Tris–HCl pH 7.5, 250 mM LiCl, 1 mM EDTA, 1% (v/v) Igepal-CA360, 1% (w/v) sodium deoxycholate.

11. TE buffer: 10 mM Tris–HCl pH 8, 1 mM EDTA.

12. DNA elution buffer: 50 mM sodium bicarbonate, 1% (w/v) SDS.

13. ZYMO DNA purification kit, Qiagen minElute kit or equivalent.

14. Qubit fluorimeter and HS dsDNA quantification kit assay (Fisher Scientific).

2.3 Biotin Pull-Down with Streptavidin Beads

1. Streptavidin C1 beads (Thermo Fisher).

2. Tween wash buffer: 5 mM Tris–HCl pH 7.5, 1 M NaCl, 0.5 mM EDTA, 0.05% (v/v) Tween 20.

3. 2× biotin binding buffer: 10 mM Tris–HCl pH 7.5, 2 M NaCl, 1 mM EDTA.

4. 2× TD buffer: 20 mM Tris–HCl pH 7.5, 10 mM MgCl$_2$, 20% dimethylformamide. Also make a 1× solution by diluting in water.

2.4 Library Preparation for Sequencing

1. Tn5 transposase (Nextera, Illumina).

2. 50 mM EDTA.

3. PCR master mix: 1× Phusion High Fidelity PCR master mix (NEB), 250 nM Nextera Ad1_noMX primer (AATGA TACGGCGACCACCGAGATCTACACTCGTCGG CAGCGTCAGATGTG), 250 nM Nextera Ad2.X primer (CAAGCAGAAGACGGCATACGA GATXXXXXXXXGTCTCGTGGGCTCGGAGATGT) (*see* **Note 5**).

4. AMPure XP beads (Beckman Coulter).

5. 80% ethanol.

6. Bioanalyzer (Agilent).

7. Illumina sequencing machine.

3 Methods

3.1 Cross-Linking, Digestion, and Ligation

1. Harvest 10–15 million cells by trypsinization to ensure single-cell suspension (*see* **Notes 6** and **7**). Centrifuge for 5 min at $500 \times g$, 4 °C and remove supernatant.

2. Resuspend cells in 10 mL 1× PBS, centrifuge for 5 min at $500 \times g$, 4 °C and remove supernatant.

3. Resuspend cells in 10 mL 1% formaldehyde in culture medium (i.e., 1–1.5 million cells/mL) and incubate at room temperature for 10 min with gentle shaking (*see* **Note 8**).

4. Add 505 μL 2.5 M glycine and incubate at room temperature for 5 min with gentle shaking.

5. Centrifuge for 5 min at $500 \times g$, 4 °C, remove supernatant and wash twice by resuspending cells in 10 mL ice-cold 1× PBS and repeating the same centrifugation and supernatant removal steps.

6. Resuspend cells in 1 mL ice-cold 1× PBS and transfer to a LoBind Eppendorf tube. Centrifuge for 5 min at $500 \times g$, 4 °C and remove supernatant (*see* **Note 9**).

7. Resuspend cells in 500 μL ice-cold Hi-C lysis buffer and rotate for 30 min at 4 °C (*see* **Notes 10** and **11**).

8. Centrifuge for 5 min at $2500 \times g$, 4 °C and remove supernatant. Resuspend in 500 μL ice-cold Hi-C lysis buffer and recentrifuge and remove supernatant to wash.

9. Resuspend gently in 100 μL 0.5% SDS, avoiding air bubbles, and incubate for 10 min at 62 °C (*see* **Note 12**).

10. Add 285 μL molecular biology-grade water and 50 μL 10% Triton X-100, mix gently to avoid air bubbles, and incubate for 15 min at 37 °C.

11. Take a 10 μL aliquot ("undigested") for quality control (*see* Subheading 3.1 **steps 19** and **20**).

12. To the remainder of the nuclei, add 50 μL 10× NEBuffer 2 and 15 μL 25 U/μL MboI (*see* **Note 13**).

13. Incubate for 4 h at 37 °C in a thermomixer (400 rpm).

14. Heat-inactivate MboI by incubating for 20 min at 62 °C (*see* **Note 14**).

15. Cool to room temperature, then take a 10 μL aliquot ("digested") for quality control (*see* Subheading 3.1 **steps 19** and **20**).

16. To the remainder of the nuclei, add 52 μL fill-in master mix and incubate for 1 h at 37 °C, 400 rpm on a thermomixer.

17. Add 948 μL ligation master mix and incubate overnight at 16 °C.

18. Add 5 μL 400 U/μL T4 DNA ligase and 40 μL 10 mM ATP and continue ligation by incubating for 1 h at room temperature.

19. Take a 30 μL aliquot ("ligated") for quality control, along with "undigested" and "digested" aliquots. Add 90 μL 10 mM Tris–HCl pH 7.5 to "undigested" and "digested" aliquots (70 μL for "ligated" aliquot) and 3 μL 20 mg/mL proteinase K to all aliquots. Incubate overnight at 65 °C.

20. Cool down samples to room temperature and perform gel electrophoresis on one sixth of the sample, with RNase A directly in the loading dye (20 μL of 10 mg/mL RNase A in 1 mL loading dye). Clear shifts in gel mobility should be visible between "undigested" and "digested" samples (Fig. 2a), and between "digested" and "ligated" samples (Fig. 2b).

3.2 Sonication and Chromatin Immunoprecipitation

1. Centrifuge nuclei for 5 min at 2500 × g, room temperature and remove supernatant. Resuspend in 400 μL fresh nuclear lysis buffer and keep on ice for 20 min (*see* **Note 15**).

2. Sonicate samples in a Bioruptor Pico (Diagenode) for 8 cycles at 4 °C on medium setting, with 30 s OFF/30 s ON pulses (*see* **Notes 16** and **17**).

3. Centrifuge for 15 min at 1600 × g, 4 °C and keep supernatant in a fresh LoBind Eppendorf tube.

4. Take a 10 μL aliquot ("sonicated") and process as in Subheading 3.1, **steps 19** and **20**, to check sonication efficiency by gel electrophoresis. Ideally, the sonicated fragments should be 200–700 bp (Fig. 2b).

5. Split fragmented chromatin into two LoBind Eppendorf tubes of 200 μL each. To each, add 800 μL ChIP dilution buffer (*see* **Note 18**).

6. Preclear lysates by adding 25 μL protein G Dynabeads to each sample (i.e., 50 μL beads for 10–15 million cells) and incubate for 1 h at 4 °C with rotation.

7. Place tubes on magnetic stand and transfer cleared supernatant to new LoBind Eppendorf tube.

8. Add 3 μg of H3K27ac antibody to each tube (for 10–15 million cells) and incubate overnight at 4 °C with rotation (*see* **Note 19**).

9. Concomitant with **step 8**, preblock 50 μL protein G Dynabeads by adding 500 μL of 0.1 mg/mL BSA in ChIP dilution buffer and incubating overnight at 4 °C with rotation.

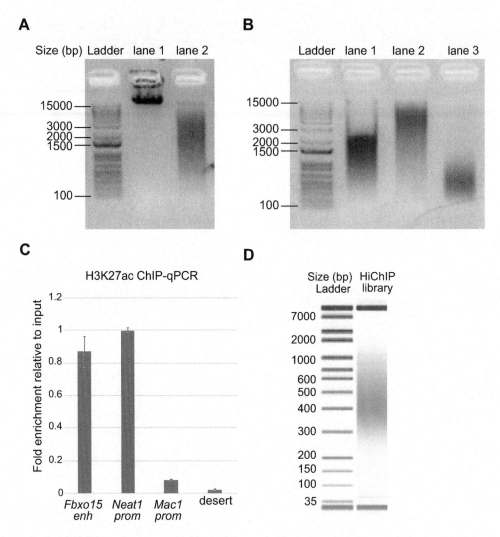

Fig. 2 (**a**) Quality Control for digestion step: Agarose gel electrophoresis (1%) of undigested DNA (lane1) and MboI digested DNA (lane2). (**b**) Quality Control for ligation and sonication: Agarose gel electrophoresis (1%) of MboI digested DNA (lane1), ligated DNA (lane2) and sonicated DNA (lane3). (**c**) Quality control for chromatin immunoprecipitation step: enrichment of immunoprecipitated DNA using the H3K27ac antibody in murine pluripotent stem cells. Primers amplifying the *Fbxo15* enhancer and *Neat1* promoters were used as positive controls while primers around the *mac* and *desert* region were used as negative control in a real-time qPCR to assess fold enrichment of immunoprecipitated DNA relative to input DNA. (**d**) Quality control for HiChIP library size: Bioanalyzer run of final HiChIP library

10. Place preblocked beads on magnetic stand, remove supernatant, and wash by resuspending in 500 μL ChIP dilution buffer. Place beads on magnetic stand, remove supernatant, and resuspend beads in 50 μL ChIP dilution buffer.

11. Transfer beads to tube containing antibody-treated fragmented chromatin and incubate for 3 h at 4 °C with rotation.

12. Wash beads by placing on a magnetic stand, removing supernatant and resuspending in 800 μL low salt wash buffer by inverting the tubes five times, then placing again on the magnetic stand. Repeat once more for a total of two washes.

13. Repeat **step 12** for two washes with high salt wash buffer.

14. Repeat **step 12** for two washes with LiCl wash buffer.

15. Resuspend beads in 500 μL TE buffer and transfer to fresh LoBind Eppendorf tube. Remove supernatant after placing beads on a magnetic stand.

16. Resuspend the beads in 150 μL DNA elution buffer and incubate for 30 min at 37 °C, 1300 rpm on a thermomixer.

17. Place beads on a magnetic stand and transfer eluted chromatin as the supernatant to a fresh LoBind Eppendorf tube.

18. Repeat **steps 16** and **17**, combining the two eluates (supernatants) together into one tube (~250 μL).

19. Add 15 μL 20 mg/mL proteinase K and incubate overnight at 65 °C.

20. Cool to room temperature and purify DNA with a ZYMO or Qiagen minElute kit, following the manufacturer's instructions and eluting DNA with 12 μL 10 mM Tris–HCl pH 7.5 (*see* **Note 20**).

21. Quantify the DNA with a Qubit dsDNA HS assay (*see* **Notes 21** and **22**). If necessary, adjust concentration to 15 ng/μL with 10 mM Tris–HCl pH 7.5.

3.3 Biotin Pull-Down

1. Wash 5 μL Streptavidin C1 beads twice with 100 μL Tween wash buffer by incubating at room temperature for 1 min, then placing on a magnetic stand and removing the supernatant.

2. Resuspend the beads in 10 μL 2× biotin binding buffer and add 10 μL post-ChIP DNA (*see* **Note 23**). Incubate for 15 min at room temperature, 800 rpm in a thermomixer.

3. Wash twice by adding 500 μL Tween wash buffer, incubating for 2 min at 55 °C, placing on a magnetic stand and discarding the supernatant.

4. Resuspend the beads in 100 μL 1× TD buffer, place on a magnetic stand and discard supernatant.

3.4 Library Preparation

1. Resuspend the beads in 25 μL 2× TD buffer, add 0.05 μL Tn5 transposase per ng input DNA, up to a maximum of 4 μL (*see* **Note 24**) and adjust volume to 50 μL with molecular biology-grade water. Incubate for 10 min at 55 °C, 700 rpm on a thermomixer.

2. Place beads on magnetic stand, remove supernatant and resuspend in 100 μL 50 mM EDTA. Incubate for 30 min at 55 °C, 700 rpm on a thermomixer. Tap tubes every ~10 min if beads settle to the bottom of the tube.

3. Wash beads twice in 100 μL 50 mM EDTA by placing on magnetic stand, removing supernatant, resuspending in wash solution and incubating for 3 min at 50 °C.

4. Transfer beads to a fresh LoBind Eppendorf tube, then wash beads twice with 100 μL Tween wash buffer by placing on a magnetic stand, removing supernatant, resuspending in wash solution and incubating for 2 min at 55 °C.

5. Place beads on a magnetic stand, remove supernatant, resuspend beads in 100 μL 10 mM Tris–HCl pH 7.5, and transfer beads to a PCR tube.

6. Place beads on a magnetic stand, remove supernatant and resuspend beads in 50 μL PCR master mix (*see* **Note 5**). Run the following PCR program.

72 °C, 5 min.

98 °C, 1 min.

$N \times$: [98 °C, 15 s; 63 °C, 30 s; 72 °C, 1 min].

N is the optimal number of cycles (*see* **Note 25** to determine optimal number).

7. Place beads on a magnetic stand and transfer the supernatant to a fresh PCR tube (*see* **Note 26**).

8. Add 25 μL (0.5 volumes) AMPure XP beads and mix by pipetting. Incubate for 10 min at room temperature (*see* **Note 27**).

9. Place tubes on magnetic stand and transfer exactly 70 μL clear supernatant to a new tube. Add 14 μL fresh AMPure XP beads, mix by pipetting and incubate for 10 min at room temperature.

10. Place tubes on magnetic stand and discard supernatant. Add 100 μL 80% ethanol to the beads without displacing them and keeping the tube on them magnetic stand. Incubate for 30 s at room temperature.

11. Remove the supernatant and wash the beads twice more with 100 μL 80% ethanol as in **step 10**.

12. Remove the supernatant and dry the beads by leaving the tubes on the magnetic stand uncapped at room temperature (*see* **Note 28**).

13. Resuspend the beads in 14 μL 10 mM Tris–HCl pH 7.5 and incubate for 10 min at room temperature.

14. Place beads on a magnetic stand and transfer 13 μL clear supernatant to a fresh Eppendorf tube. Quantify DNA with a Qubit fluorimeter and dsDNA HS assay (expected concentration is 1–5 ng/μL).

15. Run 2 μL sample on a Bioanalyzer to check the final size of the libraries (should be 300–700 bp; Fig. 2d) (*see* **Note 29**).

16. Sequence the HiChIP material on an Illumina platform, such as HiSeq 4000 or NextSeq 500, with paired-end 75 or 50 nt setting. Depending on the desired depth of sequencing (*see* Subheading 3.5), the library or pooled libraries can be run on multiple lanes (*see* **Note 30**).

3.5 Analysis

The initial steps of HiChIP and Hi-C analyses are almost identical. The most frequently used pipelines like HiC-Pro, HiCUP, HiC-Explorer, and HiC-Bench [15–19] are comprised of similar algorithms, which use the raw sequencing files as inputs (.fastq files) to perform alignment and various filtering steps in order to extract the usable reads from each Hi-C or HiChIP experiment and create the unnormalized matrices of interactions (Fig. 3). Similar to a Hi-C experiment the raw paired-end reads are initially assessed for adapter contamination and presence of low quality base pairs with FastQC (available online at: http://www.bioinformatics.babraham.ac.uk/projects/fastqc), while trimming of adapter sequences can be performed with tools such as trimmomatic, cutadapt [20, 21] and other similar tools if necessary. We describe these initial steps using the HiC-Bench pipeline up to the point of generating the "filtered.reg" file which contains all the usable filtered reads as described in (https://github.com/NYU-BFX/hic-bench/wiki).

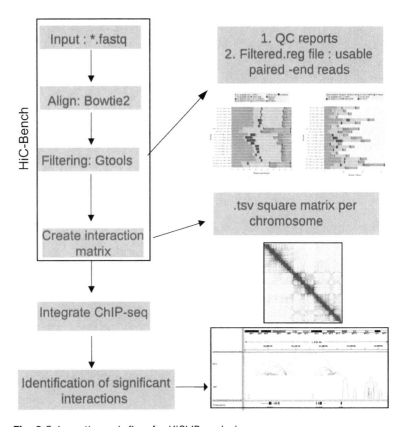

Fig. 3 Schematic work-flow for HiChIP analysis

3.5.1 Use of HiC-Bench

1. Clone repository (git clone –depth 1 https://github.com/NYU-BFX/hic-bench.git).

2. Choose pipelines (hic-bench/pipelines/hicseq-standard).

3. Set up input data directory (.fastq files).

   mkdir inputs/fastq (build directory to store .fastq files),

   mkdir sample# (build separate directory for each sample),

4. Create sample sheet in the input directory which contains all the information of the experiment in different columns which include the following.

   Sample name (ESC_hichip_replicate1).

   Group name (ESC_hichip).

   Name of Read1 from paired end .fastq file (ESC_hichip_replicate1_R1.fastq.gz).

   Name of Read2 from paired end .fastq file (ESC_hichip_replicate1_R2.fastq.gz).

   Genome version utilized for alignment and annotation (mm10).

   Restriction enzyme used for HiC/HiChIP (MboI).

   Cell Type (ESC).

5. Parameter modification: Set quality of reads, minimum distance of acceptable reads, max-offset and activate duplication filtering of the aligned data in the "align" pipeline (--mapq 20 – min-dist 10,000, --max-offset 500 –filter-dups).

6. Execute the pipeline (./run) from the main pipeline directory, *see* **step 2**).

3.5.2 Alignment, Filtering and Extraction of "Usable" Reads

1. Alignment is performed with bowtie2 algorithm (version 2.2.3) [22] and –very-sensitive-local --local option.

2. GenomicTools [23] are used to filter paired-end aligned files for duplicates; multihits; low-quality, self-ligated fragments; and short-range interactions (<10 kb distance).

3.5.3 HiChIP Stats and QCs

HiC-Bench generates reports of usable reads and filtered reads by category in a PDF format (Fig. 4). This graph provides important stats and information for the quality of two critical steps in the HiChIP protocol.

1. *Hi-C efficiency:* Similar to the Hi-C requirements, a good HiChIP library is expected to have <20–30% interchromosomal contacts and a >40–60% of far_cis contacts separated by a distance >10 kb. This group of contacts constitutes the so-called usable or valid reads.

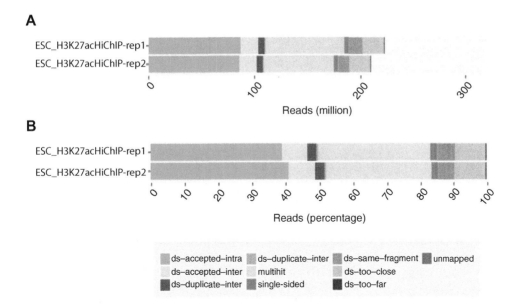

Fig. 4 QC plots of the (**a**) number and (**b**) percentage distribution of sequenced reads split in 10 different categories as generated from HiC-Bench pipeline. *ds-accepted-intra* represent the so-called usable reads (intrachromosomal contacts with >20 kb distance)

2. *ChIP efficiency*: H3K27ac HiChIP is expected to enrich for interactions between regions decorated by this mark. The degree of this enrichment depends on the efficiency and specificity of the chromatin immunoprecipitation step. To control for that, we calculate the percentage of usable reads that fall under ChIP-seq peaks as were identified by independent H3K27ac ChIP-seq experiments in the same cell type. A successful H3K27ac HiChIP usually gives ~15% enrichment, although we have observed fluctuations between 9-27% depending on the cell type.

3.5.4 Generation of Interaction Matrices

HiC-Bench pipeline provides interacting pairs in two formats (Fig. 3).

1. Square matrices in .tsv format for each chromosome with the number of raw filtered pairs for each interacting bin-pair. The first row and column contain the coordinates of each 10 kb bin (*see* **Note 31**).

2. A filtered.reg file which contains all the raw filtered cis- and trans- usable pairs in a tab delimited file that contain 8 columns: unique read ID, left pair chromosome, left pair strand, left pair start coordinate, left pair start coordinate, right pair chromosome, right pair strand, right pair start coordinate, right pair start coordinate.

D00796:294:CC5P1ANXX:5: 1206:3415:4505	chr1 + 100000000 100000000 chr1 - 102345139 102345139
D00796:294:CC5P1ANXX:5: 1105:18621:73524	chr1 + 100000332 100000332 chr1 - 110079326 110079326
D00796:294:CC5P1ANXX:5: 1312:16486:92272	chr1 + 100000515 100000515 chr1 + 98608442 98608442
D00796:294:CC5P1ANXX:5: 2301:16266:3124	chr1 + 100000531 100000531 chr1 + 97931684 97931684
D00796:294:CC5P1ANXX:5: 1311:4594:83140	chr1 + 100000556 100000556 chr1 - 100465000 100465000

3.5.5 File Transformation for Visualization in JuiceBox

Generate .hic files for visualization with the use of filtered.reg file.

1. Transform filtered.reg file into .bedpe format with the following code: awk '{print $2"\t"$4+1"\t"$4+2"\-t"$6"\t"$8+1"\t"$8+2"\t"$1"\t""."."\t"$3"\t"$7}' filtered. reg > filteredreg.bedpe.

2. Intersect the generated bedpe file with the 10 kb mm10 binned genome with the use of bedtools [24]: intersectBed -a filtered. reg.bedpe -b bin10kb -loj > filtered.reg_bin10kb.bedpe.

3. Use R to aggregate the # reads (agg_filtered.reg_bin10kb. bedpe) which will contain the bin coordinates and raw number of reads per bin and generate .txt file for juicer at 10 kb resolution: awk '{print "0" "\t"$1 "\t"$2 "\t"$2+1 "\-t" "1" "\t"$4 "\t"$5 "\t"$5+1 "\t"$7}' agg_filtered. reg_bin10kb.bedpe > juicer_agg.txt.

4. Use juicer tools (version 1.19.02) [25] to generate .hic files: java -Xmx####m -jar juicer_tools_1.19.02.jar pre juicer agg. txt.gz kr agg.txt.hic -d -k NONE -r 10,000 mm10.

3.5.6 Identification of Significant Interactions

An important step in HiChIP analyses is the identification of significant HiChIP contacts. Currently, there is no gold standard method for the detection of significant contacts using HiChIP (or PLAC-seq or ChIA-PET) data (*see* **Note 32**). Over the last years, different groups have either developed new or implemented older Hi-C pipelines for calling HiChIP interactions [5, 26–38]. Below, we describe the steps for a straightforward Mango-based approach that we applied in our recent HiChIP publication [14]. In addition, we provide a direct comparison with three other commonly used methods: HiCCUPs [5], FitHiC2 [25] and c-Loops [22] (*see* Subheading 3.5.8).

1. Apply CPM normalization across the matrices (generated in Subheading 3.5.4) for each chromosome.

2. Apply binomial test per diagonal of the counts-matrix for each interacting 10 kb bin with more than 1 raw read up to a maximum of 2 Mb distance. Interacting bins with p-value <0.10 and CPM > 3 can be considered as significant in each replicate. Only significant interacting bins that are common among replicates are used. *Stringency criteria can be modified by adjusting CPM and p-value or q-value cutoffs.*

3. Integrate H3K27ac ChIP-seq peaks (independent experiment in the same cell type) to choose contacts with at least one H3K27ac peak in one or both interacting anchors.

The identified interactions can be then integrated with other -omics datasets, such as RNA-seq or ChIP-seq for other histone marks or protein factors. Downstream analysis can be performed to (a) associate putative enhancers to target genes; (b) predict regulatory impact of interactions; (c) infer regulatory synergies or redundancies and gene coregulation; and (d) speculate on the mechanisms that mediate the observed interactions.

3.5.7 Categorization of HiChIP Loops

Though Hi-C can identify both regulatory and structural loops, H3K27ac HiChIP enriches for contacts between active (H3K27ac-occupied) enhancers and/or promoters. By default, all HiChIP loops must contain an H3K27ac peak in one or both anchors classifying the loops in single and bianchor groups. Depending on the presence of one or more transcription start sites within any of the interacting bins we classify anchors in three distinct categories. Each anchor can be considered as follows.

Promoter anchor (P): contains a transcription start site (TSS) with or without a H3K27ac peak.

Enhancer anchor (E): contains at least one H3K27ac peaks and no TSS.

Nonenhancer/no-promoter anchor (X): there is no H3K27ac or TSS.

Based on this classification of the HiChIP anchors we can identify 5 different loop classes.

Enhancer–enhancer interaction: E-E.

Enhancer–promoter interaction: E-P.

Promoter–promoter interaction: P-P.

Enhancer–nonregulatory interaction: E-X.

Promoter–nonregulatory interaction: P-X.

In our ESC H3K27ac HiChIP, almost 73% of the identified loops were enhancer-linked loops (EE, EP, EX), while 50% were promoter-related (PP, EP, PX). The majority of nonregulatory anchors within HiChIP loops (EX or PX), contained at least one accessible site and/or binding site for architectural proteins, such as CTCF or cohesin members SMC1/3, suggesting the presence of other structural or regulatory elements, such as insulators. Similar percentages were noted even when we used other loop calling methods (*see* Fig. 5 and Subheading 3.5.8).

3.5.8 Comparison of Different Loop Calling Methods

For FitHiC2 and c-Loops, both of which use binomial distribution model for p-value calculation similar to the Mango pipeline, transformation of filtered.reg file into bedpe format is necessary for identifying significant interactions, while .hic file generation is required for HiCCUPs. Following the recommendation of the developers, we used the following parameters for each method:

1. c-Loops (https://github.com/YaqiangCao/cLoops): cLoops -f chr*.bedpe -o loop_chr* -eps 5000 -minPts 10 -hic -j -s -w, where each command was used for each chromosome and replicate separately. CLoop algorithm identifies contacts with variable anchor size (median anchor size ~21 kb). In order to compare with the other methods, we transformed anchors into 10 kb interacting bins by selecting the center of the cloop anchor and fitting it into a 10 kb bin.

2. HiCCUPs (https://github.com/aidenlab/juicer/wiki/HiCCUPS): java -Xmx*m -jar */juicer_tools_1.22.01.jar hiccups --threads 10 -r 10,000 esc_rep*.hic ./loop_esc_rep*, using 10 kb resolution for each replicate.

3. FitHiC2 (https://github.com/ay-lab/fithic): fithic -i i1.bed.gz -f f1.bed.gz -o fit1 -r 10,000 -L 10000 -U 10000000, where i.bed.gz and f1.bed.gz were fixed according to the FitHiC instructions (https://github.com/ay-lab/fithic) for each chromosome and replicate separately.

For each method, only commonly identified significant interactions between replicates with CPM > 3 were used (Fig. 5a). All methods detected different numbers of significant contacts, ranging from ~2000 for HiCCUPS to ~79,000 for FitHiC2. The size distribution of the loops captured by each method was also highly variable, with the Mango-based approach showing a strong preference for short interactions (75% < 50 kb) in contrast with c-Loops, which preferentially detects longer contacts (75% > 100 kb) (Fig. 5b). For both FitHiC2 and HiCCUPS the size range of detected contacts was between 50–300 kb. Pairwise comparison showed that >50 and 75% of the loops detected by either FitHiC2 or HiCCUPs fully overlapped with the ones called by Mango. In contrast, c-Loops showed less than 10% overlap with any other method, even upon extension of the 10 kb anchors by 2 bins. Moreover, all methods (except for c-Loops) showed overall similar distribution of features within the loop anchors when tested for: (a) the ratio of single-anchored vs bi-anchored (presence of

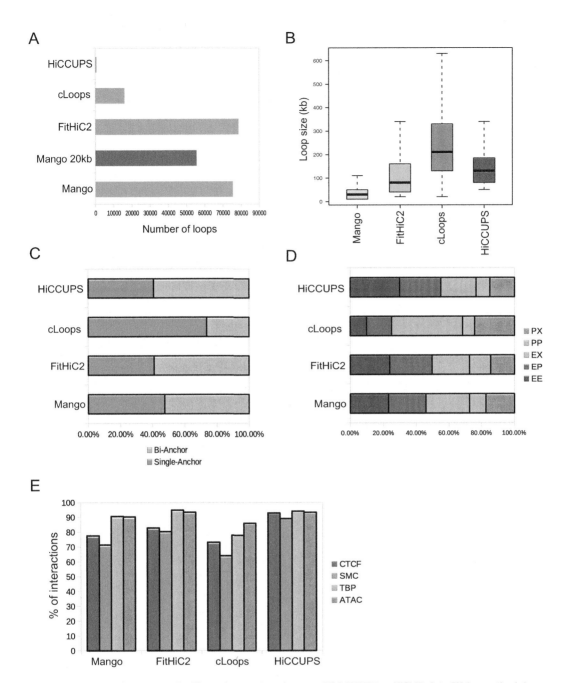

Fig. 5 (**a**) Number of detected significant interactions in our mESC H3K27ac HiChIP data [3] by method. Loop size identified by Mango-like approach extends from 10 kb up to 2 Mb while for the rest of the methods loop size begins from 20 kb. In order to fairly compare the loops we considered two loop groups generated from Mango, one containing all the loops (teal color) and the one containing all loops above 20 kb (blue color). The rest of the groups are depicted with teal color. (**b**) Boxplot of ESC loop size from 4 different loop calling methods. (**c**) Classification of HiChIP interactions as single-anchored and bi-anchored (presence of H3K27ac peak(s) on one or both loop anchors) for each loop calling method. (**d**) Classification of HiChIP interactions in 5 categories based on the presence and/or absence of H3K27ac and gene transcription start site in any of the anchors for each of the loop calling methods. (**e**) Bar plot depicting the percentage of HiChIP contacts detected by each loop-calling method that have an accessible DNA region or a CTCF, SMC1/3 or TBP binding site in any of their anchors

H3K27ac peak(s) on one or both anchors) loops (Fig. 5c), (b) the percentage of different loop subcategories (E-E, E-P, P-P as discussed in Subheading 3.5.7) (Fig. 5d), and (c) the percentage of loop anchors that overlap with ATAC-seq or ChIP-seq peaks of key different architectural factors (e.g., CTCF and cohesin) or transcriptional regulators, such as TBP (Fig. 5e).

3.5.9 Identification of Differential Interactions Between MEFs and ESCs

H3K27ac HiChIP can also be used to detect potential 3D rewiring of enhancer connectomes between two different cell types, for example MEFs and ESCs. *See* **Note 33** for limitations and important considerations and **Note 34** for alternative approaches.

1. MEF and ESC loops were merged and differential analysis was performed on the CPM signal across replicates.

2. A pseudo-count of one was added to interactions with 0 CPM in order to generate log2 fold change (log2[FC]) values between MEF and ESC.

3. R and two-tailed unpaired Student's t-test was performed between MEF and ESC replicates.

4. Loops with p-value <0.1 and log2[FC] >2 or log2[FC] < 2 were considered to be differential between MEF and ESC (21,659 MEF specific and 18,820 ESC specific loops). Loops with p-value >0.5 and absolute log2 fold change (abs(log2 [FC])) < 0.5 were considered as nondifferential (8095 loops).

There are multiple ways to visualize examples of detected loops as arcs in IGV or Epigenome browser, as annotated peaks (pixels) on heatmaps in either HiGlass or JuiceBox [5, 39] (Fig. 6a, b) or with the use of virtual-4C [14] (Fig. 6c). Heatmaps and aggregate analysis plots (APA) consist two of the most practical ways to represent the full set of differential loops (Fig. 6a) identified between two different cell types or conditions.

4   Notes

1. Mouse embryonic fibroblasts are isolated by dissection of mouse embryos at E15.5–16.5 and cultured in DMEM supplemented with 10% heat-inactivated fetal bovine serum, GlutaMAX, penicillin–streptomycin, nonessential amino acids, and β-mercaptoethanol in low oxygen condition (4% O_2). Only low-passaged cells (up to 5 passages) are used to avoid senescence.

2. Alternative restriction enzymes that are commonly used include: NlaIII, Csp6I, and DpnII. If choosing a restriction enzyme different from MboI, adapt the protocol by using a compatible restriction buffer.

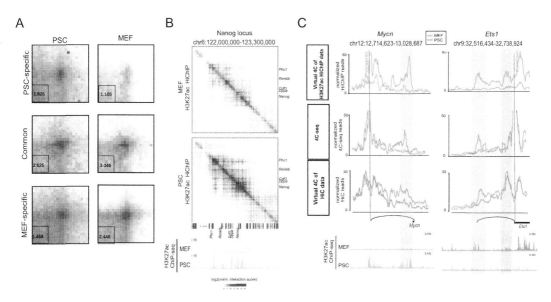

Fig. 6 (**a**) APA plots of the ESC-specific (here denoted as PSC, for pluripotent stem cells), common and MEF-specific loops in ESC and MEF H3K27ac HiChIP matrices [3]. Black box on the bottom left corner of each APA (P2LL) represents the ratio of the central pixel to the mean of the pixels in the left low corner. Knight–Ruiz balancing method and a width of 10 bins was used for generating each APA plot. (**b**) HiChIP heatmaps (generated with Juicebox) around the *Nanog* locus in MEF and ESC using the respective CPM normalized counts-matrices at 10 kb resolution. BigWig of depth-normalized H3K27ac ChIP-seq tracks of MEF and PSC are presented below. (**c**) Line plot representing the normalized signal of all interacting bins from HiChIP, 4C-seq and Hi-C, between MEF (orange) and ESC (teal) starting from the promoter regions of *Mycn* (left) and *Ets1* (right). BigWig of depth normalized H3K27ac ChIPseq tracks of MEF and PSC are presented below the Virtual 4C plots

3. Protein A and G are structurally very similar but they have slightly different affinities for IgG subclasses across different species. Please refer to the following website to choose the optimal condition depending on the antibody used for the ChIP step: https://www.neb.com/tools-and-resources/selection-charts/affinity-of-protein-ag-for-igg-types-from-different-species

4. Any antibody that has been successfully used for ChIP-seq in the cell type of interest is likely appropriate for HiChIP too. In general, we recommend using ChIP-grade antibodies and concentrations similar to the ones used for ChIP-seq. Table 1 shows a list of commercially available antibodies that have been successfully used for HiChIP experiments.

5. If handling multiple samples intended for multiplexing, a master mix can be prepared using all reagents except for the Nextera Ad2.X primers. In this case add 49 μL of master mix without the second primer to each sample, and then add 1 μL of barcoded Nextera Ad2.X primer to the individual samples. Table 2 shows a list of suitable barcoded primers.

Table 1
List of ChIP-grade antibodies that have been successfully used in HiChIP experiments

Epitope	Company	Catalog number
Smc1a [9]	Bethyl Laboratories	A300-055A
RNA Pol II [40]	Biolegend	664906
CTCF [11]	Abcam	ab70303
Oct4 [7]	Santa Cruz	sc-8628
Klf4 [14]	R&D	AF3158
H3K27ac [11, 14]	Abcam	ab4729
YY1 [15]	Abcam	ab1009237
Pax3 [13]	Santa Cruz	sc-4926

6. If mouse ESCs are cultured on feeders, make sure to preplate cells after trypsinization on gelatinized plates for 30 min in order to let feeders attach to the surface of the plate. Collect the supernatant containing mESCs and count. Alternatively, mESCs can be preplated on gelatinized plates (without feeders) the day before starting the experiment.

7. The number of cells needed to yield a sufficient amount of DNA after the ChIP step varies depending on the overall abundance of protein/mark of interest and the efficiency of the antibody. For H3K27ac or other widespread histone marks (such as H3K4me3), 5-10 million cells are recommended following this protocol, although lower numbers can also be used. For transcription factors, such as KLF4, we recommend starting with >30 million cells and splitting into two (or more) independent HiChIP assays using 10–15 million cells for each. All the steps will be carried out separately for the two samples and the DNA will be combined at Subheading 3.3, **step 2**).

8. As shown with many other C-based assays, cross-linking with 2% formaldehyde may improve final yield and efficiency, although subsequent steps, such as sonication, may need to be adjusted.

9. At this point the cell pellet can be flash-frozen in liquid nitrogen and stored at −80 °C for future use.

10. This step can be optimized based on the cell type used by either modifying the incubation time with the lysis buffer or by changing the composition of the lysis buffer (e.g., change concentration or type of detergent).

Table 2
List of suitable barcoded primers for multiplexing HiChIP experiments

Ad1_noMX	AATGATACGGCGACCACCGAGATCTACACTCGTCGGCAGCGTCAGA TGTG
Ad2.1_TAAGGCGA	CAAGCAGAAGACGGCATACGAGATTCGCCTTAGTCTCGTGGGC TCGGAGATGT
Ad2.2_CGTACTAG	CAAGCAGAAGACGGCATACGAGATCTAGTACGGTCTCGTGGGC TCGGAGATGT
Ad2.3_AGGCAGAA	CAAGCAGAAGACGGCATACGAGATTTCTGCCTGTCTCGTGGGC TCGGAGATGT
Ad2.4_TCCTGAGC	CAAGCAGAAGACGGCATACGAGATGCTCAGGAGTCTCGTGGGC TCGGAGATGT
Ad2.5_GGACTCCT	CAAGCAGAAGACGGCATACGAGATAGGAGTCCGTCTCGTGGGC TCGGAGATGT
Ad2.6_TAGGCATG	CAAGCAGAAGACGGCATACGAGATCATGCCTAGTCTCGTGGGC TCGGAGATGT
Ad2.7_CTCTCTAC	CAAGCAGAAGACGGCATACGAGATGTAGAGAGGTCTCGTGGGC TCGGAGATGT
Ad2.8_CAGAGAGG	CAAGCAGAAGACGGCATACGAGATCCTCTCTGGTCTCGTGGGC TCGGAGATGT
Ad2.9_GCTACGCT	CAAGCAGAAGACGGCATACGAGATAGCGTAGCGTCTCGTGGGC TCGGAGATGT
Ad2.10_CGAGGCTG	CAAGCAGAAGACGGCATACGAGATCAGCCTCGGTCTCGTGGGC TCGGAGATGT
Ad2.11_AAGAGGCA	CAAGCAGAAGACGGCATACGAGATTGCCTCTTGTCTCGTGGGC TCGGAGATGT
Ad2.12_GTAGAGGA	CAAGCAGAAGACGGCATACGAGATTCCTCTACGTCTCGTGGGC TCGGAGATGT
Ad2.13_GTCGTGAT	CAAGCAGAAGACGGCATACGAGATATCACGACGTCTCGTGGGC TCGGAGATGT
Ad2.14_ACCACTGT	CAAGCAGAAGACGGCATACGAGATACAGTGGTGTCTCGTGGGC TCGGAGATGT
Ad2.15_TGGATCTG	CAAGCAGAAGACGGCATACGAGATCAGATCCAGTCTCGTGGGC TCGGAGATGT
Ad2.16_CCGTTTGT	CAAGCAGAAGACGGCATACGAGATACAAACGGGTCTCGTGGGC TCGGAGATGT
Ad2.17_TGCTGGGT	CAAGCAGAAGACGGCATACGAGATACCCAGCAGTCTCGTGGGC TCGGAGATGT
Ad2.18_GAGGGGTT	CAAGCAGAAGACGGCATACGAGATAACCCCTCGTCTCGTGGGC TCGGAGATGT
Ad2.19_AGGTTGGG	CAAGCAGAAGACGGCATACGAGATCCCAACCTGTCTCGTGGGC TCGGAGATGT

(continued)

Table 2
(continued)

Ad2.20_GTGTGGTG	CAAGCAGAAGACGGCATACGAGATCACCACACGTCTCGTGGGC TCGGAGATGT
Ad2.21_TGGGTTTC	CAAGCAGAAGACGGCATACGAGATGAAACCCAGTCTCGTGGGC TCGGAGATGT
Ad2.22_TGGTCACA	CAAGCAGAAGACGGCATACGAGATTGTGACCAGTCTCGTGGGC TCGGAGATGT
Ad2.23_TTGACCCT	CAAGCAGAAGACGGCATACGAGATAGGGTCAAGTCTCGTGGGC TCGGAGATGT
Ad2.24_CCACTCCT	CAAGCAGAAGACGGCATACGAGATAGGAGTGGGTCTCGTGGGC TCGGAGATGT

11. Take a small aliquot, add Trypan Blue, and check permeabilization (blue staining) under a microscope.

12. If the samples become viscous it means the nuclei have lysed and DNA was released. In that case, this step should be optimized by trying a lower percentage or shorter incubation time with SDS.

13. For lower starting material use less restriction enzyme: 8 μL for five million cells, and 4 μL for one million cells, adjusting the amount of water to have a final reaction volume of 500 μL.

14. If using a restriction enzyme that does not require heat inactivation, ignore **step 14**.

15. Different lysis buffers can be used depending on the chromatin immunoprecipitation protocol that works best for the protein/ mark of interest and in the cell type of interest. The steps described here have been successful for H3K27ac HiChIP in various different mouse and human cell lines.

16. Alternatively, resuspend the pellet in 880 μL nuclear lysis buffer and use a Covaris E220 to sonicate the sample with the following parameters: fill level 10, duty cycle 5, PIP 140, cycle/burst 200, time 4 min.

17. Sonication time and conditions might need to be adjusted for different cell types. The ideal sonication time should be as long as needed to achieve a range of 200–700 bp fragmentation (to enable efficient immunoprecipitation and high-signal to noise ratio) and as short as possible to avoid overheating.

18. This will dilute the SDS to 0.1%. If using antibodies that require a different concentration of final SDS adjust the volume of dilution buffer accordingly.

19. The amount of antibody required for the immunoprecipitation step depends on the abundance of the protein and the efficiency of the antibody. If the antibody has been already optimized for ChIP-seq assays, we recommend using the same concentration for HiChIP. As a general guideline, we recommend 5–8 μg of antibody for transcription factors, and 2–4 μg for abundant histone modifications when starting with 10–15 million cells.

20. In order to increase yield at the elution step add 10 μL Tris–HCl pH 7.5 to the column, let stand 5 min and then spin down according to manufacturer's instructions. Use the same 12 μL to re-elute the DNA from the column.

21. 1 μL of eluted DNA can be used to check the efficiency of the ChIP step by performing RT-qPCR using primers for genomic regions known to be enriched or depleted for the protein/mark of interest (Fig. 2c).

22. Expected HiChIP yields for histone marks can range between 100–300 ng total DNA per 15 million cells. For transcription factors, post-ChIP DNA yield is much lower and can range between a total of 2–30 ng. Amounts can vary dramatically depending on cell type, abundance of the protein and efficiency of the antibody.

23. For low yield ChIPs, such as for low abundance transcription factors, it is advisable to combine two reactions of post-ChIP DNA (each coming from a 15 million cell reaction) and make up the volume to 20 μL with 10 mM Tris–HCl pH 7.5. In this case, the streptavidin beads should be resuspended in 20 μL of 2× biotin buffer instead of 10 μL.

24. Using the correct amount of Tn5 is critical to achieve an ideal size distribution for clustering on an Illumina sequencer. An overtransposed sample will exhibit lower alignment rates while an undertransposed sample will have fragments that are too large to cluster properly on the sequencer. A maximum amount of 4 μL Tn5 is recommended in order to save on the Tn5 costs, considering that a library with this much material will be amplified in 5 cycles and have enough complexity to be sequenced deeply regardless of how fully transposed the library is to achieve an ideal size distribution.

25. One of two methods can be used to estimate the optimal number of amplification cycles. First, estimate the cycle number based on the amount of material from the post-ChIP measured by Qubit (Subheading 3.2, **step 21**). As a guideline, use 8 cycles for 12.5 ng total DNA, 7 cycles for 25 ng DNA, 6 cycles for 50 ng, 5 cycles for DNA amounts greater than 50 ng. Alternatively, run 5 cycles of a regular PCR. Place samples on a magnetic rack and transfer the supernatant to a

new tube (discard the tube containing the beads). Add 0.25×
SYBR Green and then run on a qPCR machine. Identify the
total cycle number at the beginning of the exponential curve.

26. A carryover of streptavidin beads can be detrimental for the
next steps therefore make sure to wait until all the beads are
captured on the side of the tube and the supernatant is clear
(this can take several minutes). Also, make sure to measure
exactly how much supernatant is collected, since the size selection with AMPure XP beads is based on the exact initial sample
volume. The protocol assumes 50 μL of supernatant.

27. Instead of a two-sided size selection with Ampure XP beads,
the libraries can be purified using a ZYMO kit and then the
eluted libraries can be run on a 6% PAGE in order to gel-purify
DNA in the range of 200–600 bp.

28. It is critical to not overdry the beads, therefore as soon as the
edges are starting to dry, incubate one more minute and proceed with the next step. The drying step usually takes around
5 min.

29. The ideal size range is 300–700 bp. Libraries with smaller sizes
can be still sequenced but they will likely have lower alignment
rates. In our experience if the library is bigger than expected it
will not have a negative effect on the outcome of the sequencing as long as it is below 1000 bp. Much larger fragments may
indicate contamination with beads. In that case a repurification
with a ZYMO kit may help.

30. If using multiple samples that are differentially barcoded, at
this step the samples can be pooled together in equimolar
concentration and the pooled library can be sequenced in one
or multiple lanes (depending on the sequencing depth that
needs to be achieved).

31. *Sequencing depth, resolution and saturation.* The selected genomic bin can be modified to reach the desired resolution. In
Hi-C experiments, the genomic resolution (size of genomic
bin) that will be used for downstream analysis is restricted by
the sequencing depth and specifically by the number of usable
paired-end reads per bin. A minimum of 1000 reads for 80% of
the bins is recommended [5]. Therefore, to generate mouse
Hi-C data in a resolution of 10 kb, >~260 million usable reads
are required (1000 reads × 2.6 billion bp/(bin_size (bp))
10 kb windows in the mouse genome). Given that HiChIP
captures only a fraction of the genome by enriching for regions
decorated by marks or factors of interest, the resolution can be
significantly increased at a cost-effective manner. However,
calculating the expected size of the "effective genome" for
each HiChIP experiment is tricky since it depends not only
on the number and size of ChIP-seq peaks of the mark/factor

of interest, but also on their overall interactivity. Based on our experience with H3K27ac HiChIP in various mouse and human cell lines, ~100-200 M usable reads per sample are sufficient to reach 5–10 kb resolution, while deeper sequencing results in saturation. To test saturation, we applied either FitHiC2 or Mango-like method (as described in Subheading 3.5.6) on random subsets of 25, 50, 75, 100, 150, and 173 million usable reads from the merged ESC H3K27ac HiChIP experiments [14]. By plotting the percentage of the final loops (called using all usable reads) that are detected in each step, we observed a trend for saturation already around at 100 million reads (Fig. 7a). In addition, we noticed that the CPM values of significant loops captured in each consecutive step of the downsampling process were progressively lower, suggesting that the strongest contacts are efficiently detected at much lower sequencing depth (Fig. 7b).

32. *Considerations for calling and visualizing significant interactions.* Identification of significant interactions may be affected by many parameters starting from the enzyme or enzyme combinations that are used to digest the chromatin up to the selection of the parameters and cut-offs when "calling" chromatin interactions. Factors such as sequencing depth, matrix normalization (ICE, KR, distance normalization), filtering of low covered bins, modeling of the distribution of the data before differential analysis and differential analysis cut-offs such as fold, p or q-value, enrichment score, observed vs expected ratio of signal can significantly influence the outcome of an analysis and define the number and type of interactions

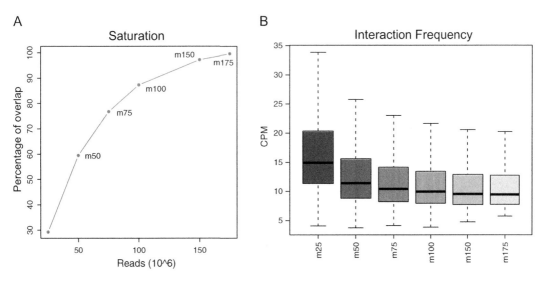

Fig. 7 (a) Line plot presenting the percentage of all interactions (as originally called using all usable reads) that are detected in each consecutive step of downsampling using FitHiC2 approach. **(b)** Strength of loops identified by FitHiC2 approach in different sequencing depths

analyzed. Use of two or more different pipelines and/or selection of deeply sequenced 3D datasets as the "gold-standard" in any analysis could reduce the false positive results, but also limit the true-positive. Though many visualization tools like Juice-Box [5] and HiGlass [39] exist and assist scientists in the systematic visualization of their data, only a small number of loops/regions can be assessed at a time. In addition to these tools virtual-4C analysis [14] of certain loci can be very helpful when genes or regions of interest are masked by the resolution of the data leading to the use of bigger bins than their actual size. By splitting the genome into smaller bins and using a sliding window covering a short area around the regions or gene of interest (+/− 1 to 2 Mb) it is easier to pick up and visualize close and strong interactions that might be missed.

33. *Limitations and considerations for calling HiChIP differential loops.* Differential HiChIP loops among different cell types or states should be interpreted with caution, since they could either represent changes in H3K27ac occupancy and/or 3D reorganization. To distinguish between these two possibilities independent validation is needed with methods that are agnostic for H3K27ac, such as locus-specific 4C-seq [41], Capture Hi-C [10], Hi-C [4], or Micro-C [42]. Alternatively, HiChIP [9] or PLAC-seq [8] against marks or factors that are minimally changing across the genome (e.g., H3K4me3, CTCF or SMC1) may also capture dynamic rearrangements. In our case, a comparison of HiChIP, 4C-seq, and Hi-C signals around selected loci in MEFs and ESCs validated a number of cell-type specific chromatin contacts as initially detected by HiChIP [14] (Fig. 6c). However, this comparison also revealed some inconsistencies, where contacts that appeared differential based on HiChIP signal, showed similar strengths in both cell types by 4C-seq and Hi-C (Fig. 6c).

34. *Alternative approaches for calling differential HiChIP contacts.* Differential HiChIP contacts can be also called by applying methods, such as Wilcoxon test, DESeq [43], and edgeR [44], which are commonly used for calling differential peaks or genes when applied on ChIP-seq or RNA-seq data. Instead of raw counts, normalized data in their distributions can be used before calling significant differences in transcripts, genomic regions or even in loops. New pipelines and methods have emerged lately focusing on differential loop strength analysis [5, 45, 46] which benefit from the high volume of deeply sequenced 3D-related data from 5C, HiC, HiChIP, and so on. A systematic comparison among all these loop-calling and differential loop calling methods is needed to appreciate strengths and limitations as well as technical biases for each approach.

Acknowledgments

We would like to thank Andreas Kloetgen and Aris Tsirigos for comments and advice regarding the HiChIP analysis and members of the Apostolou lab for critical reading of the manuscript. EA is supported by the NIH Director's New Innovator Award (DP2DA043813), the NIGMS (1R01GM138635), the NIDDK (1U01DK128852), and the Tri-Institutional Stem Cell Initiative by the Starr Foundation.

References

1. Misteli T (2020) The self-organizing genome: principles of genome architecture and function. Cell 183(1):28–45. https://doi.org/10.1016/j.cell.2020.09.014

2. Schoenfelder S, Fraser P (2019) Long-range enhancer-promoter contacts in gene expression control. Nat Rev Genet 20(8):437–455. https://doi.org/10.1038/s41576-019-0128-0

3. Di Giammartino DC, Polyzos A, Apostolou E (2020) Transcription factors: building hubs in the 3D space. Cell Cycle 19(19):2395–2410. https://doi.org/10.1080/15384101.2020.1805238

4. Lieberman-Aiden E, van Berkum NL, Williams L et al (2009) Comprehensive mapping of long-range interactions reveals folding principles of the human genome. Science 326(5950):289–293. https://doi.org/10.1126/science.1181369

5. Rao SS, Huntley MH, Durand NC et al (2014) A 3D map of the human genome at kilobase resolution reveals principles of chromatin looping. Cell 159(7):1665–1680. https://doi.org/10.1016/j.cell.2014.11.021

6. Dixon JR, Selvaraj S, Yue F et al (2012) Topological domains in mammalian genomes identified by analysis of chromatin interactions. Nature 485(7398):376–380. https://doi.org/10.1038/nature11082

7. Zhang J, Poh HM, Peh SQ et al (2012) ChIA-PET analysis of transcriptional chromatin interactions. Methods 58(3):289–299. https://doi.org/10.1016/j.ymeth.2012.08.009

8. Fang R, Yu M, Li G et al (2016) Mapping of long-range chromatin interactions by proximity ligation-assisted ChIP-seq. Cell Res 26(12):1345–1348. https://doi.org/10.1038/cr.2016.137

9. Mumbach MR, Rubin AJ, Flynn RA et al (2016) HiChIP: efficient and sensitive analysis of protein-directed genome architecture. Nat Methods 13(11):919–922. https://doi.org/10.1038/nmeth.3999

10. Mifsud B, Tavares-Cadete F, Young AN et al (2015) Mapping long-range promoter contacts in human cells with high-resolution capture Hi-C. Nat Genet 47(6):598–606. https://doi.org/10.1038/ng.3286

11. Mumbach MR, Satpathy AT, Boyle EA et al (2017) Enhancer connectome in primary human cells identifies target genes of disease-associated DNA elements. Nat Genet 49(11):1602–1612. https://doi.org/10.1038/ng.3963

12. Petrovic J, Zhou Y, Fasolino M et al (2019) Oncogenic notch promotes long-range regulatory interactions within hyperconnected 3D cliques. Mol Cell 73(6):1174–1190.e12. https://doi.org/10.1016/j.molcel.2019.01.006

13. Magli A, Baik J, Pota P et al (2019) Pax3 cooperates with Ldb1 to direct local chromosome architecture during myogenic lineage specification. Nat Commun 10(1):2316. https://doi.org/10.1038/s41467-019-10318-6

14. Di Giammartino DC, Kloetgen A, Polyzos A et al (2019) KLF4 is involved in the organization and regulation of pluripotency-associated three-dimensional enhancer networks. Nat Cell Biol 21(10):1179–1190. https://doi.org/10.1038/s41556-019-0390-6

15. Weintraub AS, Li CH, Zamudio AV et al (2017) YY1 is a structural regulator of enhancer-promoter loops. Cell 171(7):1573–1588.e28. https://doi.org/10.1016/j.cell.2017.11.008

16. Lazaris C, Kelly S, Ntziachristos P et al (2017) HiC-bench: comprehensive and reproducible Hi-C data analysis designed for parameter exploration and benchmarking. BMC Genomics 18(1):22. https://doi.org/10.1186/s12864-016-3387-6

17. Ramirez F, Bhardwaj V, Arrigoni L et al (2018) High-resolution TADs reveal DNA sequences underlying genome organization in flies. Nat Commun 9(1):189. https://doi.org/10.1038/s41467-017-02525-w

18. Wingett S, Ewels P, Furlan-Magaril M et al (2015) HiCUP: pipeline for mapping and processing Hi-C data. F1000Res 4:1310. https://doi.org/10.12688/f1000research.7334.1

19. Servant N, Varoquaux N, Lajoie BR et al (2015) HiC-pro: an optimized and flexible pipeline for Hi-C data processing. Genome Biol 16:259. https://doi.org/10.1186/s13059-015-0831-x

20. Bolger AM, Lohse M, Usadel B (2014) Trimmomatic: a flexible trimmer for Illumina sequence data. Bioinformatics 30(15):2114–2120. https://doi.org/10.1093/bioinformatics/btu170

21. Kechin A, Boyarskikh U, Kel A et al (2017) cutPrimers: a new tool for accurate cutting of primers from reads of targeted next generation sequencing. J Comput Biol 24(11):1138–1143. https://doi.org/10.1089/cmb.2017.0096

22. Langmead B, Salzberg SL (2012) Fast gapped-read alignment with Bowtie 2. Nat Methods 9(4):357–359. https://doi.org/10.1038/nmeth.1923

23. Tsirigos A, Haiminen N, Bilal E et al (2012) GenomicTools: a computational platform for developing high-throughput analytics in genomics. Bioinformatics 28(2):282–283. https://doi.org/10.1093/bioinformatics/btr646

24. Quinlan AR, Hall IM (2010) BEDTools: a flexible suite of utilities for comparing genomic features. Bioinformatics 26(6):841–842. https://doi.org/10.1093/bioinformatics/btq033

25. Durand NC, Shamim MS, Machol I et al (2016) Juicer provides a one-click system for analyzing loop-resolution Hi-C experiments. Cell Syst 3(1):95–98. https://doi.org/10.1016/j.cels.2016.07.002

26. Cao Y, Chen Z, Chen X et al (2020) Accurate loop calling for 3D genomic data with cLoops. Bioinformatics 36(3):666–675. https://doi.org/10.1093/bioinformatics/btz651

27. Carty M, Zamparo L, Sahin M et al (2017) An integrated model for detecting significant chromatin interactions from high-resolution Hi-C data. Nat Commun 8:15454. https://doi.org/10.1038/ncomms15454

28. Juric I, Yu M, Abnousi A et al (2019) MAPS: model-based analysis of long-range chromatin interactions from PLAC-seq and HiChIP experiments. PLoS Comput Biol 15(4):e1006982. https://doi.org/10.1371/journal.pcbi.1006982

29. Kaul A, Bhattacharyya S, Ay F (2020) Identifying statistically significant chromatin contacts from Hi-C data with FitHiC2. Nat Protoc 15(3):991–1012. https://doi.org/10.1038/s41596-019-0273-0

30. Phanstiel DH, Boyle AP, Heidari N et al (2015) Mango: a bias-correcting ChIA-PET analysis pipeline. Bioinformatics 31(19):3092–3098. https://doi.org/10.1093/bioinformatics/btv336

31. Bhattacharyya S, Chandra V, Vijayanand P et al (2019) Identification of significant chromatin contacts from HiChIP data by FitHiChIP. Nat Commun 10(1):4221. https://doi.org/10.1038/s41467-019-11950-y

32. Crowley C, Yang Y, Qiu Y et al (2021) FIRE-caller: detecting frequently interacting regions from hi-C data. Comput Struct Biotechnol J 19:355–362. https://doi.org/10.1016/j.csbj.2020.12.026

33. Lareau CA, Aryee MJ (2018) hichipper: a preprocessing pipeline for calling DNA loops from HiChIP data. Nat Methods 15(3):155–156. https://doi.org/10.1038/nmeth.4583

34. Li G, Chen Y, Snyder MP et al (2017) ChIA-PET2: a versatile and flexible pipeline for ChIA-PET data analysis. Nucleic Acids Res 45(1):e4. https://doi.org/10.1093/nar/gkw809

35. Mifsud B, Martincorena I, Darbo E et al (2017) GOTHiC, a probabilistic model to resolve complex biases and to identify real interactions in Hi-C data. PLoS One 12(4):e0174744. https://doi.org/10.1371/journal.pone.0174744

36. Roayaei Ardakany A, Gezer HT, Lonardi S et al (2020) Mustache: multi-scale detection of chromatin loops from Hi-C and micro-C maps using scale-space representation. Genome Biol 21(1):256. https://doi.org/10.1186/s13059-020-02167-0

37. Rowley MJ, Poulet A, Nichols MH et al (2020) Analysis of Hi-C data using SIP effectively identifies loops in organisms from C. elegans to mammals. Genome Res 30(3):447–458. https://doi.org/10.1101/gr.257832.119

38. Varani J, Hasday JD, Sitrin RG et al (1986) Proteolytic enzymes and arachidonic acid metabolites produced by MRC-5 cells on various microcarrier substrates. In Vitro Cell Dev Biol 22(10):575–582. https://doi.org/10.1007/BF02623516

39. Kerpedjiev P, Abdennur N, Lekschas F et al (2018) HiGlass: web-based visual exploration and analysis of genome interaction maps.

Genome Biol 19(1):125. https://doi.org/10.1186/s13059-018-1486-1

40. Rowley MJ, Lyu X, Rana V et al (2019) Condensin II counteracts Cohesin and RNA polymerase II in the establishment of 3D chromatin organization. Cell Rep 26(11):2890–2903.e3. https://doi.org/10.1016/j.celrep.2019.01.116

41. Simonis M, Klous P, Splinter E et al (2006) Nuclear organization of active and inactive chromatin domains uncovered by chromosome conformation capture-on-chip (4C). Nat Genet 38(11):1348–1354. https://doi.org/10.1038/ng1896

42. Hsieh TH, Weiner A, Lajoie B et al (2015) Mapping nucleosome resolution chromosome folding in yeast by micro-C. Cell 162(1):108–119. https://doi.org/10.1016/j.cell.2015.05.048

43. Anders S, Huber W (2010) Differential expression analysis for sequence count data. Genome Biol 11(10):R106. https://doi.org/10.1186/gb-2010-11-10-r106

44. Robinson MD, McCarthy DJ, Smyth GK (2010) edgeR: a Bioconductor package for differential expression analysis of digital gene expression data. Bioinformatics 26(1):139–140. https://doi.org/10.1093/bioinformatics/btp616

45. Fernandez LR, Gilgenast TG, Phillips-Cremins JE (2020) 3DeFDR: statistical methods for identifying cell type-specific looping interactions in 5C and Hi-C data. Genome Biol 21(1):219. https://doi.org/10.1186/s13059-020-02061-9

46. Sahin M, Wong W, Zhan Y et al (2020) HiC-DC+: systematic 3D interaction calls and differential analysis for Hi-C and HiChIP. bioRxiv:2020.2010.2011.335273. https://doi.org/10.1101/2020.10.11.335273

Part III

Sequencing-Based Approaches to Assess Nuclear Environment

Chapter 8

Measuring Cytological Proximity of Chromosomal Loci to Defined Nuclear Compartments with TSA-seq

Liguo Zhang, Yu Chen, and Andrew S. Belmont

Abstract

Distinct nuclear structures and bodies are involved in genome intranuclear positioning. Measuring proximity and relative distances of genomic loci to these nuclear compartments, and correlating this chromosome intranuclear positioning with epigenetic marks and functional readouts genome-wide, will be required to appreciate the true extent to which this nuclear compartmentalization contributes to regulation of genome functions. Here we present detailed protocols for TSA-seq, the first sequencing-based method for estimation of cytological proximity of chromosomal loci to spatially discrete nuclear structures, such as nuclear bodies or the nuclear lamina. TSA-seq uses Tyramide Signal Amplification (TSA) of immunostained cells to create a concentration gradient of tyramide–biotin free radicals which decays exponentially as a function of distance from a point-source target. Reaction of these free radicals with DNA deposits tyramide–biotin onto DNA as a function of distance from the point source. The relative enrichment of this tyramide-labeled DNA versus input DNA, revealed by DNA sequencing, can then be used as a "cytological ruler" to infer relative, or even absolute, mean chromosomal distances from immunostained nuclear compartments. TSA-seq mapping is highly reproducible and largely independent of the target protein or antibody choice for labeling a particular nuclear compartment. Our protocols include variations in TSA labeling conditions to provide varying spatial resolution as well as enhanced sensitivity. Our most streamlined protocol produces TSA-seq spatial mapping over a distance range of ~1 micron from major nuclear compartments using ~10–20 million cells.

Key words TSA-seq, Nuclear compartments, Nuclear bodies, Genome organization, Gene position, Proximity labeling

1 Introduction

1.1 Overview   Fluorescence in situ hybridization (FISH) methods have established a curious but highly conserved correlation between gene expression and the radial positioning of chromosomes, with gene expression generally increasing with a more interior average radial position [1, 2]. This correlation is curious because there is no single

---

Liguo Zhang and Yu Chen contributed equally as co-first authors.

Tom Sexton (ed.), *Spatial Genome Organization: Methods and Protocols*,
Methods in Molecular Biology, vol. 2532, https://doi.org/10.1007/978-1-0716-2497-5_8,
© The Author(s), under exclusive license to Springer Science+Business Media, LLC, part of Springer Nature 2022

known structure localized at the nuclear center to explain this gradient of gene expression as a function of radial position. Instead, we have proposed that this gradient in gene expression with radial position reflects a convolution of the correlations of gene expression with position relative to multiple nuclear compartments that themselves show a nonrandom, radial distribution within the nucleus [3, 4]. These correlations would include the known association of heterochromatic chromosome regions with the nuclear periphery [5–7] and active chromosome regions with the more interiorly located nuclear speckles [8, 9]. But the genome is arranged nonrandomly relative to a number of other known nuclear structures, including for example nuclear pores, nucleoli, chromocenters, paraspeckles, and PML and Cajal bodies [10, 11]. Gene positioning relative to several of these known nuclear compartments has already been tightly correlated with regulation of genes [9, 12–14]. The combined volume of just these known nuclear compartments would sum to a significant fraction of the total nuclear volume.

Nuclear bodies have been defined in the past as intranuclear, non–membrane enclosed structures that can be visualized by transmission electron microscopy (TEM) as independent domains without immunostaining [10, 11]. Emerging evidence suggests many of these known nuclear bodies, including nucleoli and nuclear speckles, may be formed through phase-separated condensates, including but not limited to liquid-liquid phase separation [15–19]. Recent studies, however, have revealed additional nuclear condensates that do not necessarily form distinct bodies as visualized by TEM without immunostaining. Examples would include condensates formed by local concentrations of Mediator [20–22], different post-translational modifications of RNA pol2 [23–25], and various factors involved in RNA processing which localize outside any of the original nuclear bodies visualized by TEM [26–28].

Thus, a large fraction of the total nuclear volume is comprised of regions close to one or more of these many distinct nuclear structures, bodies, and condensates – both known and unknown. Full exploration of the relationship between genome intranuclear positioning and genomic function awaits further mapping of the genome relative to these different nuclear compartments. Traditional technologies to study intranuclear gene position include both microscopy and genomic approaches.

Conventional microscopy approaches using FISH are very low throughput. More recent multiplexed oligo-library based methods have greatly advanced the throughput to hundreds to thousands of chromosome target loci, yet this still translates to sampling at no more than ~1 Mbp resolution. Moreover, these approaches are still relatively low throughput approaches for imaging the entire genome and require specialized equipment [29, 30]. Additionally, there remain concerns regarding possible spatial distortions in both nuclear structures and chromosomes introduced by the harsh DNA denaturation conditions used for DNA FISH [31, 32].

Genomic approaches currently include molecular proximity assays such as ChIP-seq [33] and DamID-seq [34], which until recently were the most direct ways to probe for intranuclear chromosome positioning. Newer molecular proximity methods include CUT&RUN [35, 36], CUT&Tag [37], and pA-DamID [38]. All of these methods assay the relative fractions of alleles in direct, molecular contact with their specific molecular targets. DamID and pA-DamID, especially, but also ChIP have been very successful in mapping lamina associated domains (LADs) [5–7, 38–40]. To date, however, they have not provided a straightforward, generic method for mapping chromosome contacts with any nuclear structure/body/compartment. This may be because these methods require actual molecular contact of the interphase chromosome with the target protein and therefore they will only report on the actual fraction of the target protein in molecular contact with the interphase chromosome. If the target protein is close cytologically but not in actual molecular contact, then these methods will not produce a signal. Conversely, the signal produced by these methods might be heavily skewed, or even dominated, by those chromosome loci in closest molecular contact with what might be a small fraction of the target protein located away from the actual target nuclear structure. As a specific example, one might predict for these reasons that using either DamID or ChIP to identify chromosome regions close to nuclear speckles would be problematic: First, SON and SRRM2 [41], the two most highly enriched proteins in nuclear speckles [28] that have commonly been used as markers for nuclear speckles, are both located in the interior core region of nuclear speckles [42] and are likely not in molecular contact with the vast majority of chromatin surrounding nuclear speckles. Second, most nuclear speckle proteins, including SON, are present throughout the nucleus at lower concentration bound over nascent transcripts and/or transcriptionally active genes. Indeed, previous ChIP-seq performed using antibodies to SON and a monoclonal antibody putatively targeting SC35 (SRSF2), but recently revealed actually to be SRRM2 [41], showed peaks over most active genes regardless of their distances to nuclear speckles [3]. In both cases, ChIP-seq was likely dominated by the small fraction of target proteins in molecular contact with active genes rather than the majority of target proteins within nuclear speckles.

Newer genomic methods such as SPRITE [43, 44] and MARGI [45] have inferred genomic regions associating with nuclear speckles by measuring interactions between DNA and known RNAs enriched at these nuclear speckles. MARGI, which uses a ligation-based approach to measure RNAs near DNA, presumably measures molecular-scale interactions; therefore, similar concerns raised in the preceding paragraph for molecular proximity mapping approaches may apply as well to MARGI. In contrast, SPRITE uses sonication to form complexes whose sizes can be varied according to sonication conditions. SPRITE therefore probes larger distances than typical molecular proximity methods and as a result is more likely than molecular proximity methods to capture chromosome regions adjacent to nuclear structures/bodies/compartments. However, the actual distance(s) measured by SPRITE are typically indeterminate, given that the sizes of the complexes produced by sonication are not usually measured.

We developed TSA-seq to map the relative positions of genomic loci over cytological distance scales relative to known nuclear compartments such as nuclear speckles [3, 4]; specifically, we were interested in developing a genomic method that would help bridge the gap between sequencing-based genomic mapping approaches and light microscopy. TSA-seq is based on the widely used signal amplification technique for immunochemistry called Tyramide Signal Amplification or TSA [46, 47]. The rationale of TSA-seq is to combine antibody staining with TSA to generate a diffusible label generated from and centered around a specific target protein that identifies a given nuclear structure/compartment/body. TSA uses antibodies coupled to horseradish peroxidase (HRP) to catalyze the formation of tyramide free-radicals that diffuse away from the antibody and target protein prior to reaction and covalent labeling of neighboring macromolecules. Previously, the presumed major substrates for tyramide labeling were the tyrosines and tryptophans in proteins. Recently, though, we showed that DNA was also a substrate [3]. Tyramide is typically conjugated to biotin as an epitope tag, but other haptens, including FITC, are also commonly used [28].

A similar molecular proximity labeling method, APEX (or APEX2), uses a different peroxidase that generates the same tyramide–biotin (phenol–biotin) free radical. APEX labeling has commonly been thought to extend only ~20 nm from the peroxidase free radical source [48–52]. However, this inferred distance was derived from previous transmission electron microscopy (TEM) visualization of immunostaining using HRP or APEX to catalyze the generation of free radicals that oxidize DAB (3,3'-Diaminobenzidine). This oxidation then leads to DAB polymerization and formation of a localized DAB precipitant that is then stained by osmium tetroxide for TEM visualization. However, we showed that the diffusion of the TSA-generated, tyramide free radical labeling leads to an exponential decay spreading as far as 1000 nm from the TSA antibody target [3]. Thus, in contrast to the DAB precipitant, the tyramide–biotin (phenol–biotin) free radical is soluble and diffuses over large distances. TSA labeling therefore can be used to derive information on distances of chromosome regions up to ~1 μm from target nuclear structures.

By manipulating either the viscosity of the TSA reaction buffer (using the addition of sucrose) or the lifetime of the tyramide free radical (using the addition of DTT to quench tyramide free radicals), we described how we could vary the TSA spatial resolution to produce three different TSA-seq mapping resolutions [3]. Importantly, as currently implemented, TSA-seq uses the tyramide–biotin labeling of DNA rather than chromosomal proteins for its readout of chromosome positioning. We found that histones are poorly labeled by the TSA reaction, implying that protein TSA labeling might be biased by varying local concentrations of nonhistone chromosomal proteins. The use of DNA labeling, in theory anyway, avoids possible bias introduced by varying genomic distribution of chromosomal proteins with different amino-acid composition and tyrosine-reactivity.

However, because of the low efficiency of DNA tyramide labeling, the original TSA-seq protocol (TSA-seq 1.0) requires several hundred million cells for each staining target and replicate [3]. In our newly developed TSA-seq 2.0 method, we increased the sensitivity by 10–20-fold by deliberately saturating protein labeling, leaving DNA labeling enhanced but not saturated [4]. Thus, TSA-seq 2.0 requires only ~10–20 million cells for the same nuclear compartment targets, while providing similar distance mapping capability as TSA-seq 1.0 [4].

Together, the TSA-seq 1.0 and 2.0 approaches described three "reaction" conditions (TSA-seq 1.0), for varying spatial mapping resolution, and seven "enhancement" conditions (TSA-seq 2.0), for varying mapping sensitivity. While we continue to further optimize TSA-seq, and extend it to the mapping of additional nuclear compartments, currently our general approach to applying TSA-seq to any new project involves: (1) making a choice about the reaction condition(s) appropriate for the spatial resolution range(s) we want to explore, and; (2) determining which enhancement condition will maximize DNA labeling for our particular mapping target without producing an overlabeling (a significant fraction of sonicated DNA fragments from any genomic region with more than one biotin per fragment that would introduce a nonlinearity between the amount of DNA biotinylation and the amount of biotin-DNA pulldown).

At first glance, the large number of possible combinations between reaction and enhancement conditions might appear confusing. Conceptually, however, understanding how to develop a working TSA-seq protocol is logical and straightforward. Below in Subheading 1.2, we present an overall guide toward how to tailor the TSA-seq protocol to a particular target and application. We then discuss in detail a number of specific points (*see* Subheading 1.3–1.9) to take into consideration in accomplishing this tailoring of the TSA-seq protocol, including important controls and quality checks that should routinely be included during protocol development and application.

In the remaining chapter sections, we then present Materials and one detailed TSA-seq protocol specific for mapping the genome relative to nuclear speckles using a particular antibody against the SON protein. This detailed protocol is accompanied by a number of Notes describing additional considerations.

1.2 General Guide to Designing a Tailored TSA-seq Protocol

Designing and executing a successful TSA-seq experiment conceptually involves several steps related to the immunostaining and TSA reaction itself (Fig. 1).

First, one must determine whether TSA-seq is in fact the appropriate method for the research goal(s). This requires an appreciation for the differences between the cytological distance mapping that TSA-seq provides as compared to molecular proximity assays such as ChIP-seq or Cut and Run/Tag, as discussed in Subheading 1.1. It also requires considering how well matched the intranuclear spatial distribution of the target protein is to the spatial resolution of the TSA-staining reaction.

Second, one must identify a suitable primary antibody to label the target nuclear compartment and suitable cell fixation and immunostaining protocols that work well with this primary antibody while still preserving nuclear architecture. This is done with small numbers of cells, typically growing adherent cells directly on coverslips for subsequent microscopy examination or attaching cells grown in suspension onto coverslips, either before or after immunostaining.

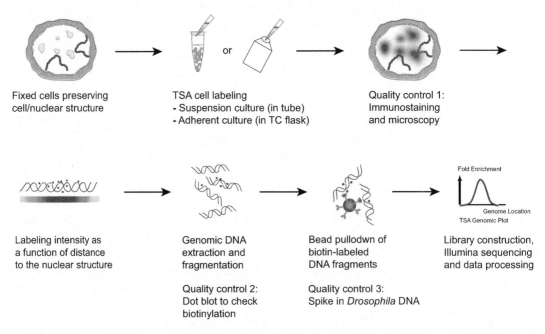

Fixed cells preserving cell/nuclear structure

TSA cell labeling
- Suspension culture (in tube)
- Adherent culture (in TC flask)

Quality control 1: Immunostaining and microscopy

Labeling intensity as a function of distance to the nuclear structure

Genomic DNA extraction and fragmentation

Quality control 2: Dot blot to check biotinylation

Bead pullodwn of biotin-labeled DNA fragments

Quality control 3: Spike in *Drosophila* DNA

Library construction, Illumina sequencing and data processing

Fold Enrichment

Genome Location

TSA Genomic Plot

Fig. 1 TSA-seq experimental workflow. Schema showing TSA-seq workflow, including quality control steps. (Reprinted and modified from Fig. 2A of Ref. [3], with reprint permission from Rockefeller University Press)

Third, one then must develop a working TSA staining protocol, including the choice of HRP-labeled secondary antibody, and confirm by microscopy the TSA staining specificity achieved by the net combination of immunostaining and TSA reaction, using the minimum enhancement condition that avoids protein saturation. TSA-seq is a "What you see is what you get" method: the labeling visualized in the microscope after the TSA reaction performed under conditions that do not saturate protein labeling should be close to the corresponding genomic labeling produced in the subsequent TSA-seq procedure.

Fourth, one next explores different TSA enhancement conditions, again using microscopy on small numbers of cells. Fifth, one now needs to scale up the antibody staining and TSA reaction, optimized in these earlier steps on small numbers of cells, to the larger numbers of cells required for the TSA-seq protocol. The most common error during this step is to underestimate the amount of total primary and secondary antibodies added to the sample needed to maintain similar levels of free, unbound antibodies during staining when cell numbers are increased. Once this is achieved, the sixth step (after TSA staining of the larger numbers of cells required for TSA-seq) is to harvest the cells and proceed to DNA isolation, sonication, and then assay of biotin labeling using dot-blot analysis.

Note that this sixth step marks the transition from cell biological immunostaining to biochemical protocols. Whereas immunostaining protocols need to be adapted and optimized to each new primary and secondary antibody, biochemical protocols instead can be standardized and applied uniformly to all TSA staining samples. Note also that we perform **steps 1–5** similarly to how we approach sample preparation and immunostaining for super-resolution light or electron microscopy, considering carefully each part of the protocol for its potential to perturb nuclear and chromosomal structure in ways that would change the results or resolution of the TSA-seq measurement. This is because our comparison of measured FISH (fluorescence in situ hybridization) distances of specific genomic regions to nuclear speckles versus distances to nuclear speckles predicted by TSA-seq revealed a typical difference of only ~50 nm, exceeding the resolution of conventional light microscopy. Thus, we avoid the type of sample treatments used commonly for many biochemically based genomic methods. This includes removing adherent cells off of the cell culture dish substrate prior to staining, isolation of nuclei, detergent treatment and exposure of permeabilized cells to buffers that can perturb large-scale chromatin structure prior to cell fixation, freezing of cells prior to staining, or any steps during staining—such as air-drying of cells—that can perturb cell structure. Once TSA staining is completed, one can use traditional biochemical procedures without concern about altering results.

Step 7 is the DNA pulldown of biotinylated DNA. Although conceptually simple, this is the most critically demanding step, given the typically low fraction of biotinylated DNA fragments, particularly using lower levels of TSA enhancement. We therefore recommend initial pilot experiments using DNA samples of unlabeled genomic DNA spiked with biotinylated fragments of known sequence to verify specificity and yield of this pulldown. We now routinely add both biotinylated and nonbiotinylated spike-in DNA fragments to TSA-seq input DNA samples prior to pulldown of the biotinylated DNA. These spike-in controls monitor the efficiency of biotinylated DNA pulldown, the enrichment of biotinylated DNA in the pulldown sample, and the depletion of nonbiotinylated DNA in the pulldown DNA sample. **Step 8** is preparation of DNA sequencing libraries. **Step 9** is mapping the raw reads back to the genome and processing these mapped reads to produce a normalized TSA-seq score.

Producing meaningful TSA-seq data for a particular research application should be considered an iterative process. Examination of the results from several quality control tests, including sequencing read numbers of the spike-in controls, plus examination of the TSA-seq score in a genome browser with comparison to other genomic data, may lead to adjustments of the overall TSA-seq protocol. These adjustments might include changes to antibody concentrations, changes in the choice of TSA reaction conditions to change the spatial resolution of the TSA-mapping, changes in the choice of TSA enhancement condition to increase labeling efficiency or to avoid overlabeling. Additional possible adjustments include repeating the DNA pull-down two or even three times to further increase the pulldown enrichment of biotinylated DNA and changes in the library construction method to minimize PCR bias. Below we discuss in greater detail some of these individual steps.

1.3 Nuclear Compartment Target and Antibody Selection

TSA-seq will be informative only if the staining target shows a spatially distinct, relatively sparse nuclear staining pattern relative to the several hundred nm or larger TSA staining radius. TSA is a "cytological" rather than molecular assay, providing spatial information over distances of hundreds but not tens of nm. Continuous immunostaining patterns throughout the nucleus, with only small variations in intensity over distances of several hundred nm and above, should not yield significant differences in TSA labeling over the genome. Similarly, staining of large numbers of distinct foci that are very closely spaced are likely only to give low spatial resolution maps if the TSA spreading of tyramide labeling creates a merged, more continuous staining between foci. An example of the latter would be the TSA-seq obtained after anti-RNA pol2 Ser5phos staining [3], which produced TSA-seq scores strikingly similar to Hi-C compartment (EV1, Eigenvector 1) scores [3].

Because TSA-seq is based on immunostaining, both the specificity of the chosen marker protein to the nuclear compartment of interest and the specificity of the antibody to the marker protein are critical for successful TSA-seq mapping. Therefore, we suggest beginning by optimizing the immunostaining of the target nuclear compartment. More specifically, we recommend starting by surveying possible primary antibodies to use for marking these target compartments. This is especially true for new or little-studied target compartments. Nowadays, this can begin by combining a literature survey with public databases, such as the Human Protein Atlas (https://www.proteinatlas.org). A few promising candidates can then be tested by initial staining, allowing selection of antibodies that produce the highest specific to nonspecific staining at the highest dilutions. Some nonspecific staining might be tolerated, if necessary, but this will reduce the final dynamic range of the TSA-seq results. However, nonspecific staining in a particular staining pattern may skew the TSA-seq results, producing systematic error in the mapping of the genome relative to the target compartment. For example, cytoplasmic staining (either specific or nonspecific) from either the primary or secondary antibody could affect the TSA-seq results given the large staining radius of the TSA reaction. A low-level cytoplasmic background staining could contribute to a TSA-seq pattern resembling anti-lamin TSA-seq due to the spreading of the biotin–tyramide generated by the TSA reaction into the nucleus, thus reacting with the chromatin at the nuclear periphery. This would be true anytime the cytoplasmic background staining was within the TSA staining radius of the nuclear periphery.

Once candidate primary antibodies are identified, we suggest identifying fixation conditions which best balance preservation of the target protein intranuclear distribution, preservation of interphase chromosome and nuclear morphology, uniform staining of the target protein distribution, and preservation of epitope recognition by the primary antibody. Ideally, we prefer antibodies that allow use of significant formaldehyde fixation time and concentration without any detergent permeabilization prior to fixation. Some proteins change their distribution with formaldehyde fixation but not, for example, after alcohol fixation. However, alcohol fixation significantly perturbs nuclear and chromatin structure, so we try to avoid these antibodies. Some antibody epitopes are destroyed by even short formaldehyde fixation times. If given the choice, we will choose antibodies that allow for stronger fixation.

Typically, we begin our survey of primary antibodies using well-tested, fluorescently labeled secondary antibodies used for other projects in the laboratory. Serial dilutions of the primary antibody are done to identify optimal antibody concentrations that maximize specific versus background staining—the TSA labeling itself amplifies the immunostaining reaction so higher specificity is the top criterion at this point. We favor low antibody concentration combined with longer incubation times at 4 °C (overnight or most of the working day). For most antibodies, we obtain brighter, more specific staining using lower concentrations of antibodies but for longer times as compared to protocols with incubation times of just a few hours but higher antibody concentrations; however, in some cases this increases background. Lower antibody concentrations reduce cost when staining is scaled to tens of millions of cells.

Once we identify a reasonable primary antibody titer, we then test HRP-labeled secondary antibodies. Any new HRP-labeled secondary antibody needs to be tested for specificity using the negative control of no added primary antibody. Secondary antibody concentrations are then tested using the candidate primary antibody and usually an additional primary antibody known to stain a well-recognized cell structure as a positive control. We test this secondary antibody staining using a fluorescently labeled tertiary antibody directed against the secondary antibody.

1.4 Testing TSA Staining Conditions

Once primary and secondary antibody titers are optimized, different reaction and enhancement TSA conditions can be explored, again using small-scale coverslip cell staining and inspection of this staining by light microscopy. Variable conditions for TSA-seq are summarized in Tables 1 and 2. Specifically, we describe three TSA "reaction" conditions, to reach different mapping spatial resolutions, and multiple TSA "enhancement" conditions, to increase the amount of TSA labeling, as tested using nuclear speckle (anti-SON) TSA-seq 1.0 mapping [3] (Fig. 2). TSA-seq reaction Condition 1, using PBS as the buffer for TSA labeling, provides the lowest spatial resolution but probes the largest volume surrounding a target nuclear structure. TSA-seq reaction Condition 2 uses 50% sucrose added to PBS to reduce diffusion of the tyramide free radicals and provide intermediate spatial resolution while probing nearly the same volume surrounding the target structure. TSA-seq reaction Condition 3 uses 50% sucrose in PBS containing an added 0.75 mM DTT to quench the tyramide free radicals, which reduces the tyramide free radical life-time, thus limiting how far the tyramide free radicals can diffuse. Condition 3 provides the highest spatial resolution but probes the smallest volume surrounding the

Table 1
TSA reaction conditions

Reaction condition	PBS	Sucrose (w/v)	DTT (mM)
1	1×	–	–
2	1×	50%	–
3	1×	50%	0.75

Table 2
TSA enhancement conditions

Enhancement condition	Tyramide–biotin	H$_2$O$_2$	Reaction time (mins)	Cell number needed for 5 ng pulldown DNA[a]
A	1:10,000	0.0015%	10	150 million
AI	1:10,000	0.0015%	20	
AII	1:5000	0.0015%	20	
B	1:3000	0.0015%	10	50 million
C	1:3000	0.0015%	30	30 million
D	1:1000	0.0015%	30	20 million
E	1:300	0.0015%	30	15 million

[a]Tested for anti-SON TSA labeling by Reaction Condition 2 (Table 1) in K562 cells

target structure. Also, the amount of tryamide–biotin labeling progressively decreases from Conditions 1 through 3, meaning an increasing number of cells are required for Conditions 1 through 3 to obtain sufficient pull-down DNA. The decreased tyramide labeling with Condition 3 is expected, given the deliberate quenching of the tyramide free radicals by DTT. The reason for the reduced tyramide reactivity using Condition 2 is unknown but may be due to tyramide reactivity with the sucrose.

All testing of TSA staining should begin using the lowest enhancement Condition A (TSA-seq 1.0) to verify the specificity of the combined immunolabeling and TSA staining in labeling of the target nuclear compartment. A low level of TSA enhancement is needed to avoid saturating the reactive groups (tyrosines and tryptophans) in the proteins surrounding the target nuclear compartment. The tyramide protein labeling dominates the light microscopy TSA signal; it should be a proxy for the actual tyramide labeling on DNA, but only under these nonsaturating conditions. Therefore, one should use enhancement Condition A (minimum enhancement) to get an appreciation for the actual intranuclear distribution of TSA labeling. In practice we like to visualize the biotin distribution using fluorescently labeled streptavidin (or anti-biotin antibody) but then also to visualize the distribution of the target protein using an additional fluorescently labeled secondary antibody against the primary antibody. This allows direct comparison in the same nucleus of the target protein immunostaining with the TSA staining.

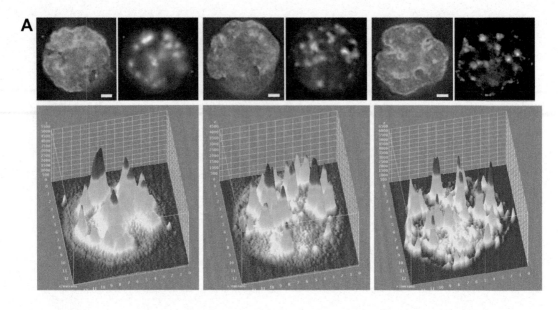

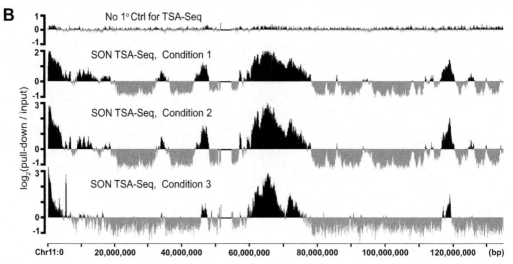

Fig. 2 Nuclear speckle (SON) TSA-seq 1.0 mapping using three reaction conditions. (**a**) Microscopy image (top) and intensity profile (bottom) of anti-SON TSA staining using reaction Condition 1 (left panels), 2 (middle panels), or 3 (right panels). Top: red (left insets)/gray (right insets): Streptavidin staining for TSA biotin labeling; green: DAPI staining of DNA; scale bars: 2 μm. Bottom: surface plots showing image intensity profiles of biotin staining (Reprinted from Fig. 1D of Ref. [3], with reprint permission from Rockefeller University Press). (**b**) SON TSA-seq 1.0 chromosome 11 profiles for (top to bottom): no primary antibody negative control (reaction Condition 1) and anti-SON antibody TSA staining using reaction Condition 1, Condition 2, or Condition 3. (Reprinted and modified from Fig. 2B of Ref. [3], with reprint permission from Rockefeller University Press)

At this stage, different TSA reaction conditions can be explored, selecting one tuned to the goals of the TSA-seq mapping. While Condition 3 provides the highest spatial resolution, it also may probe the smallest fraction of the genome. For example, whereas most of the nucleus is within the ~1-μm staining radius of nuclear speckles using Conditions 1 and 2, a large fraction of the nucleus will be outside the nuclear speckle TSA staining radius using Condition 3. Therefore, whereas Conditions 1 and 2 will provide spatial positioning information for nearly the entire genome, a significant fraction of the genome will be labeled with similar, near background levels using Condition 3. This loss of spatial information for a significant fraction of the genome is demonstrated by the conversion of small SON TSA-seq peaks or "peaks-within-valleys" visualized in Condition 1 or 2 TSA-seq maps to featureless valleys in Condition 3 TSA-seq maps [3] (Fig. 2).

In practice, we use TSA reaction Condition 2 (plus sucrose) for most applications. Reaction Condition 2 offers slightly higher spatial resolution with only several-fold reduction in TSA labeling relative to Condition 1 (buffer only). If we want to maximize either labeling intensity and/or the range of TSA staining—for instance to probe larger distances from the target compartment—then we use reaction Condition 1. Previously, we have used reaction Condition 3 rarely, due to both its restricted spatial range and its reduced labeling intensity. However, we are now exploring enhancement conditions that suitably boost the labeling efficiency using reaction Condition 3 for specific experiments where we are interested in higher spatial resolution.

Once the reaction condition is determined, different enhancement conditions can then be tested. Table 2 contains seven TSA "enhancement" conditions using progressively increasing biotin–tyramide concentration and/or reaction times; Conditions A, B, C, D, and E were tested using reaction Condition 2 for nuclear speckle (anti-SON) mapping [4] (Fig. 3). The enhancement in tyramide–DNA labeling provided by these enhancement conditions translates into smaller cell numbers required for TSA-seq experiments. Conceptually, the limit on the degree of this reaction enhancement is imposed by our goal of maintaining a linearity between the amount of tyramide–DNA labeling and the amount of biotinylated DNA pulldown. We therefore enhanced DNA tryamide–biotin labeling only up to levels that would still produce the vast majority of DNA fragments after sonication with only 0 or 1 biotin tags per fragment, even over the maximally labeled genomic regions (highest peaks visualized in the TSA-seq maps) [4]. In the particular case of SON-TSA labeling in K562 cells, we verified that the enhancement condition producing the highest tyramide labeling (Condition E) provided the same estimated distance to nuclear speckles as that obtained using the lowest enhancement condition (Condition A), corresponding to the original SON TSA-seq 1.0 reaction protocol [4].

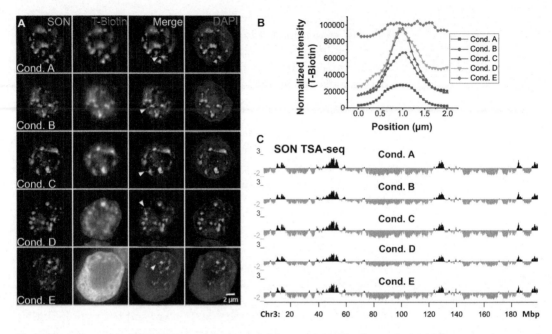

Fig. 3 Nuclear speckle (SON) TSA-seq 1.0 versus 2.0 "super-saturation" mapping. (**a**) Microscopy images of anti-SON immunostaining (left, green), streptavidin staining for TSA biotin (left middle, red), merged SON and streptavidin staining (right middle), and merged SON and streptavidin with DAPI (blue) DNA staining (right). Top to bottom: SON TSA-seq reaction Condition 2 plus enhancement Conditions A (TSA-seq 1.0), B, C, D, and E. With increasing levels of enhancement from B through E, biotin staining progressively spreads outward from nuclear speckles, eventually reaching the cytoplasm, with protein saturation also progressively spreading outward from nuclear speckles. (**b**) Streptavidin biotin staining normalized intensities along line profiles shown in panel A demonstrate the increase in nuclear speckle TSA staining intensities with increasing enhancement levels which then saturates and spreads outward from the nuclear speckle peaks at enhancement Conditions D and E. (**c**) TSA-seq chromosome 3 profiles for corresponding enhancement Conditions A–E (top to bottom) are similar, despite the obvious saturation effect visualized by immunostaining in panel (**a**) and (**b**) for Conditions D–E. This is explained by the microscopy images being dominated by protein tyramide labeling, whereas the TSA-seq exploits the DNA tyramide labeling, which remains below saturation (no more than 1 biotin label per DNA fragment for most DNA fragments). (Reprinted and modified from Fig. 1 A, B and E of Ref. [4], with author-retained copyright)

Using this minimal enhancement Condition A together with reaction Condition 2, the anti-SON TSA labeling visualized by light microscopy using anti-biotin staining shows the expected decay of labeling with distance from nuclear speckles; this anti-biotin staining visualized by microscopy is presumably dominated by tyramide labeling on proteins. With progressively increasing TSA reaction enhancement, a progressive loss of the expected decay in biotin staining with distance from nuclear speckles is seen, until with Condition E a near uniform biotin staining throughout the nucleus is observed. We interpreted this change in staining as due to the progressive saturation of tyramide labeling on proteins spreading outward from the labeling site [4]. Therefore, whereas light microscopy can be used as a quality check for both the magnitude and specificity of TSA labeling using enhancement Condition A, light microscopy is used primarily as a quality control for the intensity of TSA labeling with higher enhancement conditions [4]. This provides a useful reference for comparing to the TSA staining achieved during the actual experiment when it is scaled up to tens of millions of cells in a dish or flask.

Although for SON TSA-seq we have obtained quantitatively similar results using enhancement conditions A-E, this was not true with lamin B1 TSA-seq. Whereas enhancement Conditions A through C yield similar lamin B1 TSA-seq results, use of Condition E sometimes noticeably changes the final TSA-seq results to a pattern that does not correlate as well to Lamin B1 DamID (data not shown). We presume this reflects nonlinear effects produced by overlabeling of individual DNA fragments, particularly in LAD regions that tightly contact nuclear lamina. Therefore, we routinely use Condition AI, AII, or C (Table 2) for lamin B TSA-seq.

Based on this experience, for new targets we are conservative, by only choosing enhancement Condition E where we need to maximize yield of biotinylated DNA to reduce cell numbers, but choosing lower enhancement conditions such as Condition C if they produce sufficient biotinylated DNA to construct a sequencing library using ~10–30 million cells.

1.5 Scaling Up TSA Staining for the Larger Cell Numbers Required for TSA-seq

A very common experimental error is to use the same antibody titer as used in the original TSA pilot experiments performed on coverslips in the actual cell staining for a TSA-seq experiment without considering the staining solution volume, and, more specifically, the ratio of cell number to staining solution volume. Because antibodies are expensive, a common decision is to reduce the volume of antibody staining solution to the minimum required (for example to uniformly coat the surface area of the tissue culture flask) without considering the number of cells used for staining. The volume of the antibody staining solution relative to number of cells used for coverslip staining is typically high enough to ensure that the concentration of free antibody is nearly the same as the total antibody concentration. However, when staining in a large flask, as one reduces the staining volume, the ratio of cell number to antibody staining volume will increase and the concentration of free antibody may progressively fall. Thus, the actual titer of free, unbound antibody may be significantly lower in the bulk TSA-seq staining experiment than what was used in the initial TSA staining conducted on coverslips. This is of particular concern for TSA labeling of cells in suspension, where the cell number to volume ratio can be increased significantly beyond what was used in the coverslip cell staining. Therefore, we recommend estimating the actual ratio of cell number to volume of antibody staining solutions used in the initial test and considering this ratio in scaling up antibody solution volumes used in the TSA-seq experiments. We also recommend testing the efficiency of TSA labeling after the actual TSA staining performed in each TSA-seq experiment by removing a small aliquot of cells. This aliquot of cells can then be immunostained and studied by light microscopy, to visualize the biotin distribution produced by the TSA labeling relative to the primary antibody labeling of the target protein.

The simplest way to scale up the reaction for adherent cells is first to identify the minimum volume that can be used to safely stain coverslips at the minimum antibody total concentration needed to achieve satisfactory staining. Then one can simply scale the volume of the staining solution by the cell number or the surface area of the flasks or dishes, assuming a similar cell number per cm^2 is used for both the coverslip and flask staining.

The next step for scaling the TSA reaction up to the numbers of cells needed for a TSA-seq experiment is to estimate the number of cells that will be required. For a nuclear speckle (anti-SON) TSA-seq mapping experiment, using reaction Condition 2 and enhancement Condition A, ~100–300 million cells are required to obtain a final 5 ng of DNA for sequencing after the pulldown of biotinylated DNA. This compares to ~50–150 million cells needed for nuclear lamin (anti-Lamin B1) TSA-seq mapping using the same reaction and enhancement conditions. Fewer cells are required for lamin TSA-seq due to a higher yield of biotinylated DNA after the lamin immunostaining and TSA reaction; presumably, this is due to the larger fraction of genomic DNA within the TSA staining radius of the anti-lamin versus anti-SON staining. As the fraction of the nuclear volume stained by the TSA reaction decreases, a larger number of cells will be required to obtain the same, minimum amount of pulldown DNA needed to prepare the DNA sequencing library. Thus, targeting a protein present in just a few small nuclear bodies per nucleus will require more cells than targeting a major nuclear compartment for which a larger fraction of the genome is within the staining radius of this compartment.

Our experience has been that reaction Condition 1 requires ~four to fivefold fewer cells, as compared to that required for reaction Condition 2, while reaction Condition 3 requires ~four to fivefold increased cell numbers; this is directly proportional to the different amount of tyramide–DNA labeling generated using reaction Conditions 1–3, as discussed previously.

We can reduce the required cell numbers for SON TSA-seq from ~100–300 million cells to ~10–30 million cells using enhancement Condition E. Table 2 shows the multi-fold reduction in cell numbers required for SON TSA-seq staining at reaction Condition 2 as a function of different enhancement conditions. In general, the required number of cells for the TSA-seq experiment will be a function of the target abundance and intranuclear distribution and the choice of both "reaction" and "enhancement" TSA conditions, as described previously in this section.

1.6 Harvesting and Lysing Cells, DNA Purification and Sonication, and Assaying DNA Biotinylation

From this step on, the TSA-seq methodology changes from cell biological to biochemical protocols.

Buffers and tubes should be free of contaminants (molecular biology grade). The purified and sonicated DNA should be quantified by spectrophotometer (Nanodrop). A_{260}/A_{280} ratio should be ~1.8 and A_{260}/A_{230} ratio should be ~2.2 to ensure good DNA quality.

Fragmented DNA is subjected to dot blot analysis to estimate DNA biotinylation levels prior to streptavidin bead pulldown of biotinylated DNA. A fivefold serial dilution of 250 bp biotinylated DNA fragments prepared by PCR are used as biotinylated DNA standards. This DNA fragment was PCR amplified from a BAC (BACR48E12) containing the cloned *Drosophila* genomic DNA using primers: GAAACATCGC/iBiodT/GCCCATAAT (forward) and AGAAGCAGCTACGCTCCTCA (reverse). The use of the biotinylated forward primer results in exactly 1 biotin per 250 bp PCR fragment. These biotinylated DNA standards were combined with sonicated human genomic DNA (unbiotinylated, 100–600 bp) to produce the same final DNA concentration as the actual DNA samples isolated from the TSA-stained cells. Dilution of biotinylated PCR standards in unbiotinylated genomic DNA is meant to produce similar ratios of biotinylated versus nonbiotinylated DNA fragments in the standards as in the actual DNA sample from TSA-stained cells, to avoid the possibility of varying efficiencies for cross-linking the DNA to the dot-blot membrane. The cross-linking of the biotinylated standards to the membrane was observed to be higher without the added genomic DNA, resulting in an inaccurate comparison of biotinylation levels between sample and standards.

1.7 Pulldown of Biotinylated DNA

Since the expected amount of biotinylated DNA in the sample is very small compared to the amount of nonbiotinylated DNA, the biggest concern is minimizing the nonspecific binding of nonbiotinylated DNA to the beads during the bead pulldown such that most of the pulldown DNA will be biotinylated DNA. Too much nonspecific pulldown of nonbiotinylated genomic DNA will lead to failure of the TSA-seq profiling.

We strongly recommend testing the entire bead pulldown procedure by combining equal and small amounts of biotinylated and unbiotinylated spike-in DNA fragments with known sequence from a different genome than the tested sample (e.g., *Drosophila* sequences) together with a large amount of unlabeled genomic DNA from the same genome as the sample to mimic the DNA pulldown procedure of a real experiment. Then qPCR can be performed to test the relative enrichment and depletion ratio of the known biotinylated and unbiotinylated DNA and to verify that the biotinylated DNA makes up most of the DNA of the pulldown sample.

Similarly, in the real TSA-seq input DNA sample, we recommend to spike in small amounts of biotinylated and unbiotinylated DNA with defined sequences prior to the avidin pulldown procedure. The amounts of the biotinylated and unbiotinylated DNA spike-in sequences are adjusted to the minimum that will produce statistically significant read counts in both the input and pulldown samples. In our routine experiments conducted on human and mouse samples, we make the two spike-in DNA species from *Drosophila* DNA sequences, ~250 bp in length, that do not overlap with the target cell genome.

Initially, we had been adding both spike-in fragments at the same concentration to the fragmented genomic DNA prior to bead pulldown of biotinylated DNA. The biotinylated spike-in will be significantly enriched in the pulldown sample; therefore, it is important to keep this at a very low concentration in the input DNA sample. Typically, we obtain an enrichment ratio for the biotinylated spike-in, defined as the ratio of spike-in reads in the pulldown to total pulldown reads divided by the ratio of spike-in reads in the input to total input reads, ranging from ~1000–5000. We have been adding biotinylated spike-in at levels which produce ~100–500 reads per 30–40 million reads in the input library. At these levels of enrichment ratios, we may obtain ~1–5% of pulldown library reads corresponding to the biotinylated spike-in. Thus, it is important not to add too high a concentration of biotinylated spike-in fragments such that a large fraction of the pulldown sample will consist of this spike-in. The exact amount of biotinylated spike-in to add will depend on the estimated amount of biotinylated genomic DNA in the sample. The lower this level the less of the biotinylated spike-in should be added to avoid the spike-in from dominating the pulldown library. Empirically, we have been adding 0.25 ng of biotinylated spike-in DNA per 30 million cells for our typical SON TSA-seq experiments performed at reaction Condition 2 and enhancement condition E (assuming to recover 5 ng pulldown DNA).

For the nonbiotinylated spike-in fragment, we test depletion ratios, defined as the ratio of spike-in reads in the input to total input reads divided by the ratio of spike-in reads in the pulldown to total pulldown reads. Initially we had been adding similar amounts of nonbiotinylated to biotinylated spike-in fragments to the input DNA and observed depletion ratios of ~4–200 after one round of bead pulldown. Two or three rounds of bead pulldown increased the depletion ratio. More recently we have been adding larger amounts of the nonbiotinylated spike-in fragment to ensure that we obtain sufficient nonbiotinylated spike-in reads in the pulldown library to enable us to accurately compute a depletion ratio. In practice, we have been assuming a depletion ratio of ~100–1000, and, setting a target of ~500 nonbiotinylated reads per 30 million total reads in the pulldown library, calculating based on these depletion ratios what amount of nonbiotinylated spike-in we need to add to the input sample. Because this nonbiotinylated DNA spike-in will be substantially depleted in a successful pulldown, it is much easier (compared to biotinylated DNA spike-in) to add a "safe" amount of spike-in such that the nonbiotinylated spike-in reads will represent only a small fraction of the total input reads.

Calculation of both the enrichment and depletion ratios of the biotinylated and nonbiotinylated spike-in fragments, respectively, from the sequencing results determine how well the pulldown protocol is working.

From this step to the end of the TSA-seq protocol, we strongly recommend using DNA LoBind Eppendorf tubes to minimize the nonspecific binding of DNA to the tube. This DNA binding to the tube will increase the nonbiotinylated DNA that is carried over from the input DNA in the purified sample as well as result in a reduced yield of the very small amounts of biotinylated DNA that are obtained in the pulldown samples.

In some cases, we repeat the pulldown procedure once or even twice. We do this if we suspect a higher-than-expected carryover of nonspecific DNA—for instance if our final yield of pulldown DNA is much higher than expected. Typically, we lose ~25–50% of the pulldown DNA with repeated pulldown steps. Therefore, we do this only when we have sufficient pulldown DNA to begin with, and we will set aside several ng of the pulldown DNA from the first pulldown as a backup in case we do not recover sufficient DNA for library preparation after the second pulldown.

1.8 DNA Library Preparation, Sequencing and Data Processing

We use standard DNA library construction and sequencing kits from Illumina. PCR amplification cycles should be limited to minimize PCR bias introduced during the library construction.

As discussed previously, TSA-seq is fundamentally different from molecular proximity labeling methods such as ChIP-seq. TSA-seq produces a continuous signal related to the relative cytological distances between chromosome regions and the target nuclear compartment. Therefore, TSA-seq data normalization [3, 4] is different from commonly used "peak calling" methods in ChIP-seq. We normalize the pulldown library aligned read count by the input library aligned read count per genomic bin (typically ~20 kb), with an additional normalization that compensates for the different total read counts of the input and pulldown libraries. Normalization by the input read number compensates both for variation in read count due to cell cycle composition of the cell sample and variable timing of DNA replication across the genome as well as for the fraction of each bin that will be unmappable due to repetitive DNA reads. We then compute the log2 ratio of this normalized pulldown read count.

In some experiments, we have obtained input library reads that show clear bias in read count across the genome. Plotting the log2 ratio of input reads in each genomic bin divided by the mean input read count per bin, we observed large genomic domains showing several-fold enhanced or depleted read counts (log2 values of ~1–2 in magnitude), rather than smaller amplitude variations (log2 values of ~0.2–0.4 in magnitude) across the genome that is consistent for each cell type. Small amplitude variations across the genome in the unbiased input library are expected due to differences in DNA replication timing as well as copy number variation in different cell lines.

Differences in biased libraries relative to unbiased input libraries correlate well with large chromosome domains of varying GC-content; therefore, we believe the biased libraries may result from PCR bias during library construction. GC-content itself varies with many chromosome domain features, including Hi-C A/B compartments, DNA replication timing domains, LADs and iLADs. Therefore, this input library bias has the potential to skew measurements of chromosome domains related to nuclear positioning that are measured by TSA-seq. Figure 4 shows a comparison of

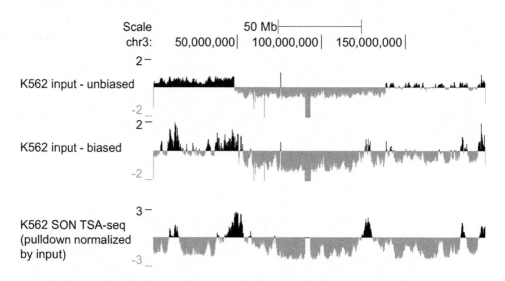

Fig. 4 Comparison of input libraries with low versus high PCR bias. Input reads were normalized using the log2 of the ratio of observed versus mean read count per bin. Chromosome 3 profiles are compared for library with low PCR bias (top) versus high PCR bias (middle). Data from K562 cells. The lower magnitude variation in normalized input reads observed in the low PCR bias track is consistent with varying DNA content due to early versus late DNA replicating genomic regions (K562 cell sample was from log-phase growth). However, the higher magnitude variation in the normalized input reads in the high PCR bias sample is far too large to be explained by variable DNA replication timing. Note also that the biased input read chromosome profile (middle) shows localized peaks that parallel the observed SON TSA-seq peaks seen in the SON TSA-seq score chromosome profile (bottom), suggesting a preferential PCR amplification of speckle-associated genomic regions, which show the highest genomic levels of GC content [3]. We now routinely examine the normalized input read chromosome profiles to check the quality of our sequencing libraries for each TSA-seq experiment

unbiased versus biased input library profiles from K562 cells. From our experience, important quality-control steps to ensure unbiased input libraries include good DNA quality after DNA purification and sonication (judged by spectrophotometry), minimizing the number of PCR cycles during the NGS library construction, as well as a consistent workflow and careful processing of samples during the library construction procedure.

1.9 Summary

The entire TSA-seq procedure includes TSA cell labeling, quality control monitoring of TSA cell labeling using light microscopy, genomic DNA extraction and fragmentation, quality control measurement of the levels of DNA tyramide–biotin labeling using dot blot analysis, preparation of spike-in DNA fragments using *Drosophila* DNA, bead purification of biotin-labeled DNA fragments, and NGS library construction, sequencing, and data processing (Fig. 1).

Below, we describe a detailed TSA-seq protocol for nuclear speckle (SON) labeling, using reaction condition 2 and enhancement condition E as an example.

2 Materials

2.1 TSA Cell Labeling

1. Cell culture medium.

2. 10× and 1× CMF-PBS (calcium, magnesium-free phosphate-buffered saline).

3. 8% (w/v) paraformaldehyde (PFA): make fresh for each experiment by dissolving 0.8 g in 7 mL ddH$_2$O at 60 °C, adding 1 N NaOH dropwise until solution is clear, cooling to room temperature, adding 1 mL 10× CMF-PBS and then adjusting pH to 7.4, making the volume up to 10 mL with ddH$_2$O, then filtering (0.22 μm). For adherent cells, dilute this stock to 1.6% PFA in 1× CMF-PBS.

4. 1.25 M (for suspension cells) and 20 mM (for adherent cells) glycine in 1× CMF-PBS.

5. 0.5% and 0.1% PBST: 0.5% or 0.1% (v/v) Triton X-100 in 1× CMF-PBS.

6. 1.5% and 0.15% stocks of hydrogen peroxide in 1× CMF-PBS, made fresh from 30% hydrogen peroxide stock for each experiment.

7. GS blocking buffer: 5% (v/v) normal goat serum in 0.1% PBST.

8. Primary antibody solution: Rabbit anti-SON polyclonal antibody (Atlas Antibodies HPA023535), diluted 1:1000 in GS blocking buffer (*see* **Note 1**).

9. Secondary antibody solution: HRP-conjugated goat anti-Rabbit IgG (H + L) antibody (Jackson ImmunoResearch 111–035-144), diluted 1:1000 in GS blocking buffer (*see* **Note 1**).

10. 60% sucrose: 60% (w/v) sucrose in $1\times$ CMF-PBS.

11. 50 mM DTT (for alternative versions of the protocol).

12. Biotin–tyramide stock solution (coupling protocol is adapted from [53, 54]; *see* **Note 2**):

    Warm bottle containing 100 mg EZ-Link Sulfo-NHS-LC-Biotin (NHS-Biotin) to room temperature in a fume hood, wrapped with foil (*see* **Note 3**).

    Weigh 30 mg tyramine-HCl in a 15 mL conical tube in a fume hood.

    Mix 5 mL N,N-dimethylformamide (DMF) and 50 μL triethylamine (TEA) in a 15 mL conical tube, and transfer 3 mL to the tube with the tyramine-HCl. Mix for several minutes to dissolve.

    Transfer 2.848 mL to a 50 mL conical tube and wrap with foil, creating solution B.

    Add 10 mL DMF to the bottle of NHS-Biotin and dissolve by mixing thoroughly. Transfer all to the tube of solution B, and keep at room temperature for 3 h in the dark.

    Add 7.18 mL ethanol, mix, and store as aliquots at −20 °C.

13. 1% biotin–tyramide solution: dilute the biotin–tyramide stock 1:100 in $1\times$ CMF-PBS, make fresh right before the step that needs it.

14. 12 mm round coverslips.

15. Humid chamber (*see* **Note 4**).

16. High TE buffer: 10 mM Tris-HCl, 10 mM EDTA, pH 8.0, filtered after preparation through a 0.22 μm filter.

2.2 Immunostaining to Check Tyramide–Biotin Labeling

1. Fluorescent staining solution: 1:200 Streptavidin–Alexa Fluor 594 (Invitrogen S11227) and 1:500 goat anti-rabbit IgG (H + L) (Jackson ImmunoResearch 111–095-144) in GS blocking buffer.

2. Mounting medium: 0.3 μg/mL DAPI, 10% (w/v) Mowiol 4–88, 1% (w/v) DABCO, 25% glycerol in 0.1 M Tris-HCl pH 8.5.

3. Fluorescent microscope with $>40\times$, high NA objective.

2.3 Genomic DNA Extraction and Sonication

1. 20% (w/v) SDS.

2. 20 mg/mL proteinase K (stock).

3. Lysis buffer 1: Make fresh (from SDS and proteinase K stocks) 0.5% (w/v) SDS, 0.2 mg/mL proteinase K in high TE buffer.

4. Lysis buffer 2: Make fresh (from SDS and proteinase K stocks) 1% (w/v) SDS, 0.2 mg/mL proteinase K in high TE buffer.

5. Hybridization chamber.

6. 4 M NaCl.

7. Phenol–chloroform–isoamyl alcohol (25:24:1).

8. Chloroform–isoamyl alcohol (24:1).

9. 100 mg/mL RNase A. Dilute to 10 mg/mL with water immediately before use.

10. 3 M sodium acetate pH 5.2.

11. 20 mg/mL glycogen.

12. 100% and 70% ethanol.

13. Molecular biology-grade water.

14. Nanodrop spectrophotometer.

15. Bioruptor Pico (Diagenode) sonicator or equivalent.

2.4 Dot Blot to Check Biotinylation of DNA

1. 1 ng/μL *Drosophila* BAC BACR48E12 (BACPAC Resources Center).

2. 10 μM biotinylated *Drosophila* DNA primers: GAAA-CATCGC/iBiodT/GCCCATAAT (forward) and AGAAG-CAGCTACGCTCCTCA (reverse).

3. 10 mM dNTPs.

4. 2 U/μL Q5 High Fidelity DNA polymerase (NEB, provided with 5× Q5 buffer).

5. QIAquick PCR Purification Kit (Qiagen).

6. Nitrocellulose membrane (0.45 μm pore).

7. UV Stratalinker 2400 crosslinker or equivalent.

8. Dot blot blocking buffer: SuperBlock blocking buffer in TBS (Thermo Fisher) with 0.05% (v/v) Tween 20.

9. Streptavidin-HRP (Invitrogen 43–4323), diluted 1:10,000 in dot blot blocking buffer.

10. 0.05% TBST: 0.05% (v/v) Tween 20 in 1× TBS (Tris-buffered saline).

11. SuperSignal West Femto chemiluminescent substrate (Thermo Fisher).

12. HyBlot CL film (Denville) or iBright machine (Invitrogen).

2.5 Biotin-Labeled DNA Fragment Pulldown and Sequencing Library Preparation

1. Nonbiotinylated *Drosophila* DNA primers: CCAATGCGATG-GATATGTCA (forward) and GCGAGGTTTCATTTTTG-GAA (reverse).

2. 2× W&B buffer: 10 mM Tris-HCl pH 7.5, 1 mM EDTA, 2 M NaCl, filtered through 0.22 μm filter. Also diluted in water to make 1× stock.

3. Dynabeads M270 streptavidin beads (Invitrogen).

4. LoBind Eppendorf tubes.

5. Magnetic stand.

6. TSE 500 buffer: 20 mM Tris-HCl pH 8, 1% (v/v) Triton X-100, 0.1% (w/v) SDS, 2 mM EDTA, 500 mM NaCl, filtered through 0.22 μm filter.

7. Bead digestion mixture: make fresh (from SDS and proteinase K stocks) 0.5% (w/v) SDS, 1 mg/mL proteinase K, 100 μM free biotin in high TE buffer.

8. 10 mM Tris-HCl pH 8.5.

9. Qubit dsDNA HS (high sensitivity) quantification kit.

10. TruSeq ChIP Sample Preparation kit.

11. AMPure XP beads (Beckman).

3   Methods

3.1   TSA Cell Labeling

For suspension cells, proceed with Subheading 3.1.1; for adherent cells, proceed with Subheading 3.1.2.

3.1.1   TSA Labeling for Cells in Suspension Culture

1. Grow cells in appropriate culture medium to log-phase growth. For K562 cells, aim for a density of ~30 million cells in 40 mL medium.

2. Transfer 40 mL cell suspension to a 50 mL conical tube, add 10 mL 8% PFA in 1× CMF-PBS, mix and incubate for 20 min at room temperature with gentle nutation. This fixation concentration and time may need to be adjusted for each target and antibody (*see* Subheading 1).

3. Add 5.6 mL 1.25 M glycine, mix and incubate for 5 min at room temperature by gentle nutation to quench free aldehyde groups.

4. Centrifuge for 5 min at $130 \times g$, room temperature (*see* **Note 5**), remove supernatant and resuspend in 10 mL 0.5% PBST by gentle pipetting with a cut 1 mL tip or a long 15 mL pipette. Incubate at room temperature for 30 min by gentle nutation to permeabilize cells.

5. Centrifuge for 5 min at $130 \times g$, room temperature, remove supernatant and resuspend in 2 mL 1× CMF-PBS by gentle pipetting with a cut 1 mL tip. Mix 0.5 mL of 30% hydrogen peroxide and 7.5 mL 1× CMF-PBS and then add to the 2 mL cell suspension to reach a final concentration of 1.5% for hydrogen peroxide. Incubate at room temperature for 1 h by gentle nutation to quench endogenous peroxidases (*see* **Note 6**).

6. Centrifuge for 5 min at $130 \times g$, room temperature (*see* **Note 5**), remove supernatant and resuspend in 10 mL 0.1% PBST by gentle pipetting with a cut 1 mL tip or a long 15 mL pipette. Repeat twice for a total of three washes.

7. Centrifuge for 5 min at 130 × *g*, room temperature, remove supernatant and resuspend in 1 mL per ten million cells of GS blocking buffer by gentle pipetting with a cut 1 mL tip or a long 15 mL pipette. Transfer to a 15 mL conical tube and incubate at room temperature for 1 h by gentle nutation.

8. Centrifuge for 5 min at 130 × *g*, room temperature, remove supernatant and resuspend in 1 mL per ten million cells of anti-SON primary antibody solution by gentle pipetting with a cut 1 mL tip or a long 15 mL pipette. Incubate at 4 °C for 10–12 h by gentle nutation.

9. Centrifuge for 5 min at 130 × *g*, room temperature, remove supernatant and resuspend in 10 mL 0.1% PBST by gentle pipetting with a cut 1 mL tip or a long 15 mL pipette, then incubate at room temperature for 5 min by gentle nutation. Repeat twice for a total of three washes.

10. Centrifuge for 5 min at 130 × *g*, room temperature, remove supernatant and resuspend in 1 mL per ten million cells of HRP-conjugated secondary antibody solution by gentle pipetting with a cut 1 mL tip or a long 15 mL pipette. Incubate at 4 °C for 10–12 h by gentle nutation.

11. Wash cells as in **step 9** (*see* **Note 7**).

12. TSA-seq 1.0 (enhancement Condition A) with reaction conditions 1, 2 or 3 are shown in Table 3. Here we describe the TSA-seq 2.0 protocol for nuclear speckle (anti-SON) labeling with enhancement condition E, reaction condition 2 (Table 4). Prepare Buffer 1 and Buffer 2 according to Table 4 during the

Table 3
TSA labeling 30 million suspension cells with reaction condition 1, 2, or 3, enhancement condition A

TSA condition (reaction condition + enhancement condition)		1 + A	2 + A	3 + A
Buffer 1 (buffer to resuspend cells, half of total volume)	PBS (μL)	1500	250	205
	60% sucrose (μL)	–	1250	1250
	50 mM DTT (μL)	–	–	45
Mix reagents following the order above and use Buffer 1 to resuspend cells				
Buffer 2 (buffer with tyramide–biotin and H₂O₂, half of total volume)	PBS (μL)	1440	190	190
	1% biotin–tyramide (μL)	30	30	30
	0.15% H₂O₂ (μL)	30	30	30
	60% sucrose (μL)	–	1250	1250
Mix reagents following the order above, add Buffer 2 to cell suspension, mix, and start timer for 10 min nutating				
	Total volume (μL)	3000	3000	3000

Table 4
TSA labeling 30 million suspension cells with reaction condition 2, enhancement condition E

TSA condition (reaction condition + enhancement condition)		2 + E
Buffer 1 (buffer to resuspend cells, half of total volume)	PBS (μL)	250
	60% sucrose (μL)	1250
Mix reagents following the order above and use Buffer 1 to resuspend cells		
Buffer 2 (buffer with tyramide–biotin and H_2O_2, half of total volume)	PBS (μL)	210
	Tyramide–biotin (μL)	10
	0.15% H_2O_2 (μL)	30
	60% sucrose (μL)	1250
Mix reagents following the order above, add Buffer 2 to cell suspension, mix, and start timer for 30 min nutating		
	Total volume (μL)	3000

washes of **step 11**. Centrifuge cells for 5 min at $130 \times g$, room temperature, remove supernatant and resuspend in 1.5 mL Buffer 1 by gentle pipetting with a cut 1 mL tip or a long 15 mL pipette. Add Buffer 2 to the cell suspension and incubate at room temperature for 30 min by gentle nutation.

13. Wash cells as in **step 9** (*see* **Note 8**).

14. Centrifuge for 5 min at $130 \times g$, room temperature, remove supernatant and resuspend cells in 10 mL $1\times$ CMF-PBS by gently pipetting with cut 1 mL tip or a long 15 mL pipette. Incubate at room temperature for 5 min by gentle nutation.

15. Transfer 10–20 μL cell suspension onto a 12 mm round coverslip placed in a humid chamber (*see* **Note 4**) and proceed to Subheading 3.2. Store another 50 μL aliquot in a microtube at 4 °C as backup material for immunostaining.

16. Centrifuge for 5 min at $700 \times g$, room temperature (*see* **Note 9**) and remove supernatant. Store the cell pellets at -80 °C or immediately proceed to Subheading 3.3.

3.1.2 TSA Labeling for Cells in Adherent Culture

Adherent cells are fixed and stained within the flasks to minimize perturbation of cell and nuclear architecture.

1. Culture cells in the appropriate medium in a T75 flask to log-phase growth. For HCT116 cells, aim for ~75% confluency.

2. Remove medium and quickly add 7.5 mL freshly made 1.6% PFA in $1\times$ CMF-PBS. Incubate at room temperature for 20 min with gentle rocking. This fixation concentration and time may need to be adjusted for each target and antibody (*see* Subheading 1).

3. Remove liquid and rinse cells with 7.5 mL 1× CMF-PBS.

4. Wash/permeabilize cells by removing liquid, adding 7.5 mL 0.5% PBST, and incubating at room temperature for 5 min with gentle rocking. Repeat twice for a total of three wash/permeabilization incubations.

5. Remove liquid and add 7.5 mL 20 mM glycine in 1× CMF-PBS, then incubate at room temperature for 5 min with gentle rocking to quench free aldehyde groups. Repeat twice for a total of three glycine treatments.

6. Remove liquid, rinse cells with 7.5 mL 1× CMF-PBS, remove liquid, then add 7.5 mL 1.5% hydrogen peroxide prepared in 1× CMF-PBS and incubate at room temperature for 1 h with gentle rocking to quench endogenous peroxidases (*see* **Note 10**).

7. Wash cells by removing liquid and rinsing three times with 7.5 mL 1× CMF-PBS.

8. Remove liquid and add 3 mL GS blocking buffer (1 mL per 25 cm^2 flask surface is needed to prevent air drying of cells). Incubate at room temperature for 1 h with gentle rocking.

9. Remove liquid and add 3 mL anti-SON primary antibody solution. Incubate at 4 °C for 10–12 h with gentle rocking.

10. Remove liquid and wash cells three times in 7.5 mL 0.1% PBST, incubating at room temperature for 5 min with gentle rocking each time.

11. Remove liquid and add 3 mL HRP-conjugated secondary antibody solution. Incubate at 4 °C for 10–12 h with gentle rocking.

12. Remove liquid and wash cells three times in 7.5 mL 0.1% PBST, then once in 7.5 mL 1× CMF-PBS, incubating at room temperature for 5 min with gentle rocking each time (*see* **Note 7**).

13. TSA-seq 1.0 (enhancement Condition A) reaction conditions 1, 2 or 3 are shown in Table 5. Here we describe the TSA-seq 2.0 protocol for nuclear speckle (anti-SON) labeling with enhancement Condition E and reaction Condition 2 (Table 6). Prepare the TSA reaction buffer according to Table 6 during the washes of **step 12**. Add the TSA reaction buffer to cells and incubate at room temperature for 30 min by gentle rocking.

14. Wash cells as in **step 12**.

15. Remove all but ~0.5 mL liquid and scrape off a small portion of the attached cells with a cell scraper or cell lifter (*see* **Note 11**). Transfer the cell suspension to a 1.5 mL microtube and then add 7.5 mL high TE buffer to the flask, incubating at room temperature for 5 min with gentle rocking and proceed to Subheading 3.3 (*see* **Note 12**).

Table 5
TSA labeling attached cells in a T75 flask with reaction condition 1, 2, or 3, enhancement condition A

TSA condition (reaction condition + enhancement condition)		1 + A	2 + A	3 + A
Buffer 1 (half of total volume)	PBS (μL)	–	–	410
	60% sucrose (μL)	–	–	2500
	50 mM DTT (μL)	–	–	90
Mix reagents following the order above and add Buffer 1 to cells in flask. Only required for condition 3- need to soak cells in DTT buffer				
Buffer 2 (buffer with tyramide–biotin and H$_2$O$_2$, half of total volume)	PBS (μL)	5880	880	380
	1% tyramide–biotin (μL)	60	60	60
	0.15% H$_2$O$_2$ (μL)	60	60	60
	60% sucrose (μL)	–	5000	2500
Mix reagents following the order above, add Buffer 2 to cells in flask, mix, and start timer for 10 min incubation				
	Total volume (μL)	6000	6000	6000

Table 6
TSA labeling attached cells in a T75 flask with reaction condition 2, enhancement condition E

TSA condition (reaction condition + enhancement condition)		2 + E
TSA reaction buffer (6 mL for a T75 flask)	PBS (μL)	920
	Tyramide–biotin (μL)	20
	0.15% H$_2$O$_2$ (μL)	60
	60% sucrose (μL)	5000
Mix reagents following the order above, add to cells in flask, and start timer for 30 min nutating		

16. Disperse the scraped cell clumps to a single-cell suspension by pipetting with a cut micropipette tip (*see* **Note 13**). Transfer 10 μL cell suspension onto a 12 mm round coverslip placed in a humid chamber (*see* **Note 4**) and proceed to Subheading 3.2. Store the remainder of the cells at 4 °C as backup material for immunostaining.

3.2 Quality Control 1: Immunostaining and Microscopy

1. Briefly air-dry the cells on the coverslip (*see* **Note 14**).

2. Add 50 μL fluorescent staining solution (*see* **Note 15**) and incubate at room temperature for 2 h or at 4 °C for 10–12 h.

3. Transfer coverslips to wells of a 24-well plate and wash three times by removing liquid and incubating for 5 min with 1 mL 0.1% PBST at room temperature.

4. Mount coverslips in mounting medium and check TSA staining by fluorescence microscopy.

3.3 Genomic DNA Extraction and Sonication

1. For suspension-culture cells (Subheading 3.1.1), resuspend cell pellets in 1 ml per ten million cells lysis buffer 1 (*see* **Note 16**). For adherent cells (Subheading 3.1.2), remove liquid and lyse cells by adding 1.5 mL per T75 flask lysis buffer 2, incubating for 5 min at room temperature (*see* **Note 17**) and transferring lysate to a 15 mL conical tube (*see* **Note 18**).

2. Incubate at 55 °C for 12–16 h in a hybridization chamber with constant rotation of the conical tube (*see* **Note 19**).

3. Add 158 μL 4 M NaCl (assuming 3 mL/30 million cells) to a final concentration of 0.2 M and incubate at 65 °C for 20–24 h in a hybridization chamber with constant rotation (*see* **Note 19**).

4. Add equal volume (~3.16 mL) phenol–chloroform–isoamyl alcohol (25:24:1), vortex vigorously for 20 s, then centrifuge for 5 min at >3260 × g, room temperature. Carefully transfer the upper aqueous phase to a new 15 mL conical tube. Repeat extraction one more time.

5. Add equal volume (~3.16 mL) chloroform–isoamyl alcohol (24:1), vortex vigorously for 20 s, then centrifuge for 5 min at >3260 × g, room temperature. Carefully transfer the upper aqueous phase to a new 15 mL conical tube.

6. Incubate at 55 °C for 1 h with the cap off to evaporate phenol residue.

7. Cool solution to room temperature, add 8 μL 10 mg/mL RNase A and incubate at 37 °C for 1 h or overnight with constant rotation.

8. Repeat **steps 4–6** to remove all RNase A and further purify the genomic DNA from protein contaminants.

9. Cool solution to room temperature. Transfer the ~3 mL DNA solution to a 15 mL conical tube suitable for high-speed centrifugation. Add 300 μL (1/10 volume) 3 M sodium acetate pH 5.2, ~8.3 μL 20 mg/mL glycogen and ~ 6.6 mL (2 volumes) 100% ethanol. Mix well and keep at −20 °C for 20 min-2 h (*see* **Note 20**).

10. Centrifuge for 20 min at 18,000 × g, 4 °C, discard supernatant and rinse pellet with 5 mL prechilled (−20 °C) 70% ethanol.

11. Centrifuge for 5 min at 18,000 × g, 4 °C, discard supernatant, and air-dry pellet until transparent (do not overdry pellet).

12. Dissolve DNA pellet in 100 μL (assuming 30 million cells) molecular biology-grade water and quantify with a Nanodrop spectrophotometer. DNA can be stored at −20 °C or immediately processed by sonication (**steps 13** and **14**).

13. Dilute and aliquot the genomic DNA into microtubes (100 μL per tube; optimally at 1 mg/mL).

14. Sonicate to ~100–600 bp using a Bioruptor Pico (Diagenode) with a mode of 30 s ON/30 s OFF. Sonicated DNA can be stored at −20 °C or immediately processed in Subheading 3.4 (*see* **Note 21**).

3.4 Quality Control 2: Assaying DNA Biotinylation Levels with Dot Blots

1. To make biotinylated *Drosophila* DNA for dot blot standards, mix in a PCR tube 31.5 µL nuclease-free water, 1 µL 10 mM dNTPs, 2 µL 1 ng/µL *Drosophila* BAC BACR48E12, 2.5 µL 10 µM forward biotinylated Drosophila DNA primer, 2.5 µL 10 µM reverse biotinylated Drosophila DNA primer, 10 µL 5× Q5 buffer and 0.5 µL 2 U/µL Q5 High Fidelity DNA polymerase.

2. Run the following program in a thermocycler:

   98 °C, 30 s;

   35× [98 °C, 20 s; 57 °C, 20 s; 72 °C, 30 s];

   72 °C, 5 min;

   4 °C, hold.

   This amplifies a 250 bp fragment, containing exactly 1 biotin moiety per DNA fragment ("Drosophila48E12_bio"):

   GAAACATCGCTGCCCATAATGGTTGCTATATATCC ACGATCGCTAAGGGAACGCTTTCCGTTGGCCTTCGC AAATCTTGAAGCCAACGCCCATCGCCTAGTGACCACC TACGATCTGCACGAGACCCTTAAAGATGTGGTCGATC TGGAAAACCTTAGTGACGAGCGGATTTTGAATCGTAC TCTGCGACTGCGAAACAATCACAACGTTTCATTATTC CTGCCCATTCCTGAGGAGCGTAGCT GCTTCT.

3. Run 3 µL of PCR product on an agarose gel to confirm a single 250 bp band.

4. Purify the rest of the PCR product with a QiaQuick PCR purification kit, following the manufacturer's instructions.

5. Make a serial dilution of the purified biotinylated *Drosophila* DNA and combine with sonicated unbiotinylated human cell (e.g., K562) genomic DNA (100–600 bp) to reach a final standard DNA concentration of 750 ng/µL (*see* **Note 22**) (Table 7).

Table 7
Dot blot standard DNA concentrations

Standard DNA	S0	S1	S2	S3	S4	S5	S6
Biotinylated *Drosophila* DNA (ng/µL)	100	20	4	0.8	0.16	0.032	0.0064
Unbiotinylated human cell genomic DNA (100–600 bp) (ng/µL)	650	730	746	749.2	749.8	750	750
Total DNA concentration (ng/µL)	750	750	750	750	750	750	750

6. Combine the sonicated TSA-labeled DNA from multiple tubes, take a 5 µL aliquot and adjust to the same concentration as the standard DNA (750 ng/µL, *see* **Note 22**).

7. Spot 1 µL standards and sonicated DNA samples onto a nitrocellulose membrane and UV cross-link DNA to the membrane after drying (0.24 J). Keep volume and concentration the same for all standards and samples (*see* **Note 23**).

8. Block the membrane with dot blot blocking buffer at room temperature for 1 h (*see* **Note 24**).

9. Incubate the membrane in 1:10,000 streptavidin-HRP in dot blot blocking buffer overnight at 4 °C.

10. Wash the membrane six times by vigorously shaking in 0.05% TBST for 5 min at room temperature.

11. Treat the membrane with SuperSignal West Femto chemiluminescent substrate and develop with HyBlot CL film or on an iBright machine.

12. Calculate the sample average biotinylation level (kilobases of DNA per one biotin) using the following equation:

$$\frac{1}{l_{std}} \times c_{std} = \frac{1}{l_{smp}} \times c_{smp}$$

l_{std}: Biotinylated *Drosophila* DNA length per biotin (0.25 kb). l_{smp}: average sample DNA length (kb) per biotin (to be calculated from sonication profile). c_{std}: concentration of the 0.25 kb biotinylated *Drosophila* DNA in the standards (Table 7) that has the same signal intensity with the tested sample DNA. c_{smp}: concentration of the tested sample DNA.

Figure 5 shows an example of dot blot for assaying biotinylation level of DNA from HCT116 cells that have been labeled by anti-SON TSA with reaction Condition 2 plus enhancement Condition E. In this example, three dilutions of biotinylated *Drosophila* DNA have been used (S0, S1, S2 in Table 7) with l_{std} of 0.25 kb. The to-be-tested sample DNA concentration (c_{smp}) is 750 ng/µL. From the dot blot, the to-

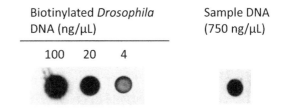

Biotinylated *Drosophila*   Sample DNA
DNA (ng/µL)   (750 ng/µL)

100   20   4

Fig. 5 Dot blot estimation of DNA biotinylation level. Sonicated genome DNA (100–600 bp) was subjected to dot blot analysis using chemiluminescence detection of biotin by streptavidin-HRP staining. Comparison with standards of biotinylated DNA is used to estimate average biotinylation level of sample DNA

be-tested sample DNA shows chemiluminescent intensity similar to the S1 standard DNA (c_{std}: 20 ng/µL). Solving the equation given above, predicts the sample DNA has an average biotinylation level of 9.375 kb (l_{smp}) per biotin.

3.5 Drosophila Spike-In DNA Preparation

1. Prepare biotin spike-in DNA by repeating Subheading 3.4, steps 1–4.

2. Prepare nonbiotin spike-in DNA by repeating Subheading 3.4, steps 1–4, except using nonbiotinylated Drosophila DNA primers and a 55 °C PCR annealing temperature.
   ("Drosophila48E12_non-bio")
   CCAATGCGATGGATATGTCAAGATACCGTGGAAT
   GTGCGTAGTATCCAGCGGATTGCCAAGCTCGCTGTG
   GCCAACATCAATCGACTGCTTGCACCCTATCCACAGT
   GTGAACAGCTGGAATTGCTCAATGTGGAGGATGCCTA
   TCTGAGAAAACAACACAAGCTCAATAAAACGATCATT
   GTGCGACTG GTGACGCAACCAGGAAATGGCCATTTT
   GATGCCACGGTC CTTTCCAAAAATGAAACCTCGC.

3. Confirm the two purified spike-in DNA sequences by Sanger sequencing.

3.6 Bead Pulldown of Biotin-Labeled DNA Fragments

1. Add 0.25 ng biotinylated and 0.25 ng (or higher amount) of nonbiotinylated spike-in DNA to the sonicated TSA-labeled DNA. This is assuming 5 ng biotinylated DNA will be recovered by the bead pulldown from 30 million initial cells (refer to Subheading 1.7 for guideline of the amount of spike-in to add.

2. Take 10 µL of the pooled DNA after adding the spike-in DNA, transfer to a microtube and store at −80 °C as the input DNA for preparing sequencing library.

3. Add an equal volume of 2× W&B buffer to the remaining DNA and mix well.

4. Place 150 µL Dynabeads M270 streptavidin beads in a LoBind Eppendorf tube. Place the tube on a magnetic stand for 1–2 min to immobilize the beads. Carefully remove the liquid with a 1 mL pipette tip, without disturbing the beads. Wash beads in 1 mL 1× W&B buffer by rotating for 1 min at room temperature. Repeat for a total of three washes.

5. Bring the DNA solution to 4 times the original bead volume (600 µL) by adding 1× W&B buffer. Remove the buffer from the beads after immobilization on a magnetic stand, and resuspend beads in the 600 µL DNA solution. Rotate for 30 min at room temperature, then overnight at 4 °C to bind biotinylated DNA to beads (*see* **Note 25**).

6. Wash beads four times with 1× W&B buffer, four times with TSE 500 buffer, four times with 1× W&B buffer, then four

times with high TE buffer, each time removing the previous wash after immobilization of the beads on a magnetic stand (*see* **Note 26**).

7. Resuspend the beads in 900 μL bead digestion mixture and aliquot to two LoBind Eppendorf tubes at 450 μL each. Mix well and rotate at 55 °C for 12–16 h.

8. Add ~24 μL 4 M NaCl (to final 0.2 M), mix and add ~474 μL (equal volume) phenol–chloroform–isoamyl alcohol (25:24:1), vortex vigorously for 20 s, then centrifuge for 5 min at 16,000 × *g*, room temperature. Carefully transfer upper aqueous phase to a new microtube.

9. Add ~450 μL (equal volume) chloroform/isoamyl alcohol (24:1), vortex vigorously for 20 s, then centrifuge for 5 min at 16,000 × *g*, room temperature. Carefully transfer upper aqueous phase to a new microtube.

10. Incubate tube at 55 °C for 1 h with the cap off to evaporate phenol residues.

11. Cool the solution down to room temperature. Resplit the solution to three Eppendorf tubes with ~300 μL each tube. Add 30 μL (1/10 volume) 3 M sodium acetate pH 5.2, 0.83 μL 20 mg/mL glycogen and ~ 662 μL (2× volume) 100% ethanol. Mix well and keep at −20 °C for 8–10 h (*see* **Note 27**).

12. Centrifuge for 30 min at 16,000 × *g*, 4 °C, discard supernatant and briefly rinse pellet with 1 mL prechilled (−20 °C) 70% ethanol (*see* **Note 28**).

13. Remove all supernatant without disturbing the pellet, and air-dry the pellet until it is transparent (do not overdry pellet). Dissolve the DNA pellet in 17.4 μL 10 mM Tris-HCl pH 8.5. Pool the dissolved DNA from three tubes to have a total of 52 μL.

14. Use a 2 μL aliquot for quantification with a Qubit HS assay. Store the remaining DNA at −80 °C as the pulldown DNA for preparing sequencing library.

3.7 DNA Library Preparation, Sequencing, and Data Processing

1. Construct sequencing libraries from the input and pulldown DNA, each with 5–10 ng, using the Illumina TruSeq ChIP Sample Prep Kit, following the manufacturer's instructions with 8–10 cycles of PCR (*see* **Note 29**).

2. Purify libraries with 1:1 (volume) AMPure XP beads, elute with 23 μL 10 mM Tris-HCl pH 8.5. Repeat the bead purification once (*see* **Note 30**).

3. Perform high-throughput sequencing in an Illumina system (*see* **Note 31**).

Table 8
**Examples for successful experiments showing spike-in DNA read number in input and pulldown
libraries (anti-SON TSA-seq, reaction Condition 2 and enhancement Condition E, one round bead
pulldown, K562 cells in example 1, RPE1 cells in example 2)**

Example	Library	Total read number	Biotin spike-in DNA read number	Nonbiotin spike-in DNA read number
1	Input	29,086,615	266	104
	Pulldown	34,534,667	1,170,396	28
2	Input	37,897,282	547	345
	Pulldown	36,032,736	465,137	2

4. Map the raw sequencing reads from the pulldown and input libraries to the sequences of the biotinylated spike-in DNA and nonbiotinylated spike-in DNA fragments with Bowtie2 [55]. Check read numbers from the two libraries mapped to the two species of spike-in DNAs. See Table 8 for examples of read numbers of spike-in DNAs after successful experiments involving a single DNA bead pulldown.

5. Map the raw sequencing reads to the human genome (hg38) with Bowtie2 [55], remove PCR duplicates and normalize the pulldown library with the input library using the TSA-seq 2.0 processing software [4].

4   Notes

1. One can use 0.1% PBST instead of the GS blocking buffer during the primary and secondary antibody incubations.

2. This reaction works best in an anhydrous environment. Make fresh DMF solution, control the room humidity, and do all steps as quickly as possible to avoid water absorption.

3. The NHS ester is very unstable. Once a vial is opened, dissolve the entire contents in DMF and use at once. It is not recommended to store this solution.

4. The humid chamber can be a 15 cm tissue culture petri dish wrapped with aluminum foil. Lay a thin layer of wet tissue paper on bottom of the dish. To create a support for the coverslips you can use the plastic divider that is used for holding pipette tips from a tip box. Alternatively use any object with a flat surface and an ~0.5 cm height that will fit within the dish with the petri dish lid in place. Cover the top surface of this plastic divider or other object using a layer of parafilm and place it in the dish as a staining platform for the coverslips. The fluid surface tension should keep the 50 µL antibody solution as a single drop/layer covering the coverslips but be careful when moving the chamber.

5. Ensure a low centrifugation speed for **steps 4** through **14** to avoid perturbing the nuclear morphology. **Steps 4** and **6** may require a second centrifugation because of the large volume.

6. In this step, there may be many bubbles forming as a result of the H_2O_2 treatment. The amount of these bubbles varies for different cell lines. Seal the conical tubes with parafilm to prevent liquid leaking from the tubes as a result of high pressure generated from this bubbling; release the generated gas 2–3 times during the incubation by uncapping and then resealing the tubes. Also, do not directly resuspend cells in diluted 1.5% H_2O_2 because the cells become very sticky in H_2O_2 solution. As a result, the cells cannot be resuspended thoroughly and many cells will be lost due to their adhering to pipette tips.

7. We recommend preparing the TSA-staining reagents immediately before the TSA reaction during these preceding wash (and centrifugation) steps.

8. Higher centrifugation speeds ($1000 \times g$) are required to pellet the cells in the 50% sucrose buffer after the TSA reaction and before the first washing step. After removing the sucrose buffer, return to normal centrifuge speed ($130 \times g$) during the cell washing steps.

9. Since the TSA labeling is already completed, and an aliquot of cells have been removed for validating the TSA labeling by immunostaining, at this point there is no longer a concern about perturbing nuclear morphology. Therefore, one can apply higher centrifugation speeds ($600–700 \times g$) at this step to improve recovery of cells for downstream application. Stop point: one can store the cell pellet at -80 °C. Again, the staining is now complete and there is no longer a concern about perturbing nuclear morphology by freeze-thawing. We do not recommend freezing cell pellets prior to antibody and TSA staining, even though this seems to be common in many molecular proximity assays. Since TSA-seq measures distances on a cytological size scale, perturbing nuclear morphology by freezing would also change the final TSA-seq results. This is different from molecular proximity assays such as ChIP-seq, DamID, or Hi-C, in which one would expect the results to be less dependent on perturbation of nuclear morphology introduced by cell freezing and thawing.

10. In this step, there may be many bubbles forming because of the H_2O_2 treatment; the amount of bubbling varies for different cell lines. TSA staining of cell lines with weak attachment to the substrate (e.g., HCT116) needs extra care at this step because these bubbles may peel the cells off the plastic of the tissue culture flask. The cell density is an important variable

modifying the severity of this problem. For example, we typically stain HCT116 cells at ~70%–80% confluency. At lower cell densities than this we have observed a larger number of cells not in contact with neighboring cells detaching at an increased frequency from the plastic. Conversely, at cell densities higher than ~70–80%, we have observed cells detaching and peeling off the plastic substrate as a single, connected cell sheet.

11. We recommend that you scrape sufficiently that you actually can visualize the cell material in the media that you collect.

12. After removing the PBS containing the scraped cells from the flask, immediately add the high TE buffer to the flask and continue with cell lysis, which should be done in parallel with the remaining steps of the quality-control immunostaining procedure.

13. It may be difficult to do this for some cell lines. In this case, try to disperse the cell pellet onto the coverslip so that cell clumps are as small and flat as possible for the subsequent microscopy.

14. *Do not allow the coverslips to dry!* The goal is that no water drop should be left on the coverslip but there should still be a slight layer of liquid over the coverslip to prevent drying of the coverslip. Drying of the cells is both destructive of cell shape, as visualized by light microscopy, as well as disruptive of cell ultrastructure in ways which may not be visible at light microscopy resolution but will produce nonspecific binding of antibodies. Cell drying is the most likely cause of irreproducible variation in antibody binding specificity.

15. With typical polyclonal secondary antibodies, after the HRP-secondary antibody incubation there remain epitopes available on the primary antibody for a different secondary antibody to bind. The FITC-conjugated secondary antibody against the primary antibody is for checking the target protein distribution. The streptavidin-Alexa Fluor 594 staining, which marks the biotin distribution, should overlap but extend beyond the FITC staining of the primary antibody. The degree of this spreading beyond the staining of the primary antibody will depend on the TSA staining conditions. Other fluorescent combinations can also be used.

16. Resuspend cells by vortexing. Do not use pipettes, especially after −80 °C storage, because cells will stick tightly to the pipette. Even if the cells are not totally resuspended, they will be lysed very quickly in the lysis buffer.

17. The lysis buffer volume is based on the expected DNA amount but needs to be enough to lyse cells off the entire growth area of the tissue culture flask. Usually this would be ~1–2 mL per ten million cells. The attached cells should be lysed off the flask in minutes and one can observe this process under a tissue-culture microscope.

18. Rinse the flask with more lysis buffer and transfer it into the same conical tube. If you are dealing with multiple flasks for the same experimental staining, and the cell number is small per flask, you can do the lysis in multiple flasks sequentially, transferring the same lysis buffer from one flask to the next, thus minimizing the total lysis buffer volume and concentrating the DNA within the cell lysate. This cell lysate solution should be immediately subjected to genomic DNA extraction (no stopping point here to prevent Proteinase K activity loss due to freeze-thaw cycle).

19. Choose a trusted conical tube brand and seal the cap well to make sure there is no leaking. Tubes are fragile if they have been stored at −80 °C and therefore we recommend that you transfer solution to new tubes for incubation. Stop point: one can store the solution at −80 °C after this step (**Steps 2 or 3**) is completed.

20. If 15 mL conical tubes suitable for high-speed centrifugation are not available, one can aliquot the DNA solution into 1.5 mL or 2 mL Eppendorf tubes. Adding glycogen aids DNA precipitation when DNA concentration is low (for some adherent cell types with low cell numbers such as HFF cells).

21. We have found the following added step tremendously useful in facilitating DNA fragmentation and increasing the reproducibility of this fragmentation when using the Bioruptor Pico (Diagenode) sonicator. Cut a 20 μL pipette tip and keep the lower tip portion with a length of ~2 cm. Use a flame to melt the plastic at the tip end to seal this end. Then, after allowing the tip to cool, drop the tip into the tube and DNA solution (only tested in Bioruptor Pico (Diagenode)). We acknowledge Dr. Jiang Xu, University of Southern California (ORCID: 0000-0002-0509-1250), for suggesting this step. The rationale is that the pipette tip acts as a stir bar during sonication that prevents small drops of liquid from sticking to different parts of the tube, thereby sonicating at a different rate than the remaining volume in the tube. We recommend running 1 μL of the DNA solution on a 2% agarose gel every 5 cycles of sonication to check the extent of DNA fragmentation.

22. We have also used concentrations of 400 ng/μL or 600 ng/μL for the dot blot standard DNA and the sonicated sample DNA to be tested. It is important to use the same concentrations for the standard DNA and the sample DNA in one dot blot experiment.

23. Keep the volume and DNA concentration the same for all standards and samples to be blotted so that the size of the dot and crosslinking efficiency can be kept consistent across samples. Optional: for all standards and samples, spot a second

1 μL volume of DNA onto the previously spotted position after drying to have more DNA for detection.

24. Do not use milk or any other blocking reagent that contains biotin.

25. Optional: one can save 5–10 μL of the DNA solution in 1× W&B buffer immediately before binding to the beads for dot blot analysis (Subheading 3.4) as an additional troubleshooting step, if necessary. The overnight 4 °C incubation can be as short as 2 h. For other TSA labeling conditions (such as enhancement Condition A), if the DNA solution volume is greater than four times the bead volume, do the bead binding in sequential steps, adding four times the bead volume in each step. In each of these sequential steps, repeat the 30 min binding at RT until all the DNA solution has gone through the binding process. Extend the last binding batch to 4 °C for 2 h or overnight. Alternatively, add all the DNA solution to the beads, incubate with rotation of the tubes at RT for 1 h and then for 8 h at 4 °C.

26. For each wash, rotate the tubes at room temperature for 1–2 min. For washing steps except for the TSE 500 buffer wash, resuspend the beads by inverting and clicking the tubes. Do not pipette the beads since they will stick to the pipette tips and cause severe sample loss. It is normal that the beads will bind to the tubes. In the detergent-containing TSE 500 buffer washes, one can pipette beads and easily get the bound beads off the tube. We strongly recommend that you change tubes during these detergent-containing washes to reduce carryover of nonbiotinylated DNA. Save the solutions following the binding to the beads and the last wash of the beads in case later troubleshooting is needed. One can measure DNA concentration in the last step washing buffer with a Qubit assay and no DNA should be detectable if the washes are done well. When removing buffer from the tube after each wash, one can remove most of the buffer with 1 mL pipette tips while keeping the tube on magnet stand, and then wait for beads that were dislodged to become immobilized again before removing the remaining buffer with 20 μL pipette tips. But do not allow the beads to dry out.

27. Glycogen is necessary to aid DNA precipitation because of low DNA concentration. The prolonged incubation at −20 °C improves precipitation and recovery of the pulldown DNA which will be at a low concentration. Avoid incubations longer than this to reduce salt precipitation.

28. Quickly rinse, do not leave the pellet in 70% ethanol for very long, and avoid a second centrifugation, especially when the DNA yield is low. Prolonged incubation in 70% ethanol will partially redissolve the DNA, resulting in even lower yield.

29. Use a minimum of 5 ng of DNA for library construction. If more pulldown DNA is available, using more DNA helps to reduce PCR cycles and increasing library quality. Omit the step of gel purification of ligation product in the kit manual. To avoid PCR bias introduced by overamplification, start with 8 cycles of PCR and run 8 μL (of 50 μL) PCR product on gel. If no PCR product is visible, perform an additional two PCR cycles.

30. Measure the DNA library concentration using the Qubit HS and check the library quality using a Bioanalyzer. Confirm that there are no adapter-dimers left in the libraries. Otherwise, further purify libraries with AMPure XP bead until no adapter-dimers are left.

31. Library concentrations are measured by qPCR and pooled (multiplexing) at equimolar concentration for each lane prior to submission of the final pooled sample for sequencing (Illumina). Normally, we have been using sequencing at a depth of ~30 million 100 bp single end reads per sample. Each pulldown sample (~30 million reads) should have a corresponding input sample (also ~30 million reads) for normalization.

Acknowledgments

We thank Drs. William Brieher, Brian Freeman, K.V. Prasanth, and Lisa Stubbs (UIUC, Urbana, IL) for helpful suggestions in developing these protocols. We also thank Belmont laboratory members for reagents and suggestions. We thank the UIUC Biotechnology center for guidance with DNA sonication and sequencing library preparation. We thank Drs. Jian Ma, Bas van Steensel, David Gilbert, and Huimin Zhao from the Belmont 4DN NOFIC U54 Center and other members of the 4D-Nucleome Consortium for helpful suggestions and feedback. This work was supported by National Institutes of Health grants R01GM58460 (ASB) and U54 DK107965 (ASB).

Author contributions: ASB conceptualized the TSA-seq idea and supervised the development of TSA-seq 1.0 and 2.0. YC developed TSA-seq 1.0 with protocols for tyramide–biotin labeling, TSA cell labeling, genomic DNA purification and fragmentation, dot blot and biotinylated DNA bead pulldown, contributed by LZ for Condition 3 (DTT). LZ developed TSA-seq 2.0, added protocols for TSA adherent cell labeling and *Drosophila* DNA spike in controls, and optimized protocols/methods for TSA labeling, genomic DNA purification and fragmentation, dot blot, and sequencing library construction. LZ, ASB, and YC prepared the manuscript.

References

1. Takizawa T, Meaburn KJ, Misteli T (2008) The meaning of gene positioning. Cell 135(1): 9–13. https://doi.org/10.1016/j.cell.2008. 09.026

2. Bickmore WA (2013) The spatial organization of the human genome. Annu Rev Genomics Hum Genet 14:67–84. https://doi.org/10. 1146/annurev-genom-091212-153515

3. Chen Y, Zhang Y, Wang Y et al (2018) Mapping 3D genome organization relative to nuclear compartments using TSA-Seq as a cytological ruler. J Cell Biol 217:4025–4048. https://doi.org/10.1083/jcb.201807108

4. Zhang L, Zhang Y, Chen Y et al (2020) TSA-seq reveals a largely conserved genome organization relative to nuclear speckles with small position changes tightly correlated with gene expression changes. Genome Res 31: 251–264. https://doi.org/10.1101/gr. 266239.120

5. Guelen L, Pagie L, Brasset E et al (2008) Domain organization of human chromosomes revealed by mapping of nuclear lamina interactions. Nature 453(7197):948–951. https:// doi.org/10.1038/nature06947

6. Kind J, Pagie L, de Vries SS et al (2015) Genome-wide maps of nuclear lamina interactions in single human cells. Cell 163:134–147. https://doi.org/10.1016/j.cell.2015.08.040

7. van Steensel B, Belmont AS (2017) Lamina-associated domains: links with chromosome architecture, heterochromatin, and gene repression. Cell 169(5):780–791. https://doi. org/10.1016/j.cell.2017.04.022

8. Hall LL, Smith KP, Byron M et al (2006) Molecular anatomy of a speckle. Anat Rec A Discov Mol Cell Evol Biol 288(7):664–675. https://doi.org/10.1002/ar.a.20336

9. Chen Y, Belmont AS (2019) Genome organization around nuclear speckles. Curr Opin Genet Dev 55:91–99. https://doi.org/10. 1016/j.gde.2019.06.008

10. Spector DL (2001) Nuclear domains. J Cell Sci 114(16):2891

11. Spector DL (2006) SnapShot: cellular bodies. Cell 127(5):1071. https://doi.org/10.1016/ j.cell.2006.11.026

12. Ferrai C, de Castro IJ, Lavitas L et al (2010) Gene positioning. Cold Spring Harb Perspect Biol 2(6):a000588. https://doi.org/10. 1101/cshperspect.a000588

13. Geyer PK, Vitalini MW, Wallrath LL (2011) Nuclear organization: taking a position on gene expression. Curr Opin Cell Biol 23(3): 354–359. https://doi.org/10.1016/j.ceb. 2011.03.002

14. Feuerborn A, Cook PR (2015) Why the activity of a gene depends on its neighbors. Trends Genet 31(9):483–490. https://doi.org/10. 1016/j.tig.2015.07.001

15. Feric M, Vaidya N, Harmon Tyler S et al (2016) Coexisting liquid phases underlie nucleolar subcompartments. Cell 165: 1686–1697. https://doi.org/10.1016/j.cell. 2016.04.047

16. Yamazaki T, Souquere S, Chujo T et al (2018) Functional domains of NEAT1 architectural lncRNA induce Paraspeckle assembly through phase separation. Mol Cell 70(6): 1038–1053e1037. https://doi.org/10.1016/ j.molcel.2018.05.019

17. Rai AK, Chen JX, Selbach M et al (2018) Kinase-controlled phase transition of membraneless organelles in mitosis. Nature 559(7713): 211–216. https://doi.org/10.1038/s41586-018-0279-8

18. Strom AR, Brangwynne CP (2019) The liquid nucleome - phase transitions in the nucleus at a glance. J Cell Sci 132(22):jcs235093. https:// doi.org/10.1242/jcs.235093

19. Hondele M, Sachdev R, Heinrich S et al (2019) DEAD-box ATPases are global regulators of phase-separated organelles. Nature 573(7772):144–148. https://doi.org/10. 1038/s41586-019-1502-y

20. Sabari BR, Dall'Agnese A, Boija A et al (2018) Coactivator condensation at super-enhancers links phase separation and gene control. Science 361(6400):eaar3958. https://doi.org/ 10.1126/science.aar3958

21. Cho W-K, Spille J-H, Hecht M et al (2018) Mediator and RNA polymerase II clusters associate in transcription-dependent condensates. Science 361(6400):412. https://doi.org/10. 1126/science.aar4199

22. Boija A, Klein IA, Sabari BR et al (2018) Transcription factors activate genes through the phase-separation capacity of their activation domains. Cell 175(7):1842–1855e1816. https://doi.org/10.1016/j.cell.2018.10.042

23. Boehning M, Dugast-Darzacq C, Rankovic M et al (2018) RNA polymerase II clustering through carboxy-terminal domain phase separation. Nat Struct Mol Biol 25(9):833–840. https://doi.org/10.1038/s41594-018-0112-y

24. Lu H, Yu D, Hansen AS et al (2018) Phase-separation mechanism for C-terminal

hyperphosphorylation of RNA polymerase II. Nature 558(7709):318–323. https://doi.org/10.1038/s41586-018-0174-3

25. Guo YE, Manteiga JC, Henninger JE et al (2019) Pol II phosphorylation regulates a switch between transcriptional and splicing condensates. Nature 572(7770):543–548. https://doi.org/10.1038/s41586-019-1464-0

26. Saitoh N, Spahr CS, Patterson SD et al (2004) Proteomic analysis of interchromatin granule clusters. Mol Biol Cell 15(8):3876–3890. https://doi.org/10.1091/mbc.E04-03-0253

27. Galganski L, Urbanek MO, Krzyzosiak WJ (2017) Nuclear speckles: molecular organization, biological function and role in disease. Nucleic Acids Res 45(18):10350–10368. https://doi.org/10.1093/nar/gkx759

28. Dopie J, Sweredoski MJ, Moradian A et al (2020) Tyramide signal amplification mass spectrometry (TSA-MS) ratio identifies nuclear speckle proteins. J Cell Biol 219(9):e201910207. https://doi.org/10.1083/jcb.201910207

29. Su J-H, Zheng P, Kinrot SS et al (2020) Genome-scale imaging of the 3D organization and transcriptional activity of chromatin. Cell 182(6):1641–1659.e1626. https://doi.org/10.1016/j.cell.2020.07.032

30. Takei Y, Yun J, Ollikainen N et al (2020) Global architecture of the nucleus in single cells by DNA seqFISH+ and multiplexed immunofluorescence. bioRxiv:2020.2011.2029.403055. https://doi.org/10.1101/2020.11.29.403055

31. Robinett CC, Straight A, Li G et al (1996) In vivo localization of DNA sequences and visualization of large-scale chromatin organization using lac operator/repressor recognition. J Cell Biol 135(6):1685–1700. https://doi.org/10.1083/jcb.135.6.1685

32. Hepperger C, Otten S, von Hase J et al (2007) Preservation of large-scale chromatin structure in FISH experiments. Chromosoma 116(2):117–133. https://doi.org/10.1007/s00412-006-0084-2

33. Landt SG, Marinov GK, Kundaje A et al (2012) ChIP-seq guidelines and practices of the ENCODE and modENCODE consortia. Genome Res 22(9):1813–1831. https://doi.org/10.1101/gr.136184.111

34. Vogel MJ, Peric-Hupkes D, van Steensel B (2007) Detection of in vivo protein-DNA interactions using DamID in mammalian cells. Nat Protoc 2(6):1467–1478. https://doi.org/10.1038/nprot.2007.148

35. Skene PJ, Henikoff S (2017) An efficient targeted nuclease strategy for high-resolution mapping of DNA binding sites. Elife 6:e21856. https://doi.org/10.7554/eLife.21856

36. Skene PJ, Henikoff JG, Henikoff S (2018) Targeted in situ genome-wide profiling with high efficiency for low cell numbers. Nat Protoc 13(5):1006–1019. https://doi.org/10.1038/nprot.2018.015

37. Kaya-Okur HS, Wu SJ, Codomo CA et al (2019) CUT&tag for efficient epigenomic profiling of small samples and single cells. Nat Commun 10(1):1930. https://doi.org/10.1038/s41467-019-09982-5

38. van Schaik T, Vos M, Peric-Hupkes D et al (2020) Cell cycle dynamics of lamina-associated DNA. EMBO Rep 21(11):e50636. https://doi.org/10.15252/embr.202050636

39. Pickersgill H, Kalverda B, de Wit E et al (2006) Characterization of the Drosophila melanogaster genome at the nuclear lamina. Nat Genet 38(9):1005–1014. https://doi.org/10.1038/ng1852

40. Briand N, Collas P (2020) Lamina-associated domains: peripheral matters and internal affairs. Genome Biol 21(1):85. https://doi.org/10.1186/s13059-020-02003-5

41. Ilik İA, Malszycki M, Lübke AK et al (2020) SON and SRRM2 are essential for nuclear speckle formation. Elife 9:e60579. https://doi.org/10.7554/eLife.60579

42. Fei J, Jadaliha M, Harmon TS et al (2017) Quantitative analysis of multilayer organization of proteins and RNA in nuclear speckles at super resolution. J Cell Sci 130(24):4180. https://doi.org/10.1242/jcs.206854

43. Quinodoz SA, Ollikainen N, Tabak B et al (2018) Higher-order inter-chromosomal hubs shape 3D genome organization in the nucleus. Cell 174(3):744–757e724. https://doi.org/10.1016/j.cell.2018.05.024

44. Quinodoz SA, Bhat P, Ollikainen N et al (2020) RNA promotes the formation of spatial compartments in the nucleus. bioRxiv:2020.2008.2025.267435. https://doi.org/10.1101/2020.08.25.267435

45. Chen W, Yan Z, Li S et al (2018) RNAs as proximity-labeling media for identifying nuclear speckle positions relative to the genome. iScience 4:204–215. https://doi.org/10.1016/j.isci.2018.06.005

46. Bobrow MN, Harris TD, Shaughnessy KJ et al (1989) Catalyzed reporter deposition, a novel method of signal amplification: application to immunoassays. J Immunol Methods 125(1–2):279–285

47. Raap AK, van de Corput MPC, Vervenne RAM et al (1995) Ultra-sensitive FISH using peroxidase-mediated deposition of biotin- or

fluorochrome tyramides. Hum Mol Genet 4(4):529–534. https://doi.org/10.1093/hmg/4.4.529

48. Gao XD, Tu LC, Mir A et al (2018) C-BERST: defining subnuclear proteomic landscapes at genomic elements with dCas9-APEX2. Nat Methods 15(6):433–436. https://doi.org/10.1038/s41592-018-0006-2

49. Myers SA, Wright J, Peckner R et al (2018) Discovery of proteins associated with a predefined genomic locus via dCas9–APEX-mediated proximity labeling. Nat Methods 15(6):437–439. https://doi.org/10.1038/s41592-018-0007-1

50. Fazal FM, Han S, Parker KR et al (2019) Atlas of subcellular RNA localization revealed by APEX-Seq. Cell 178(2):473–490e426. https://doi.org/10.1016/j.cell.2019.05.027

51. Kurihara M, Kato K, Sanbo C et al (2020) Genomic profiling by ALaP-Seq reveals transcriptional regulation by PML bodies through DNMT3A exclusion. Mol Cell 78(3): 493–505.e498. https://doi.org/10.1016/j.molcel.2020.04.004

52. Tran JR, Paulson DI, Moresco JJ et al (2021) An APEX2 proximity ligation method for mapping interactions with the nuclear lamina. J Cell Biol 220(1):e202002129. https://doi.org/10.1083/jcb.202002129

53. Hopman AHN, Ramaekers FCS, Speel EJM (1998) Rapid synthesis of biotin-, Digoxigenin-, Trinitrophenyl-, and Fluorochrome-labeled Tyramides and their application for in situ hybridization using CARD amplification. J Histochem Cytochem 46(6):771–777. https://doi.org/10.1177/002215549804600611

54. http://wiki.xenbase.org/xenwiki/index.php/Flourescin_Tyramide_Synthesis

55. Langmead B, Salzberg SL (2012) Fast gapped-read alignment with bowtie 2. Nat Methods 9(4):357–359. https://doi.org/10.1038/nmeth.1923

Chapter 9

The High-Salt Recovered Sequence-Sequencing (HRS-seq) Method: Exploring Genome Association with Nuclear Bodies

Cosette Rebouissou, Marie-Odile Baudement, and Thierry Forné

Abstract

Recent works indicate that, at specific loci, interactions of chromatin with membrane-less organelles self-assembled through mechanisms of phase separation, like nuclear bodies, are crucial to regulate genome functions, and in particular transcription. Here we describe the protocol of the high-salt recovered sequence sequencing method whose principle relies on high-throughput sequencing of genomic DNA trapped into large RNP complexes that are made insoluble by high-salt treatments.

Key words High-order chromatin organization, Nuclear bodies, Quantitative PCR

1 Introduction

The importance of genome organization at the supranucleosomal scale in the control of gene expression is increasingly recognized today [1, 2]. In mammals, Topologically Associating Domains (TADs) and the active/inactive chromosomal compartments are two of the main nuclear structures that contribute to this organization level. However, recent works indicate that, at specific loci, chromatin interactions with nuclear bodies could also be crucial to regulate genome functions, in particular transcription [3]. They, moreover, suggest that these nuclear bodies are membrane-less organelles dynamically self-assembled and disassembled through mechanisms of phase separation. Based on previous experimental approaches developed in our laboratory [4, 5], we have recently developed a novel genome-wide experimental method, high-salt recovered sequences sequencing (HRS-seq), which allows the identification of chromatin regions associated with large ribonucleoprotein (RNP) complexes and nuclear bodies [6]. Transcriptionally active nuclei preparations are treated at a high salt concentration and the insoluble fraction is purified and sequenced (Fig. 1). We argue that the physical nature of such RNP complexes and nuclear

Tom Sexton (ed.), *Spatial Genome Organization: Methods and Protocols*,
Methods in Molecular Biology, vol. 2532, https://doi.org/10.1007/978-1-0716-2497-5_9,

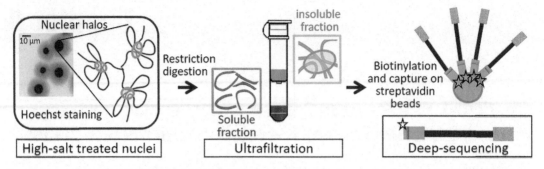

Fig. 1 Principle of the high-salt recovered sequence (HRS) assay. The principle of the HRS technique relies on purification of large RNP complexes that are made insoluble by high-salt treatments. Accurate quantifications of target sequences in the insoluble vs soluble fractions by quantitative PCR (HRS-assay) or by high-throughput sequencing (HRS-seq method) are then performed. Significant overrepresentation of a restriction fragment in the insoluble fraction reflects its retention into large RNP complexes

bodies appears to be central in their ability to promote efficient interactions between distant genomic regions. The development of novel experimental approaches, including our HRS-seq method described in the protocol below, is opening new avenues to understand how self-assembly of phase separated nuclear bodies contributes to mammalian genome organization and gene expression.

2   Materials

1. Cell culture medium (appropriate for cells of interest) and trypsin-EDTA (for adherent cells).

2. Phosphate-buffered saline (PBS).

3. Buffer 1: 0.3 M sucrose, 60 mM KCl, 15 mM NaCl, 5 mM MgCl$_2$, 0.1 mM EGTA, 15 mM Tris-HCl pH 7.5, 0.5 mM DTT, 0.1 mM PMSF, 3.6 ng/mL aprotinin, 5 mM sodium butyrate (*see* **Note 1**).

4. Buffer 2: Buffer 1 with 0.8% (v/v) NP40.

5. Buffer 3: 1.2 M sucrose, 60 mM KCl, 15 mM NaCl, 5 mM MgCl$_2$, 0.1 mM EGTA, 15 mM Tris-HCl pH 7.5, 0.5 mM DTT, 0.1 mM PMSF, 3.6 ng/mL aprotinin, 5 mM sodium butyrate (*see* **Note 1**).

6. Glycerol buffer: 40% (v/v) glycerol, 50 mM Tris-HCl pH 8.3, 5 mM MgCl$_2$, 0.1 mM EDTA.

7. Triton buffer: 10 mM Tris-HCl pH 7.5, 100 mM NaCl, 0.3 M sucrose, 3 mM MgCl$_2$, 0.5% (v/v) Triton X-100.

8. Sterile 2 mL centrifugal filter units with 0.22 μm pore and 12 mL collection tubes (e.g., Millipore Ultrafree-CL GV 0.22 μm).

9. Homemade pistons: pistons of 2 mL syringe modified by adding a seal of interior diameter = 6 mm/exterior = 9 mm or home-made apparatus (HMA) for pressurizing filtration units (*see* **Note 2**).

10. Extraction buffer: 20 mM Tris-HCl pH 7.5, 2 M NaCl, 10 mM EDTA, 0.125 mM spermidine.

11. 1× StyI enzyme buffer: 50 mM Tris-HCl pH 7.5, 100 mM NaCl, 10 mM $MgCl_2$, 0.2 mM spermidine (*see* **Note 3**).

12. 10 U/μL StyI enzyme.

13. 10× PK buffer: 100 mM Tris-HCl pH 7.5, 50 mM EDTA, 5% (w/v) SDS. Also diluted to a 2× stock.

14. 20 mg/mL proteinase K.

15. NucleoSpin Gel and PCR Cleanup kit compatible with SDS-containing samples (e.g., Macherey-Nagel, with buffer NTB).

16. Vacuum manifold.

17. Qubit fluorimeter and dsDNA assay, or equivalent.

18. 15 μM biotinylated adaptor 1 with complementarity for StyI restriction sites:

    1R: 5′P- CWWGTCGGACTGTAGAACTCTGAACCTGTC CAAGGTGTGA-Biotin-3′ and

    1F: 3′- AGCCTGACATCTTGAGACTTGGACA -5′ (*see* **Note 4**).

19. Quick Ligation Kit (NEB), containing Quick DNA ligase and 10× buffer.

20. Dynabeads MyOneTM Streptavidin C1.

21. 2× "Binding and Washing" BW buffer: 10 mM Tris-HCl pH 7.5, 1 mM EDTA, 2 M NaCl. Also diluted to a 1× stock.

22. Magnetic support for Eppendorf tubes.

23. TE buffer: 10 mM Tris-HCl pH 7.5, 1 mM EDTA.

24. MmeI (2 U/μL) and 10× buffer (e.g., NEBuffer 4) provided by the enzyme supplier.

25. 10× SAM: 5 μL 32 mM S-adenosyl-methionine diluted in water to a final volume of 325 μL (*see* **Note 5**).

26. Second adaptor ligation mix (50 μL per reaction): 8 U/μL T4 DNA ligase (NEB) in 1× supplied T4 DNA ligase buffer, 0.6 μM GEX adaptor 2:

    2F: 5′-CAAGCAGAAGACGGCATACGANN-3′

    2R: 3′-GTTCGTC TTCTGCC GTATGCT-P5′ (*see* **Note 4**).

27. PCR master mix (48 μL, added to 2 μL template, per reaction): 0.024 U/μL Phusion High Fidelity DNA polymerase in 1.04×

HF Phusion buffer, 260 nM GEX PCR primer 1 (5′-CAAGCA GAAGACGGCATACGA -3′), 260 nM GEX PCR primer 2, (5′- AATGATACGGCGACCACCGACAGGTTCA GAGTTCTACAGTCCGA-3′), 260 μM dNTPs.

28. 1× NEBuffer 2 (10× stock from NEB): 10 mM Tris-HCl pH 7.9, 50 mM NaCl, 10 mM MgCl$_2$.

29. Costar Spin-X centrifuge tube filters (45 μm).

30. 20 mg/mL glycogen.

31. 3 M sodium acetate pH 5.2.

32. 100% ethanol and 70% ethanol.

33. 2100 BioAnalyzer (Agilent) with high sensitivity DNA Chips.

34. Next-generation sequencing machine (e.g., Illumina Hi-Seq 2000).

3 Methods

3.1 Cell Nuclei Preparation (See Note 6)

1. If working with nonadherent cultured cells, proceed with **step 2** below. If working with adherent cell cultures, trypsinize, wash, and filter through 40 μm cell strainer to make a single-cell suspension before proceeding with **step 2** below.

2. Wash cells in PBS.

3. Resuspend cells in 1.5 mL Buffer 1.

4. Add 0.5 mL Buffer 2 and mix gently. Put on ice for 3 min.

5. Split to two 14 mL tubes (1 mL each) containing 4 mL of Buffer 3.

6. Centrifuge for 20 min at 11,300 × g, 4 °C and remove supernatant (use several tips to avoid NP40 contamination of the nucleus pellet).

7. Resuspend pellets in 50 μL of glycerol buffer and take a 4 μL aliquot to count nuclei in a Thoma cell. Adjust to a final concentration of ~20 million nuclei in 100 μL and, from this stock, make 100 μL aliquots containing 150,000 nuclei diluted in glycerol buffer. Freeze into liquid nitrogen and store at −80 °C.

3.2 Generation of Nuclear Halos by High-Salt Treatment

1. Thaw on ice an aliquot containing 150,000 nuclei in a 1.5 mL Eppendorf tube (important: do not increase the number of nuclei treated during these steps; *see* **Note 7**).

2. Add 900 μL Triton buffer to the nuclei and homogenize by pipetting.

3. Incubate for 15 min on ice (mix gently by pipetting 3 times during incubation), then incubate for 20 min at 37 °C in a water bath or equivalent apparatus.

4. Transfer solution to the filter of a centrifugal filtration unit equipped with a 12-mL collector tube below.

5. Push 2 drops out with the help of the piston or Home-Made Apparatus (HMA) (*see* **Note 2**).

6. Apply again a pressure with the help of the piston or HMA to push the liquid through the filter at a drop-by-drop rhythm (no more than 1 drop per second) until as little liquid remains as possible above the filter (do not dry the filter).

7. Add 1 mL extraction buffer, homogenize gently by pipetting and push 6 drops out (dead volume of the column). Incubate for 5 min at room temperature.

8. Use the piston (or HMA) to pass the entire volume of extraction buffer through the filter at a drop-by-drop rhythm.

9. Add 950 μL enzyme buffer. Pass all the liquid through the filter at a drop-by-drop rhythm. Repeat three more times, passing all the liquid through the filter and emptying the collection tube to the waste (these washings allow removal of high salt concentration present in the extraction buffer).

3.3 Digestion of Nuclear Halos

1. For each filtration unit, add a mix containing 960 μL of enzyme buffer and 20 μL 10 U/μL StyI, and homogenize gently by pipetting (*see* **Note 8**).

2. Add a band of Parafilm under the filter part to obstruct the filtration unit and place it back on the collector tube.

3. Incubate the filtration unit for 30 min at 37 °C (take care that all the liquid in the filtration unit should be immersed into the water bath and check regularly that there is no leakage due to a lack of Parafilm obstruction).

4. Collect evaporation drops in the tube cap and around tube walls and place them back in the liquid onto the filter. Add 20 μL 10 U/μL StyI, mix gently by pipetting and incubate for another 30 min.

3.4 DNA Purification of StyI-Digested DNA from Soluble and Insoluble Fractions

1. Remove tubes from the water bath. Collect evaporation drops in the tube cap and around the tube walls, and place them back with the remaining liquid on the filter. Homogenize gently by pipetting.

2. Remove the Parafilm, collect residual drops on the Parafilm and place them back in the liquid on the filter. Place the filter part on a clean collector tube and pass all the liquid through the filter.

3. Add 1 mL of enzyme buffer to rinse the nuclear halos and pass it all through the filter in the same collector tube as **step 2**. Save this eluate, corresponding to the **soluble fraction** (2 mL).

4. Add 500 μL enzyme buffer to the filter, place a Parafilm to obstruct the filtration unit and put the filtration unit on a clean collector tube (12 mL).

5. Add 500 μL 2× PK buffer and 10 μL 20 mg/mL proteinase K to the filter. Homogenize gently by pipetting.

6. Add 200 μL 10× PK buffer and 10 μL 20 mg/mL proteinase K to the **soluble fraction** from **step 3**. Homogenize gently by pipetting.

7. Incubate the filtration unit and the tube containing the soluble fraction for 30 min at 50 °C in a water bath (take care to keep them well immersed).

8. Pass all the liquid through the filter of the filtration unit. Save this eluate in a clean collection tube, corresponding to the **insoluble fraction** (1 mL).

   Steps 9–13 refer to the Macherey-Nagel NucleoSpin Clean-up protocol.

9. Add 5 mL of NTB buffer to the insoluble fraction and 11 mL to the soluble fraction (5 volumes). The soluble and insoluble fractions are then processed in parallel.

10. Add 700 μL of one fraction to a NucleoSpin column and pass all the liquid by aspiration on a vacuum manifold (or by centrifugation for 30 s at $11,000 \times g$, room temperature). Repeat these 700 μL loadings until all of one fraction has passed through the same column.

11. Pass 700 μL of NT3 buffer through the column using the vacuum manifold (or centrifuge for 30 s at $11,000 \times g$, room temperature) and empty collector tube in the waste.

12. Repeat **step 11** one more time, then place the column on a clean Eppendorf tube.

13. Add 50 μL NE buffer, incubate for 1 min minimum at room temperature, then centrifuge for 1 min at $11,000 \times g$, room temperature.

14. Quantify the DNA within the fractions with a Qubit fluorometric assay or equivalent.

15. Repeat Subheadings 3.2–3.4 on further aliquots until at least 150 ng DNA is obtained for each fraction. Add milli-Q water to each fraction to make a final volume of 500 μL (*see* **Note 9**).

3.5 Redigestion of Genomic DNA (See Note 10)

1. Add 55 μL 10× digestion buffer (*see* **Note 3**) and 1 μL 10 U/μL StyI to each fraction from Subheading 3.4. Incubate for 30 min at 37 °C.

2. Purify the DNA with NucleoSpin columns as in Subheading 3.4, **steps 9–13**, except that 2.5 mL NTB buffer (5 volumes) is first added.

3. Quantify the DNA within the fractions with a Qubit fluorometric assay or equivalent. If required, concentrate each fraction by ethanol precipitation and dissolve the DNA in NE buffer to reach a final concentration of at least 10 ng/μL (*see* **Note 11**).

3.6 Sequencing Library Preparation

1. To 150 ng of StyI-digested genomic DNA (soluble or insoluble fraction), add 2 μL 15 μM biotinylated adaptor 1, 25 μL 2× Quick Ligation Reaction Buffer, 6.66 μL of 400 U/μL Quick DNA Ligase, and milli-Q water to a final reaction volume of 50 μL (*see* **Note 12**). Incubate for 15 min at room temperature.

2. Resuspend the stock of Dynabeads MyOne™ Streptavidin C1 beads by vortexing, then transfer 10 μL to a 1.5 mL Eppendorf tube.

3. Add 50 μL 1× BW buffer to the beads, vortex and briefly centrifuge the tube, then place the beads on a magnetic support for 2 min until the solution becomes clear.

4. Remove the supernatant by pipetting, remove the tube from the magnetic support.

5. Repeat **steps 3** and **4** two more times for a total of three washes. Remove the supernatant and add 50 μL 2× BW buffer to the beads.

6. Add 50 μL of DNA ligated with the first adaptor (from **step 1**). Incubate for 15 min at room temperature on a rotator wheel.

7. Place the tubes on a magnetic support for 2 min, until the solution becomes clear.

8. Wash three times with 100 μL (1 volume) 1× BW buffer, then two times with 100 μL (1 volume) TE buffer (as described in **steps 3** and **4**).

9. Remove the supernatant, add 10 μL 10× NEBuffer 4, 10 μL 10× SAM and 76 μL milli-Q water to the beads (*see* **Note 5**).

10. Add 4 μL 2 U/μL MmeI and incubate for 90 min on a thermomixer at 37 °C, 1400 rpm.

11. Remove supernatant and wash beads as in **step 8**.

12. Remove supernatant and add 50 μL second adapter ligation mix to the beads. Incubate for 2 h at 20 °C in a thermomixer with 15 s of agitation every 2 min.

13. Remove supernatant and wash beads as in **step 8**, then remove the supernatant and resuspend beads in 10 μL milli-Q water.

14. Transfer 2 µL of beads with the ligated DNA (from **step 13**) to PCR tubes with 48 µL PCR master mix per tube. Run the following program in a thermal cycler.

98 °C, 30 s.

15× [98 °C, 10 s; 60 °C, 30 s; 72 °C, 15 s].

72 °C, 10 min.

15. Load the PCR reaction on to a 6% 1× TBE acrylamide gel and perform electrophoresis.

16. Visualize the gel with ethidium bromide or equivalent and UV, and excise the main band (expected size 95–97 bp), placing into a 0.5 mL tube that has been perforated 5 times with a needle and placed itself in a 2 mL Eppendorf tube.

17. Centrifuge for 2 min at 20,000 × g, room temperature and discard the 0.5 mL tube.

18. Add 100 µL 1× NEBuffer2 to the acrylamide in the 2 mL tube. Incubate overnight at room temperature on the rotator wheel.

19. Transfer the buffer/acrylamide mix to a Spin-X-filter column and centrifuge for 2 min at 20,000 × g, room temperature.

20. Transfer the buffer containing the DNA into a 1.5 mL tube. Add 1 µL 20 mg/mL glycogen, 10 µL 3 M sodium acetate pH 5.2 and mix by shaking.

21. Add 325 µL 100% ethanol and shake. Centrifuge for 30 min at 19,000 × g, 4 °C.

22. Remove the supernatant and wash the pellet with 500 µL 70% ethanol. Centrifuge for 15 min at 14,000 × g, 4 °C.

23. Remove the supernatant, dry the pellet at 37 °C, then add 10 µL milli-Q water and dissolve the DNA for 15 min at 37 °C.

24. Verify on Agilent Bioanalyzer that the samples provide a single DNA band at the expected size (95–97 bp).

25. Load sequencing flow cells according to the recommendation of the supplier (e.g., for Illumina HiSeq 2000 apparatus, use 50-nt single reads, and load 6 pM of DNA library per flow cell and the following primer for cluster generation: 5′-CCACCGA CAGGTTCAGAGTTCTACAGTCCGAC-3′).

3.7 Bioinformatic Analyses of HRS-seq Data (See Note 13)

1. Trim HRS-seq raw data (fastq files) for primer sequences.

2. Align reads to the reference genome, removing those mapping to multiple positions or with more than two mismatches.

3. Remove reads shorter than 18 nt or longer than 20 nt (*see* **Note 14**).

4. Remove reads that do not end with the sequence WWGG (*see* **Note 15**).

5. Count the total number of reads obtained for each StyI restriction fraction in the reference genome.

6. Use appropriate statistical tools (e.g., DEseq or edgeR) to determine which StyI restriction fragments display a statistically significant overrepresentation of read counts in the insoluble fraction compared with the soluble fraction. These fragments are likely those most associated with nuclear bodies, and are those used for further bioinformatic analyses (e.g., comparison between different experimental conditions or cell types).

4   Notes

1. DTT, PMSF and aprotinin should be added fresh to this buffer just before use.

2. A home-made apparatus was developed for pressurizing filtration units. Such apparatus uses a hydrostatic pump. It allows a fixed and constant pressure to be exerted on the liquid above the filter, which can be easily set up by controlling the speed of the hydrostatic pump (*see* Fig. 2).

3. We recommend to use homemade restriction buffer since commercial buffers usually contain reducing agents like DTT that may alter the insoluble fraction thus affecting DNA retention.

4. The two primers used to make the adaptors are mixed in equimolar amount and annealed together on a PCR block by heating to 100 °C and then decreasing the temperature to 25 °C in 0.1 °C steps, incubating for 1 min at each intermediate temperature. Oligonucleotides with "N" should be degenerate, containing all combinations of A, T, G, and C, to make cohesive ends compatible with any generated by the MmeI digestion. Most commercial suppliers cater for this option.

5. Some enzyme suppliers provide MmeI directly supplemented with SAM (e.g., NEB). In that case SAM is replaced by 10 μL of milli-Q water (Subheading 3.6, **step 9**).

6. The present protocol is intended for cultured cells as previously described [7]. If working with tissues, prepare nuclei as previously described in references [8, 9] and proceed with Subheading 3.1, **step 2**.

7. **Important note:** do not increase the number of nuclei treated during Subheadings 3.2–3.4. It is extremely important to keep a small number of nuclei (150,000) diluted in a big volume during these steps because nuclear halos tend to aggregate together and such aggregation could generate in vitro artifacts. For the same reason, do not pellet nuclear halos by centrifugation, but, as stated in this protocol, always use filtration units and apply gentle pressures on the liquids using a piston or HMA [10].

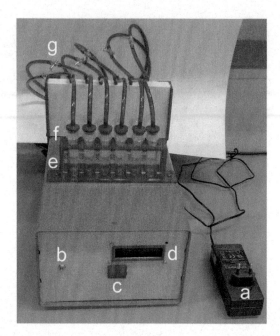

Fig. 2 View of the homemade apparatus (HMA) mounted with ultrafiltration units: (**a**) electrical plug, (**b**) switch, (**c**) pressure setting, (**d**) pressure display, (**e**) racks mounted with filtration units, (**f**) air pipes, (**g**) tap to close air pipe when not used

8. If using a different enzyme to StyI, the enzyme buffer and the biotinylated adaptor 1 provided in this protocol will need to be modified accordingly.

9. Quality controls can be made at this stage by qPCR to check for enrichment of specific sequences [4, 6].

10. Partial digestion is a technical concern; Subheading 3.5 ensures full digestion before sequencing library preparation.

11. The minimal concentration for DNA samples after Subheading 3.5 is 10 ng/μL. If less than 500 ng has been obtained, the sample will need to be concentrated to at least 10 ng/μL by ethanol precipitation (to a 50 μL sample add 2.5 μL 5 M NaCl, 1 μL 20 mg/mL glycogen, and 150 μL 100% ethanol, mix, keep for 45 min at −80 °C, and centrifuge for 15 min at 15,700 × g, room temperature). If more than 500 ng has been obtained, the sample can be used directly.

12. **Critical step:** 150 ng corresponds to the amount recommended to ensure a good quality of the final libraries. Add the biotinylated adaptor and the other reagents to the genomic DNA as specified. Do not make a "premix" of the regents with the ligase as this may lead to the formation of adaptor dimers and would reduce the quality of the sequencing library.

13. This section provides guidelines to data processing as used in our previous work [6]. The choice of software used to perform these tasks are largely dependent on available resources.

14. MmeI digestion should provide reads of 18–20 nucleotides.

15. The reads should have one extremity corresponding to the StyI cohesive end. This expected sequence terminus will be different if a different restriction enzyme is used.

Acknowledgments

We thank Mathilde Tancelin for help in improving the HRS assay, Jean Casanova for developing the HMA system, and Laurent Journot and all the engineers of the MGX platform (Montpellier) who helped to design HRS-seq DNA libraries. This work was supported by grants from the Agence Nationale de la Recherche (CHRODYT, ANR-16-CE15-0018-04), the AFM-Téléthon (N° 21024), and the Centre National de la Recherche Scientifique (CNRS).

References

1. Court F, Miro J, Braem C et al (2011) Modulated contact frequencies at gene-rich loci support a statistical helix model for mammalian chromatin organization. Genome Biol 12:R42

2. Ea V, Baudement MO, Lesne A, Forné T (2015) Contribution of topological domains and loop formation to 3D chromatin organization. Genes (Basel) 6:734–750

3. Lesne A, Baudement MO, Rebouissou C, Forné T (2019) Exploring mammalian genome within phase-separated nuclear bodies: experimental methods and implications for gene expression. Genes (Basel) 10:1049. https://doi.org/10.3390/genes10121049

4. Braem C, Recolin B, Rancourt RC et al (2008) Genomic matrix attachment region and chromosome conformation capture quantitative real time PCR assays identify novel putative regulatory elements at the imprinted Dlk1/Gtl2 locus. J Biol Chem 283:18612–18620

5. Weber M, Hagège H, Lutfalla G et al (2003) A real-time polymerase chain reaction assay for quantification of allele ratios and correction of amplification bias. Anal Biochem 320:252–258

6. Baudement MO, Cournac A, Court F et al (2018) High-salt-recovered sequences are associated with the active chromosomal compartment and with large ribonucleoprotein complexes including nuclear bodies. Genome Res 28:1733–1746

7. Ea V, Court F, Forné T (2017) Quantitative analysis of intra-chromosomal contacts: the 3C-qPCR method. Methods Mol Biol 1589:75–88

8. Milligan L, Antoine E, Bisbal C et al (2000) H19 gene expression is up-regulated exclusively by stabilization of the RNA during muscle cell differentiation. Oncogene 19:5810–5816

9. Milligan L, Forné T, Antoine E et al (2002) Turnover of primary transcripts is a major step in the regulation of mouse H19 gene expression. EMBO Rep 3:774–779

10. Weber M, Hagège H, Murrell A et al (2003) Genomic imprinting controls matrix attachment regions in the Igf2 gene. Mol Cell Biol 23:8953–8959

Part IV

Single-Cell Approaches

Chapter 10

High-Throughput Preparation of Improved Single-Cell Hi-C Libraries Using an Automated Liquid Handling System

Wing Leung and Takashi Nagano

Abstract

Hi-C is recognized as a gold standard approach to analyze the three-dimensional (3D) organization of chromatin or chromosomes on a genome-wide scale. It has revealed many characteristic features of structural organization and contributed to our understanding of how gene expression is related to the 3D organization of chromatin. However, the original Hi-C is designed to analyze the average structure across millions of cells, which makes the method unsuitable if the cell population of interest is not homogeneous or the purpose is to pursue the dynamic aspects of the structural features in individual cells. To overcome such limitations, we established single-cell Hi-C and have improved the method further in terms of data quality and throughput. Here we describe the revised single-cell Hi-C protocol, including the settings of the liquid handling system essential for increased throughput.

Key words Hi-C, Chromosome conformation capture (3C), Single-cell analysis, Chromatin interactions, Genome organization, Liquid handling system

---

1 Introduction

Chromosome conformation capture (3C)-based methodologies have revolutionized the way the three-dimensional (3D) organization of the genome is studied [1]. In particular, Hi-C, a nonbiased and high-throughput version of the original 3C method, has enabled us to analyze the 3D organization on a genome-wide scale for the first time [2]. Since then, it has contributed to the systematic identification of characteristic structures such as compartments [2], topologically associated domains [3, 4], and chromatin loops [5], and is regarded as the gold standard technique to study chromatin in 3D. However, one limitation of Hi-C is that it only gives averaged information across thousands or millions of cells, like other ensemble-cell analyses such as RNA sequencing (RNA-Seq) or chromatin immunoprecipitation sequencing (ChIP-seq). Such ensemble-cell approaches are

Tom Sexton (ed.), *Spatial Genome Organization: Methods and Protocols*,
Methods in Molecular Biology, vol. 2532, https://doi.org/10.1007/978-1-0716-2497-5_10,

unsuitable for the analysis of a heterogeneous cell population and the dynamic aspects of the epigenome in individual cells.

In contrast, the emergence of single-cell epigenomic analyses does work to overcome the limitations of ensemble-cell analyses. For example, single-cell RNA-Seq is capable of identifying a unique group of cells and inferring a dynamic differentiation trajectory among a large cell population [6], which is not possible with ensemble-cell RNA-Seq data. To make the most of such advantages of single-cell analysis in the 3D organization of chromatin, we developed a single-cell version of Hi-C in 2013 [7], which showed cell-to-cell variability in the 3D organization to some extent with 10 cells. However, analyzing a larger number of cells obviously enhances the capabilities of the method. In fact, collecting and comparing single-cell Hi-C data from thousands of cells has revealed dynamic reorganization of chromosome architecture during cell-cycle progression [8].

Thus, it is important to collect data from as many cells as possible when planning a single-cell Hi-C experiment. To achieve such an experimental throughput, one of the advantages of our second single-cell Hi-C study [8] compared to the first one [7] is the use of an automated liquid handling system during library preparation. In this article, we revise our previous single-cell Hi-C protocol [9] with extra attention to this point. Other technical improvements with the protocol revision include employing a frequent-cutting restriction enzyme (MboI) for Hi-C digestion and simplified library preparation with tagmentation (Tn5-mediated transposition of adapter sequences). In addition to increased throughput, we can get more information with better specificity from each of thousands of single cells thanks to these improvements. For more details of how single-cell Hi-C technique was developed by modifying the ensemble-cell Hi-C, see the previous version of the protocol [9].

2 Materials

1. Cell culture medium and reagents to grow cells of interest.

2. 16% paraformaldehyde (methanol-free).

3. 2 M glycine.

4. 1× phosphate-buffered saline (PBS) pH 7.4 (calcium- and magnesium-free).

5. Permeabilization buffer: 10 mM Tris-HCl pH 8, 10 mM NaCl, 0.2% (w/v) Igepal CA-630, 1× complete EDTA-free protease inhibitors (Roche).

6. 25 U/μL MboI, with manufacturer-supplied 10× restriction buffer.

7. 20% SDS.

8. 20% Triton X-100.

9. Fill-in mix: 288 μM dCTP, 288 μM dGTP, 288 μM dTTP, 288 μM biotin-14-dATP, 1 U/μL DNA polymerase I large (Klenow) fragment, made fresh for each experiment.

10. Ligation mix: 0.1 mg/mL BSA, 0.01 U/μL T4 DNA ligase in 1.05× manufacturer-supplied T4 DNA ligase reaction buffer.

11. PBT: 0.001% BSA, 0.05% (v/v) Tween 20 in 1× PBS.

12. Cell strainer (30 μm mesh).

13. Cell sorter compatible with 96-well plate format.

14. Bravo automated liquid handling system fitted with 96LT head (Agilent), or equivalent.

15. Nextera XT DNA library preparation kit (Illumina), includes Tagment DNA buffer, Amplicon Tagment mix, Neutralize Tagment buffer, and Nextera PCR master mix.

16. Dynabeads M-280 streptavidin beads (Invitrogen).

17. Magnetic stands for 1.5 mL tubes and 96-well plates (e.g., DynaMag-2 and DynaMag-96 side magnet).

18. Bead washing buffer: 5 mM Tris-HCl pH 7.5, 0.5 mM EDTA, 1 M NaCl.

19. Bead suspension buffer: 20 mM Tris-HCl pH 7.5, 2 mM EDTA, 4 M NaCl.

20. 10 mM Tris-HCl pH 7.5.

21. Nextera XT Index Kit v2 (Illumina).

22. AMPure XP beads (Beckman Coulter).

23. 80% ethanol, made fresh for each purification with AMPure XP beads.

24. 10 mM Tris-HCl pH 8.5.

25. Qubit fluorimeter and dsDNA HS assay kit (Invitrogen).

26. KAPA Library Quantification Kit Illumina/Universal (Roche), containing KAPA SYBR FAST qPCR master mix with primer premix.

27. Bioanalyzer 2100 (Agilent) and high-sensitivity chips.

28. Next-generation sequencer, Illumina-compatible.

3   Methods

3.1   Hi-C Preparation

1. Prepare a single-cell suspension of the cells of interest (1×10^5–3×10^7 cells in 21 mL) in appropriate room-temperature medium and transfer to a 50 mL centrifuge tube.

2. Add 3 mL 16% paraformaldehyde and fix for 10 min at room temperature with continuous mixing by inversion.

3. Add 1.632 mL 2 M glycine, mix well by inversion and incubate on ice for 5 min.

4. Centrifuge for 8 min at $300 \times g$, 4 °C and remove supernatant (*see* **Note 1**).

5. Resuspend cells with 50 mL ice-cold PBS, centrifuge for 8 min at $300 \times g$, 4 °C and remove supernatant (*see* **Notes 1** and **2**).

6. Resuspend cells with 50 mL permeabilization buffer (*see* **Note 3**) and incubate on ice for 30 min, intermittently mixing by inverting the tubes ~50 times every 5 min (*see* **Note 4**).

7. Centrifuge for 5 min at $300 \times g$, 4 °C and discard all but ~0.5 mL supernatant. Resuspend the cells in residual liquid and transfer to a 1.5 mL microtube.

8. Centrifuge for 5 min at $300 \times g$, 4 °C and discard all supernatant without touching the cell pellet.

9. Add 1 mL 1.24× restriction buffer gently, without dispersing the cell pellet (*see* **Note 5**). Centrifuge for 5 min at $300 \times g$, 4 °C and remove all supernatant without touching the cell pellet.

10. Repeat **step 9** with 500 μL 1.24× restriction buffer.

11. Add 400 μL 1.24× restriction buffer gently, without dispersing cell pellet. Add 6 μL 20% SDS and mix gently by pipetting, avoiding making air bubbles. Incubate for 1 h at 37 °C with gentle agitation on a rotator (2 rpm).

12. Add 40 μL 20% Triton X-100, mix gently by pipetting while avoiding air bubbles, and incubate for 1 h at 37 °C with gentle agitation on a rotator (2 rpm).

13. Add 50 μL 25 U/μL MboI, mix gently by pipetting, and incubate overnight at 37 °C with gentle agitation on a rotator (2 rpm).

14. Add 54.08 μL fill-in mix, mix gently by pipetting, and incubate for 1 h at 37 °C with gentle agitation on a rotator (2 rpm).

15. Centrifuge for 10 min at $300 \times g$, 4 °C and discard all but 50 μL supernatant (*see* **Note 6**).

16. Add 950 μL ligation mix, mix gently by pipetting, and incubate overnight at 16 °C (*see* **Note 7**).

3.2 Single-Cell Isolation and Library Preparation

1. Centrifuge ligation reaction for 10 min at $300 \times g$, 4 °C and carefully remove supernatant.

2. Resuspend cells in 1 mL PBT, centrifuge for 10 min at $300 \times g$, 4 °C and remove supernatant.

3. Resuspend cells in 1 mL PBT and pass cell suspension through a 30 μm cell strainer into a cell sorter tube.

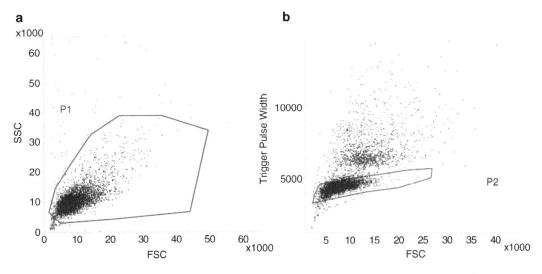

Fig. 1 Examples of scatter plots acquired during single-cell isolation by cell sorter. (**a**) The area of side scatter (SSC, *y*-axis) by the area of forward scatter (FSC, *x*-axis) plot from the whole cell suspension after Hi-C processing. (**b**) The width of FSC (Trigger Pulse Width, *y*-axis) by the area of FSC (FSC, *x*-axis) plot from the encircled population (P1) in panel **a**. The encircled population in panel **b** (P2) is subject to single-cell isolation into 96-well plates

4. Using a cell sorter, discriminate single cells (singletons) from debris and doublets by scatter plots (Fig. 1) and sort single cells into individual wells of 96-well plates. Seal each plate with sealing film and briefly centrifuge (*see* **Note 8**).

5. Add 5 μL 1× PBS to each well, seal plates with film and centrifuge briefly. Incubate overnight at 65 °C.

6. Cool plates to room temperature and add 10 μL Tagment DNA buffer to each well (*see* **Note 9** and Fig. 2a for automation of this step).

7. Add 5 μL Amplicon Tagment Mix to each well and mix by pipetting (*see* **Note 10** and Fig. 2b for automation of this step). Seal plates and centrifuge briefly.

8. In a thermal cycler, incubate for 5 min at 55 °C, followed by 10 s at 10 °C.

9. Add 5 μL Neutralize Tagment buffer and mix by pipetting (*see* **Note 11** and Fig. 2c for automation of this step). Incubate for 5 min at room temperature.

10. Transfer two aliquots of 1 mL Dynabeads M-280 streptavidin into 1.5 mL microtubes, place on magnetic stand for 1 min and remove supernatants.

11. Wash beads three times with 1 mL bead washing buffer by resuspending, placing on magnetic stand for 1 min and removing supernatant.

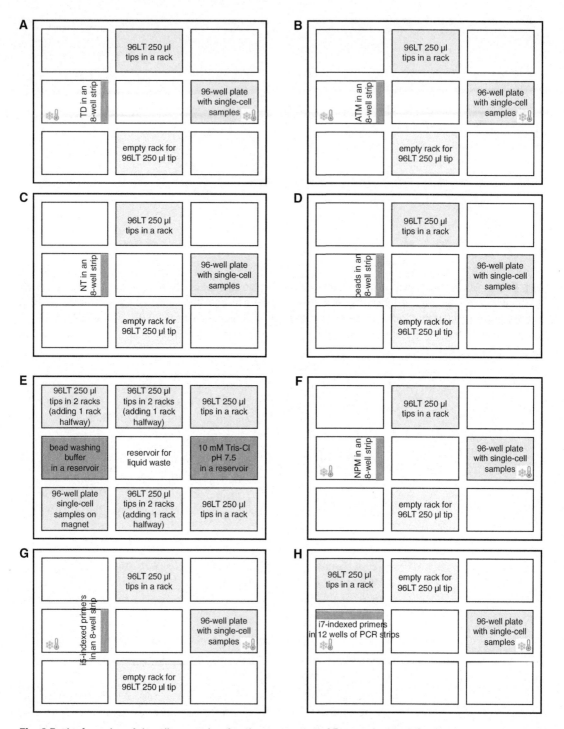

Fig. 2 Each of panels **a–h** is a diagram showing the preparation of Bravo robot deck for the step as indicated in Subheading 3.2. The top of each diagram corresponds to the rear of the deck, and the positions at the top, middle, and bottom rows of each diagram are described in the Notes as 1–3, 4–6, and 7–9, respectively

12. Resuspend each bead pellet with 420 µL bead suspension buffer.

13. Add 8 µL bead suspension to each well and mix by pipetting (*see* **Note 12** and Fig. 2d for automation of this step).

14. Put a cap on each well and incubate plates overnight at room temperature with gentle agitation on a rotator (2 rpm) (*see* **Note 13**).

15. Place plates on compatible magnetic stand for 2 min and remove supernatants (*see* **Note 14** and Fig. 2e for automation of **steps 15–17**).

16. Wash beads four times with 150 µL per well bead washing buffer, then twice with 150 µL per well 10 mM Tris-HCl pH 7.5 (*see* **Note 15**).

17. Resuspend beads in each well with 25 µL 10 mM Tris-HCl pH 7.5.

18. Add 15 µL Nextera PCR master mix to each well (*see* **Note 16** and Fig. 2f for automation of this step).

19. Add 5 µL of each i5- and i7-indexed primers of choice from Nextera XT Index kit v2, and mix by pipetting (*see* **Notes 17–19** and Figs. 2g, h for automation of this step).

20. Run the following program on a thermal cycler:

    72 °C, 3 min;

    95 °C, 30 s;

    12 or 18 cycles (*see* **Note 20**): 95 °C, 10 s; 55 °C, 30 s; 72 °C, 30 s;

    72 °C, 5 min.

    At this stage, individual single-cell libraries can be purified and sequenced individually (Subheading 3.3), or if they have been individually indexed, purified together as one pool (Subheading 3.4) (*see* **Note 20**).

3.3 Individual Library Purification (See Note 21)

1. Place plates on compatible magnetic stand for 1 min and transfer individual supernatants to a fresh 96-well plate, PCR strips or 1.5 mL microtubes, depending on the number of samples.

2. Add 30 µL AMPure XP beads to each sample, mix well and incubate for 5 min at room temperature (*see* **Note 22**).

3. Place samples on a compatible magnetic stand for 5 min and remove supernatants.

4. While keeping the samples on the magnetic stand, wash the beads twice with 200 µL 80% ethanol, removing supernatant each time.

5. Air-dry the beads on the magnet with the tube lid open (*see* **Note 23**).

6. Wet the bead pellet by adding 32 μL 10 mM Tris-HCl pH 8.5 to each sample, remove the samples from the magnet, mix well by pipetting, then incubate for 5 min at room temperature.

7. Place samples on magnetic stand for 5 min, then collect 30 μL of each supernatant into new 96-well plates, PCR strips or 1.5 mL microtubes (*see* **Note 24**).

8. Add 30 μL AMPure XP beads to each sample, mix well by pipetting and incubate for 5 min at room temperature.

9. Repeat **steps 3–5** to wash and dry beads (*see* **Note 23**).

10. Wet the bead pellet by adding 11 μL 10 mM Tris-HCl pH 8.5 to each sample, remove the samples from the magnet, mix well by pipetting, then incubate for 5 min at room temperature.

11. Place samples on magnetic stand for 5 min, then collect 10 μL of each supernatant into new 96-well plates, PCR strips or 1.5 mL microtubes (*see* **Note 24**). Proceed to Subheading 3.5 with the individual single-cell Hi-C libraries (*see* **Note 25**).

3.4 Pooled Library Purification (See Note 21)

1. Place plates on compatible magnetic stand for 1 min and pool all supernatants into a single tube (~4.8 mL from one 96-well plate).

2. Add 2.88 mL AMPure XP beads, mix well and incubate for 5 min at room temperature (*see* **Note 22**).

3. Transfer 1.3 mL to a new 1.5 mL microtube placed on a magnetic stand, leave on magnet for 3 min and remove supernatant.

4. Repeat **step 3** in same tube until entire pool has been processed.

5. While keeping the tube on the magnetic stand, wash the bead pellet four times with 1 mL 80% ethanol, removing supernatant each time.

6. Air-dry the beads on the magnet with the tube lid open (*see* **Note 26**).

7. Wet the bead pellet by adding 103 μL 10 mM Tris-HCl pH 8.5 to each sample, remove the samples from the magnet, mix well by pipetting, then incubate for 5 min at room temperature.

8. Place beads on magnetic stand for 3 min, then collect 100 μL of supernatant into a new 1.5 mL microtube (*see* **Note 24**).

9. Add 100 μL AMPure XP beads, mix well by pipetting, and incubate for 5 min at room temperature.

10. Place sample on magnetic stand for 3 min and remove supernatant.

11. Repeat **steps 5** and **6** to wash and dry beads (*see* **Note 27**).

12. Wet the bead pellet by adding 11 µL 10 mM Tris-HCl pH 8.5 to each sample, remove the samples from the magnet, mix well by pipetting, then incubate for 5 min at room temperature.

13. Place samples on magnetic stand for 5 min, then collect 10 µL of each supernatant into new 1.5 mL microtube (*see* **Note 24**). Proceed to Subheading 3.5 with the pooled single-cell Hi-C library (*see* **Note 25**).

3.5 Library Quality Control and Sequencing

1. Take 1 µL library (or libraries) and measure concentration by Qubit dsDNA HS assay.

2. If library concentration is above 2 ng/µL, prepare a dilute library around 1 ng/µL using 1 µL of the original and 10 mM Tris-HCl pH 8.5, and use this dilute library instead of the original for qPCR and Bioanalyzer analyses in steps below.

3. Set up qPCR with 4 µL 1:1000 dilution of the library added to 6 µL KAPA SYBR FAST qPCR master mix containing primer premix, following the manufacturer's instructions (*see* **Note 28**).

4. Use 1 µL of library for Bioanalyzer High Sensitivity DNA analysis (*see* **Note 29**).

5. For each library, correct the molar concentration from the qPCR measurement with the average fragment length from the Bioanalyzer results to determine the loading amount for sequencing (*see* **Note 30**).

6. Perform paired-end sequencing (2×50 or 2×100 bp) (*see* **Note 31**).

7. After de-multiplexing raw sequencing results to each single cell, process the fastq files through FastQC [10], then put the two read files through HiCUP [11] to map the reference genome, pair the two reads and filter for valid Hi-C read pairs (*see* **Note 32**).

4   Notes

1. To remove the supernatant completely, spin the tube briefly after removing most of the supernatant to collect the remainder at the bottom and remove it with a fine pipette tip without touching the cell pellet.

2. Cell pellet can be flash frozen and stored at $-80\,°C$ for several months if not immediately continuing to Hi-C steps.

3. First resuspend the cell pellet with a small volume (1–5 mL) of the permeabilization buffer by gentle vortexing or pipetting and make sure the suspension is homogenous.

4. Some cells need homogenization by Dounce homogenizer during permeabilization, which should be optimized for each cell type. A typical condition of homogenization is 2×10 strokes with a tight pestle with a 5 min interval on ice.

5. If starting with a small number of cells (e.g., less than one million) and the cells are dispersed at this point, they may not form a clear pellet after the next spin. If the cells are resuspended in $1.24\times$ restriction buffer, a clear pellet can be obtained by adding Igepal CA-630 to a final concentration of 0.02% and centrifuging for 5 min at 300 g, 4 °C.

6. To do this reproducibly, prepare a tube with 50 μL of liquid and use this tube as a guide to show how 50 μL looks like in the tube.

7. After overnight incubation cells can be stored at 4 °C for several days until single-cell isolation.

8. Sealed plates can be stored at −80 °C until Hi-C library preparation.

9. This step can be automated using a Bravo robot fitted with 96LT head and 96-filtered tips after the following preparation: Tagment DNA buffer stock is dispensed into each well of a new 8-tube PCR strip, 128 μL per tube (which is sufficient for 12 wells including the additional void volume required for automated pipetting) and placed in a PCR plate insert at position 4 (Fig. 2a). The 96-well plate with single-cell samples is set in another PCR plate insert and placed at position 6. Positions 4 and 6 are set to 4 °C.

10. This step can be automated as in **Note 9** after the following preparation: Amplicon Tagment mix stock is dispensed into each well of a new 8-tube PCR strip, 73 μL per well, and placed in a PCR plate insert at position 4 (Fig. 2b). The 96-well plate with single-cell samples is set in another PCR plate insert and placed at position 6. Positions 4 and 6 are set to 4 °C.

11. This step can be automated as in **Note 9** after the following preparation: Neutralize Tagment buffer is dispensed into each well of a new 8-tube PCR strip, 68 μL per well, and placed in a PCR plate insert at position 4 (Fig. 2c). The 96-well plate with single-cell samples is set in another PCR plate insert and placed at position 6.

12. This step can be automated as in **Note 9** after the following preparation: the bead suspension from the **step 12** in Subheading 3.2 is dispensed into each well of a new 8-tube PCR strip, 105 μL per well, and placed in a PCR plate insert at position 4 (Fig. 2d). The 96-well plate with single-cell samples is set in another PCR plate insert and placed at position 6.

13. In our experience, caps (8-cap strip) work better than a sealing film to prevent leakage during agitation. We use an empty case of pipette tips and rubber bands to set a 96-well plate on a rotator. Check the mixing status of each sample after 1–2 h.

14. These steps can be automated as in **Note 9** after the following preparation: place an automation reservoir containing 70 mL of bead washing buffer at position 4, place a second automation reservoir containing 40 mL of 10 mM Tris-Cl pH 7.5 at position 6, and set the 96-well plate containing sample bead suspensions on the magnet at position 7 (Fig. 2e). The sample bead suspensions in a 96-well plate should be on the magnet for 1–2 min at position 7, then the sealing film should be removed before starting the automation program to execute **steps 15–17** of Subheading 3.2. The used pipette tips are returned to the same racks as they are picked from, and the racks with used tips at positions 1, 2, and 8 should be replaced with racks with new tips during the program.

15. Single wash comprises adding the buffer, waiting for 1 min and removing the supernatant, while keeping the plate on the magnet.

16. This step can be automated as in **Note 9** after the following preparation: Nextera PCR master mix stock is dispensed into each well of a new 8-tube PCR strip, 190 μL per tube, and placed in a PCR plate insert at position 4 (Fig. 2f). The 96-well plate with single-cell samples (bead suspensions) is set in another PCR plate insert and placed at position 6. Positions 4 and 6 are set to 4 °C.

17. Set the i5 and i7 indices so that each library to be sequenced in the same lane can be discriminated. With the setting described here, the 12 wells in the same row of the single-cell sample plate have the same i5-indexed primer, and the 8 wells in the same column of the single-cell sample plate have the same i7-indexed primer. This makes each of the 96 wells has a distinct combination of i5 and i7 indices with eight i5- and twelve i7-indexed primers; sixteen i5 and twenty-four i7 indices enable up to 384 wells to be discriminated from each other.

18. Adding i5-indexed primers can be automated as in **Note 9** after the following preparation: the i5-index primers are dispensed into each well of a new 8-tube PCR strip, 65 μL per tube, and placed in a PCR plate insert at position 4 (Fig. 2g). The 96-well plate with single-cell samples (bead suspensions) is set in another PCR plate insert and placed at position 6. Positions 4 and 6 are set to 4 °C.

19. Adding i7-indexed primers can be automated as in **Note 9** after the following preparation: the i7-index primers are dispensed into each of 12 wells of new PCR strips, 45 μL per well, and

placed in a PCR plate insert at position 4 (Fig. 2h). The 96-well plate with single-cell samples (bead suspensions) is set in another PCR plate insert and placed at position 6. Positions 4 and 6 are set to 4 °C.

20. Choose 18 cycles to purify each single-cell library individually through Subheading 3.3 after PCR. Choose 12 cycles to purify 96 single-cell libraries as a single pool through Subheading 3.4 after PCR.

21. To avoid contamination into pre-PCR samples, the liquid handling system should not be used for post-PCR samples.

22. Make sure that the AMPure XP bead suspension to add is at room temperature and homogeneous.

23. This usually takes ~5 min. Bead pellets become nonglossy with cracks as they dry.

24. Be careful not to take beads with the supernatant.

25. Supernatants (Hi-C libraries) are stored at −20 °C until Subheading 3.5 library quality analysis and sequencing.

26. This usually takes ~60 min. Bead pellets become nonglossy with cracks as they dry.

27. The drying of the bead pellet takes 10–15 min.

28. Set up triplicate quantification for each library or library pool and omit an outlier well if the standard deviation of Ct (threshold cycle) from the triplicate is more than 0.3. This also applies to DNA standard wells.

29. Make sure that the library or pool is free from primer dimers (see Fig. 3). If it contains an obvious primer dimer peak, repeat the purification with AMPure XP beads following **steps 9–13** in Subheading 3.4, but after adjusting the library volume to 30 μL with 10 mM Tris-Cl pH 8.5 and use 24 μL of AMPure XP bead suspension instead of 100 μL.

30. Multiply average qPCR concentration by the following factor: size of DNA standard in bp (452 for the standard that comes with KAPA Library Quantification Kit) divided by average fragment length of library in bp (determined by bioanalyzer analysis).

31. We load two pools of 96 single-cell Hi-C libraries into a lane with the capacity of 200–400 million read pairs, and this essentially allowed us to sequence each library to saturation. The Nextera XT Index Kit v2 allows the discrimination of up to 384 libraries in the same sequencing run.

32. FastQC will tell if the sequencing is successful or not. HiCUP will tell if the Hi-C experiment is successful or not, most typically with the number of "unique di-tags" (for data coverage; usually more than 10,000 per singe-cell dataset) and the percentage of "unique trans-chromosomal contacts" (for

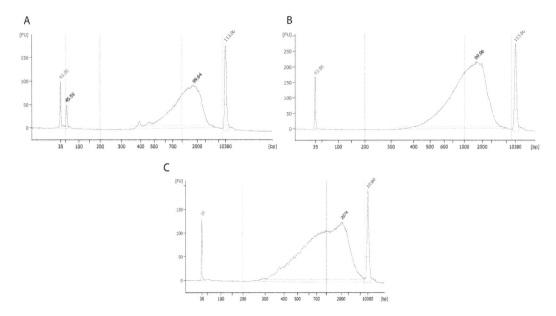

Fig. 3 Example Bioanalyzer profiles of Hi-C libraries. (**a**) A sharp peak around 50–70 bp indicates contamination of primer dimers. (**b**) The same library as **a** that has been repurified to remove primer dimers as in **Note 29**. (**c**) An example Bioanalyzer profile from 96 single-cell Hi-C libraries pooled following the steps in Subheading 3.4. Note that the sharp peaks at 35 and 10,380 bp are size markers (not from the libraries)

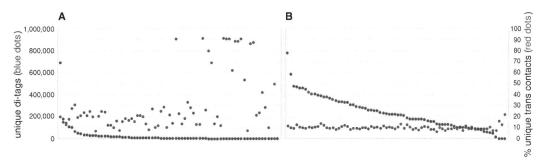

Fig. 4 Comparison of unique di-tags (blue dots; scaled on the left) and % unique trans contacts (red dots; scaled on the right) between 74 single-cell Hi-C datasets prepared from the previous [9] (**a**) and current (**b**) protocols. The 74 datasets in each plot are sorted by the unique di-tags and shown on a descending order on the x-axis. The datasets prepared on the current protocol have generally more unique di-tags with stably low % unique trans contacts

specificity of interactions detected by the Hi-C experiment; usually less than 15%). See Fig. 4 for how these two aspects have been improved from the previous version of single-cell Hi-C protocol [9]. Note that the information from FastQC and HiCUP will not tell the "single-cell-ness" of the data (for example, we cannot necessarily notice by FastQC or HiCUP if a dataset derives from multiple cells).

References

1. Dekker J, Rippe K, Dekker M et al (2002) Capturing chromosome conformation. Science 295:1306–1311

2. Lieberman-Aiden E, van Berkum NL, Williams L et al (2009) Comprehensive mapping of long-range interactions reveals folding principles of the human genome. Science 326: 289–293

3. Dixon JR, Selvaraj S, Yue F et al (2012) Topological domains in mammalian genomes identified by analysis of chromatin interactions. Nature 485:376–380

4. Sexton T, Yaffe E, Kenigsberg E et al (2012) Three-dimensional folding and functional organization principles of the drosophila genome. Cell 148:458–472

5. Rao SS, Huntley MH, Durand NC et al (2014) A 3D map of the human genome at kilobase resolution reveals principles of chromatin looping. Cell 159:1665–1680

6. Stegle O, Teichmann SA, Marioni JC (2015) Computational and analytical challenges in single-cell transcriptomics. Nat Rev Genet 16: 133–145

7. Nagano T, Lubling Y, Stevens TJ et al (2013) Single-cell Hi-C reveals cell-to-cell variability in chromosome structure. Nature 502:59–64

8. Nagano T, Lubling Y, Várnai C et al (2017) Cell-cycle dynamics of chromosomal organization at single-cell resolution. Nature 547: 61–67

9. Nagano T, Wingett SW, Fraser P (2017) Capturing three-dimensional genome organization in individual cells by single-cell Hi-C. Methods Mol Biol 1654:79–97

10. FastQC. https://www.bioinformatics.babraham.ac.uk/projects/fastqc/

11. HiCUP (Hi-C User Pipeline). http://www.bioinformatics.babraham.ac.uk/projects/hicup/

Chapter 11

Simultaneous Quantification of Spatial Genome Positioning and Transcriptomics in Single Cells with scDam&T-Seq

Silke J. A. Lochs and Jop Kind

Abstract

Spatial genome organization is considered to play an important role in mammalian cells, by guiding gene expression programs and supporting lineage specification. Yet it is still an outstanding question in the field what the direct impact of spatial genome organization on gene expression is. To elucidate this relationship further, we have recently developed scDam&T-seq, a method that simultaneously quantifies protein–DNA interactions and transcriptomes in single cells. This method efficiently combines two preexisting methods: DamID for measuring protein–DNA contacts and CEL-Seq2 for quantification of the transcriptome in single cells. scDam&T-seq has been successfully applied to measure DNA contacts with the nuclear lamina, while at the same time revealing the effect of these contacts on gene expression. This method is applicable to many different proteins of interest and can thereby aid in studying the relationship between protein–DNA interactions and gene expression in single cells.

Key words Protein–DNA interactions, Single-cell genomics, DamID, CEL-Seq2, Lamina-associated domains

1 Introduction

The spatial organization of the genome plays an important role in the regulation of gene expression and lineage specification in mammals (reviewed in [1, 2]). There are many layers of spatial genome organization, such as the localization of entire chromosomes into distinct chromosome territories [3] as well as smaller DNA loop structures, facilitated by the cohesin complex and insulator protein CTCF [4]. These layers of organization are driven by protein–DNA interactions, which therefore play a key role in regulating chromatin domains. However, it still remains largely unclear to which extent spatial genome organization directly impacts gene expression and consequently how it is involved in cellular decision-making, for example during differentiation (reviewed in [5, 6]).

There are already numerous methods available to study spatial genome organization, such as Chromosome Conformation

Tom Sexton (ed.), *Spatial Genome Organization: Methods and Protocols*,
Methods in Molecular Biology, vol. 2532, https://doi.org/10.1007/978-1-0716-2497-5_11,

Capture methods (e.g., 4C (Chapter 2) like Hi-C [7–9] (Chapter 3)), DamID [10], and Genome Architecture Mapping (GAM) [11]. Besides sequencing-based methods, super-resolution microscopy approaches have aided in understanding the three-dimensional (3D) organization of the genome [12]. Recently, a study using MERFISH combined with STORM microscopy [13] visualized the large amount of cell-to-cell variation in the spatial organization of chromatin domains (see also Chapter 12). To capture cell heterogeneity, a lot of effort has been directed towards adapting the existing methods, making them suitable for single-cell measurements at a genome-wide resolution. Examples of these include scHi-C [14] (Chapter 10), scDamID [15], and the recently developed CUT&RUN [16] method. These applications have provided the research field with many new possibilities, especially in research areas where the amount of available material is limited [17].

Nevertheless, none of these methods allows a direct comparison between protein–DNA interactions and transcription on a genome-wide scale, as they are unable to measure both outputs in the same cell. Without this combined measurement, it is difficult to disentangle the connection between spatial genome organization and gene expression patterns. To overcome this limitation, we have recently developed scDam&T-seq [18, 19], a method that combines the quantification of protein–DNA interactions and the transcriptome in the same single-cell. To measure protein–DNA interactions, we have adapted single-cell DamID [15], which uses the DNA adenine methylase (Dam) as a molecular stamp to mark all the DNA it comes into contact with. Dam specifically methylates adenines in a GATC sequence context and, when fused to a protein of interest (POI), it will leave a stable ^{m6}A mark on the DNA in the proximity of the POI. During sample preparation, the $G^{m6}ATC$ sequence is digested by the restriction enzyme DpnI, thereby allowing for directed amplification of the Dam-methylated DNA. To simultaneously measure mRNA transcripts, we have adapted the existing CEL-Seq2 protocol [20, 21] to be implemented together with the DamID approach. Subsequent amplification and sequencing of material is accomplished in a combined fashion. Using a barcoding strategy for both the DamID and CEL-Seq2 derived material, one can link both measurements for each single-cell sample during data analysis.

We have successfully applied scDam&T-seq to examine different aspects of genome organization. By fusing Dam to the Lamin B1 protein, we have been able to study the spatial segregation of Lamina-associated domains (LADs) [22] at the nuclear periphery alongside transcription [18]. In the same study, we have fused Dam to the Polycomb Repressive Complex 1 (PRC1) member Ring1B [23]. This allowed us to study X-chromosome inactivation dynamics and the correlation with transcriptional changes during differentiation of mouse embryonic stem cells (mESCs) [18]. Finally, leveraging the intrinsic preference of the Dam enzyme for methylating accessible chromatin regions, we have expressed untethered Dam in cells to perform chromatin accessibility measurements and directly link these to expression levels of genes [18]. scDam&T-seq can thus be applied to a range of different POIs and research questions.

In this chapter, we describe the scDam&T-seq protocol in detail (Fig. 1), as well as the design of the required adapter and primer sequences. The computational workflow has previously been described in Markodimitraki et al. [19]. Since the expression levels of the Dam-POI fusion proteins are essential for the successful implementation of scDam&T-seq, we first describe the procedure for the generation and optimization of a Dam-POI expressing cell line. These steps aim at identifying clones with the highest signal-to-noise ratio. To this end, we specifically present a detailed protocol for generating mESC lines with CRISPR knock-ins of the Tir1 protein into the TIGRE locus and the mAID-Dam construct in the Lamin B1 locus. Considerations on adapting this procedure for alternative cell lines or Dam-POI fusion proteins are provided at the end of the chapter. We conclude with reflections on the general application and practical issues of the method.

2 Materials

2.1 Establishing Clonal Line

1. Sterile 0.1% gelatin solution.

2. mESC culture medium: Glasgow's MEM (Gibco) supplemented with 10% fetal bovine serum (Sigma), 1× GlutaMAX (Gibco), 1× MEM non-essential amino acids (Gibco), 100 U/mL penicillin–streptomycin solution (Gibco), 1 mM sodium pyruvate (Gibco), 0.1 mM β-mercaptoethanol (Sigma), and 1000 U/mL ESGROmLIF (EMD Millipore).

3. Sterile 1× phosphate-buffered saline (PBS).

4. TrypLE Express enzyme (Gibco).

5. Lipofectamine 3000 tranfection reagent (Invitrogen).

6. Opti-MEM reduced serum medium (Gibco).

7. Plasmids Tir1 CRISPR knock-in TIGRE locus: pX330-EN1201 (Addgene 92144 [24]), pEN396-pCAGGS-Tir1-V5-2A-PuroR TIGRE (Addgene 91242 [24]).

8. 10 mg/mL puromycin dihydrochloride solution (Sigma).

9. Dimethylsulfoxide (DMSO), filter-sterile.

10. PCR-lysis mix: 10 mM Tris-acetate pH 7.5, 10 mM magnesium acetate, 50 mM potassium acetate, 0.67% (v/v) Igepal CA-630, 0.67% (v/v) Tween 20, 100 μg/mL proteinase K (Roche).

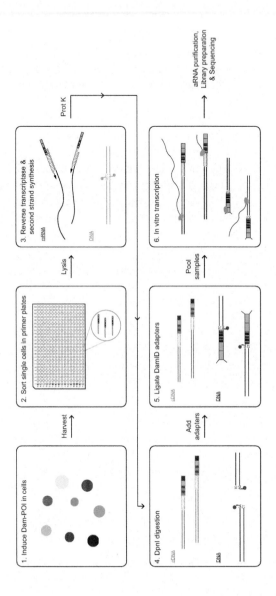

Fig. 1 A schematic overview of the scDam&T-seq procedure. (1) Induce the Dam-POI in cells for the appropriate amount of time before harvesting the cells. (2) Sort cells into 384-well plates that contain CEL-Seq2 primers. (3) Lyse cells and perform reverse transcription followed by second strand synthesis, to convert the mRNA to cDNA. (4) After Proteinase K treatment, digest ^{m6}-A methylated genomic DNA using the DpnI restriction enzyme. (5) Ligate DamID adapters to the DpnI-digested DNA fragments. (6) Pool samples and perform in vitro transcription for linear amplification of cDNA and DamID-DNA fragments. After completing these steps, one can proceed with purification of amplified RNA (aRNA), subsequent library preparation and next-generation sequencing

11. Nuclease-free water.

12. 100 μM Tir1-PCR primers.

    Tir1-PCR_primer_fw
    TTTCAGGAGGCAATGCTTGGC
    Tir1-PCR_primer_rev
    ACGCGGAACTCCATATATGGGCT

13. Tir1-PCR mix: 0.03125 U/μL MyTaq DNA polymerase, 625 nM Tir1-PCR forward primer, 625 nM Tir1-PCR reverse primer in 1.25× MyTaq Red buffer (GC Biotech).

14. Plasmids Dam CRISPR knock-in Lamin B1 locus: pUC57-Hom-BSD-P2A-mAID-Dam-Hom donor vector (Homology Chr18 59113-61113:, GRCm38/mm10), pSpCas9(BB)-2A-GFP (PX458) (Addgene 48138 [25]; sgRNA 5′ GCACGGG GGTCGCGGTCGCCA 3′). *See* **Note 1**.

15. 10 mg/mL blasticidin solution (Thermo Fisher).

16. 250 mM indole-3-acetic acid (IAA or auxin, Sigma), filter-sterile in water.

17. 100 μM Dam-PCR primers.

    Dam-PCR_primer_fw
    AACGAACCACGACGAGAGTTTATCT
    Dam-PCR_primer_rev
    TTTATACATCCTCAAATCGATTTTC

18. Dam-PCR mix: 0.03125 U/μL MyTaq DNA polymerase, 625 nM Dam-PCR forward primer, 625 nM Dam-PCR reverse primer in 1.25× MyTaq Red buffer (GC Biotech).

19. Genomic DNA isolation kit (e.g., Wizard® Genomic DNA purification kit, Promega).

20. NanoDrop spectrophotometer.

21. 20 U/μL DpnI, with manufacturer-supplied 10× CutSmart buffer (NEB).

22. Annealing buffer: 10 mM Tris-HCl pH 7.5, 50 mM NaCl, 1 mM EDTA.

23. 50 μM double-stranded ^{m6}A-adapter. Top and bottom ^{m6}A-adapter are dissolved to 50 μM each in annealing buffer, after which equal volumes of both oligonucleotides are mixed, heated to 94 °C for 2 min, then allowed to cool slowly to room temperature in a water bath or thermal cycler.

    ^{m6}A-adapter_top
    C T A A T A C G A C T C A C T A T A G G G
    CAGCGTGGTCGCGGCCGAGGA
    ^{m6}A-adapter_bottom
    TCCTCGGCCGCG

24. ^{m6}A-PCR ligase mix: 0.5 U/μL T4 DNA ligase, 2.5 μM ^{m6}A-adapter in 5× T4 DNA ligase buffer (Roche).

25. 50 μM ^{m6}A-PCR primer: NNNNNNTTACTCGTGGTCGCGGCCG AGGATC ('N' bases can be ordered as degenerate bases from an equimolar mix of A, T, G and C bases).

26. ^{m6}A-PCR mix: 0.056 U/μL MyTaq DNA polymerase, 1.38 μM ^{m6}A-PCR primer in 1.1× MyTaq Red buffer (GC Biotech).

2.2 Bulk DamID2

1. Bulk DamID2 ligase mix: 0.625 U/μL T4 DNA ligase in 6.25× T4 DNA ligase buffer (Roche).

2. 50 μM DamID double-stranded adapters. Adapters have the following design, from 5′ to 3′: 6-nt fork, T7 promoter, P5 Illumina adapter, 3-nt Unique Molecular Identifier (UMI), 4-nt barcode, 3-nt UMI and 4-nt barcode. Thus, the total barcode sequence is 8-nt and the total UMI sequence is 6-nt. The UMI sequence is split to facilitate the hybridization between the top and bottom strand and thus prevent large segments of single-stranded DNA in the adapter. The 6-nt fork is not complementary between the top and bottom adapters and prevents the formation of adapter concatemers during ligation. The bottom adapter has a 5′ phosphate group to facilitate ligation to the genomic DNA. Sequences for a set of 384 unique DamID adapters are available in Markodimitraki et al. [19].

   Example adapters are shown with UMIs indicated as 'NNN' and the split 8-nt barcodes as '*BARC*', '*ODES*'.

   Example DamID-adapter_top:
   GGTGATCCGGTAATACGACTCACTATAGGGGTT CAGAGTTCTACAGTCCGACGATCNNN*BARC* NNN*ODES*GA

   Example DamID-adapter_bot:
   /5Phos/TC*SEDO*NNN*CRAB*NNNGATCGTCG-GACTGTAGAACTCTGAACCCCTATAGTGAGTCGTATT ACCGGGAGCTT

   Top and bottom adapters are ordered as standard desalted oligonucleotides and dissolved at 500 μM in nuclease-free water. Top and bottom adapters are mixed at 50 μM each in annealing buffer in Eppendorf tubes. The tube is heated to 94 °C for 2 min in a water bath and then slowly cooled to room temperature, to allow hybridization between the top and bottom adapter. Tubes with double-stranded adapter can be stored at −20 °C.

3. SPRI magnetic beads (NGS).

4. Magnetic stand.

5. 80% ethanol.

6. MEGAscript T7 Transcription Kit (Thermo Fisher) or equivalent.

7. Fragmentation buffer: 200 mM Tris-acetate pH 8.1, 500 mM potassium acetate, 150 mM magnesium acetate.

8. Stop buffer: 0.5 M EDTA pH 8.

9. Bioanalyzer with Pico RNA 6000 chips and High Sensitivity DNA chips (Agilent).

2.3 scDam&T-seq

1. UV-PCR workstation.

2. Liquid-handling robot.

3. Mineral oil.

4. 1500 nM CEL-Seq2 primers. Primers have the following design, from the 5′ to 3′ direction: T7 promoter, P5 Illumina adapter, 3-nt UMI, 4-nt barcode, 3-nt UMI, 4-nt barcode and a poly-dT tail. Thus, the total barcode sequence is 8-nt and the total UMI sequence is 6-nt. The 3′ end contains a 'V' base, which is G, C or A. This degenerate base prevents slipping of the polymerase over the poly-A sequence and aims at annealing the primer directly upstream of the poly-A tail in the mRNA. Sequences for a set of 384 unique and DamID adapter compatible CEL-Seq2 primers are available in Markodimitraki et al. [19].

   Example primer is shown from 5′ to 3′ with UMIs indicated as 'NNN' and the split 8-nt barcodes as 'BARC', 'ODES'.

   Example CELSeq2-primer
   CCGGTAATACGACTCACTATAGGGAGTTCTA CAGTCCGACGATCNNNBARCNNNODE STTTTTTTTTTTTTTTTTTTTTTTTTTV

   Primers are ordered as standard desalted oligonucleotides, diluted in nuclease-free water to 500 μM and stored at −80 °C. To make a 'source' plate with CEL-Seq2 primers, primers are diluted to 1500 nM in nuclease-free water in a 384-well PCR plate. Each position in the plate should contain a unique CEL-Seq2 primer with a known barcode. Source plates are used to prepare 'primer' plates and can be stored at −20 °C.

6. Aluminium seals.

7. Hemocytometer or automated cell counter.

8. 1 mg/mL Hoechst 34580 (Sigma), filter-sterile in water.

9. Cell strainer cap.

10. 1 mg/mL propidium iodide (PI) (Invitrogen), filter-sterile in water.

11. FACS sorter.

12. ERCC RNA spike-in mix (Ambion). Dilute 1:1000 in nuclease-free water and store as aliquots at −20 °C, avoiding freeze-thaw cycles. Just before use, thaw an aliquot and dilute 50-fold in nuclease-free water to a 1:50,000 stock solution.

13. 10 mM dNTPs (equal mix of dATP, dGTP, dTTP, dCTP).

14. Lysis mix: 0.15% (v/v) Igepal CA-630, 1:250,000 ERCC3 RNA spike-in, 2 mM dNTPs.

15. Reverse transcriptase mix: 4.67 U/μL RNaseOUT (Invitrogen), 23.33 U/μL SuperScript II reverse transcriptase (Thermo Fisher) in 2.33× manufacturer-supplied first strand buffer and 23.33 mM DTT.

16. Second strand mix: 270 μM dNTPs, 0.08 U/μL *E. coli* DNA ligase (NEB), 0.27 U/μL DNA polymerase I (Invitrogen), 0.016 U/μL RNase H (Invitrogen) in 1.19× second strand buffer (Thermo Fisher).

17. Proteinase K mix: 5.2 mg/mL proteinase K (Roche) in 5.4× CutSmart buffer (NEB).

18. DpnI digestion mix: 2 U/μL DpnI in 1× CutSmart buffer (NEB).

19. 1 μM DamID double-stranded adapters in 384-well PCR plate. *See* Subheading 2.2 for design. Top and bottom adapters are ordered as standard desalted oligonucleotides and dissolved at 500 μM in nuclease-free water. To first make a 'mother' plate with DamID adapters, top and bottom adapters are mixed at 100 μM each in annealing buffer in a 384-well PCR plate. The plate is heated to 94 °C for 2 min in a thermal cycler and then slowly cooled to room temperature, to allow hybridization between the top and bottom adapter. Each position in the plate should contain a unique DamID adapter with a known barcode. The mother plate can be used to make working-stock plates with adapters at 0.5–2 μM concentration in nuclease-free water, plates can be stored at −20 °C.

20. scDam&T ligase mix: 0.56 U/μL T4 DNA ligase in 4.44× T4 DNA ligase buffer (Roche).

21. Bead binding buffer: 10 mM Tris-HCl pH 8, 2.5 M NaCl, 1 mM EDTA, 20% (v/v) PEG 8000, 0.05% (v/v) Tween 20.

2.4 Sequencing Library Preparation

1. 20 μM Random hexRT primer. The random hexRT primer is ordered with hand-mixed 'N' bases and has the following sequence: GCCTTGGCACCCGAGAATTCCANNNNNN.

2. NEBNext High Fidelity Master Mix (NEB) or equivalent high-fidelity PCR enzyme system.

3. 10 μM RNA PCR Index TruSeq Small RNA primers. RNA PCR Index (RPI) primers are used according to Illumina guidelines, at 10 μM concentration. See the Illumina website for more information [26]. A different indexed RPi primer is used for amplification of each individual library, together with the universal RP1 primer. Example primers are shown from 5′ to 3′, in which * indicates a phosphorothioate bond that protects the DNA from exonuclease activity:

Example RPi
CAAGCAGAAGACGGCATACGAGATCGTGATGTG
ACTGGAGTTCCTTGGCACCCGAGAATTCC*A
RP1
AATGATACGGCGACCACCGAGATCTACACG
TTCAGAGTTCTACAGTCCG*A

4. Qubit fluorimeter and dsDNA HS assay reagents (Invitrogen).

5. Illumina Sequencer or DNA sequencing facility.

3   Methods

3.1 Creating a Clonal mESC Line Expressing Tir1 and the mAID-Dam-Lamin B1 Fusion Protein

In this section, we describe how to generate a clonal mESC line with a CRISPR knock-in of the mAID-Dam construct in the Lamin B1 locus. To control the expression of the Dam-Lamin B1 fusion protein, we make use of the Auxin-Inducible Degron (AID) system [27], which requires both the expression of the Tir1 protein and a mAID module fused to Dam. We therefore first present how to integrate Tir1 into the TIGRE locus. This procedure can be adapted to other cell lines, induction systems or Dam-POI fusions of choice. *See* **Notes 1–4**.

1. Take mES cells in culture on gelatin-coated 6-well plates—to coat with gelatin, plates should be incubated for at least 30 min with a 0.1% gelatin solution at room temperature. Remove the gelatin just before seeding cells. The seeding density of cells depends on the cell line; in our hands this should be ~1 × 10^5 cells/cm^2, with 3 mL mESC culture medium per well (10 cm^2 per well, 1 × 10^6 cells).

2. Passage the cells after 2 days by washing once with 3 mL 1× PBS and trypsinizing with 0.5 mL TrypLE Express Enzyme. Incubate for 4 min at 37 °C, resuspend and inactivate with 2.5 mL mESC culture medium. Seed cells to at least 2 wells of a gelatin-coated 6-well plate at ~0.8 × 10^5 cells/cm^2.

3. The next day, perform transfections with Lipofectamine 3000 following the manufacturer's instructions using Opti-MEM medium. Transfect the plasmids for Tir1 CRISPR knock-in in the TIGRE locus in a 1:1 respective ratio in one well, using 3 μg plasmid DNA in total per transfection (1.5 μg of each plasmid). The DNA–Lipofectamine ratio should be 1:3, meaning that for 3 μg plasmid, 9 μL of Lipofectamine is used. One of the wells should be the untransfected control. Refresh cells the next day. *See* **Note 5**.

4. 48 h after transfection, passage the cells and seed each well (10 cm^2) over three 100 mm gelatin-coated dishes with 8 mL mESC culture medium per dish (60 cm^2 each, 180 cm^2 total). Start selection by adding 0.64 μL 10 mg/mL puromycin to the

8 mL volume (0.8 μg/mL final) concentration. Grow cells for 5–7 days until all cells on the untransfected control are dead and colonies have formed on the transfected plates. Refresh medium every 2 days. *See* **Note 5**.

5. Pick colonies by hand with a p20 pipette and transfer each one to a well of a 96-well round-bottom plate containing 5 μL 1× PBS. Add 25 μL TrypLE with a p200 multi-channel pipette and incubate for 5–10 min at 37 °C. Add 75 μL mESC culture medium, resuspend and transfer to a gelatin-coated flat-bottom 96-well plate, already containing 100 μL mESC culture medium per well. Puromycin selection can be removed at this point. Grow cells until confluent, refreshing medium every 2 days. *See* **Note 6**.

6. After cells are confluent, passage 1 in 2 to two gelatin-coated flat-bottom 96-well plates. Grow cells until confluent, refresh every 2 days.

7. Freeze one of the two plates in mESC culture medium supplemented with 10% DMSO. Freezing is done by washing once with 150 μL PBS and adding 25 μL TrypLE with a p200 multi-channel pipette. Incubate for 5 min at 37 °C, add 75 μL mESC culture medium and resuspend. Add 100 μL ice-cold mESC culture medium supplemented with 10% DMSO and quickly resuspend. Wrap plates in towels to freeze slowly and store at −80 °C.

8. The other 96-well plate is used to screen the clones for genomic integration of Tir1 in the TIGRE locus, by performing a PCR over the integration site. The forward primer anneals upstream of the left homology arm and the reverse on the CAGG promoter sequence. Wash once with 150 μL PBS, remove all supernatant and add 25 μL PCR-lysis mix. Incubate at 65 °C for 4 h, and transfer 5 μL to a 96-well PCR-plate containing 45 μL nuclease-free water.

9. Heat-inactivate the proteinase K by incubating the plate at 80 °C for 20 min. Transfer 2 μL of the material to a new 96-well PCR-plate already containing 8 μL Tir1-PCR mix.

10. Run the following program in a thermal cycler.

94 °C, 2 min.

30× [94 °C, 30 s; 58 °C, 30 s; 72 °C, 90 s].

72 °C, 5 min.

12 °C, hold.

11. Load 8 μL of sample on a 1% agarose gel. The expected size of the band in positive clones is ~1200 bp; no band should be present in negative clones. Select four positive clones for subsequent integration of the mAID-Dam construct in the Lamin B1 locus. *See* **Note 7**.

12. Take the selected mESC clones in culture from the frozen 96-well plate. Seed each clone in one well of a gelatin-coated 24-well plate. Grow cells until confluent before passaging. Seed cells to at least 3 wells of a gelatin-coated 6-well plate at ~0.8 × 10⁵ cells/cm²—two will be used for transfections and one can be used for further expansion and freezing.

13. The next day, perform transfections with Lipofectamine 3000 following the manufacturer's instructions using Opti-MEM medium. Transfect the plasmids for Dam CRISPR knock-in in the Lamin B1 locus in a 1:1 respective ratio in one well, using 3 μg plasmid DNA in total per transfection (1.5 μg of each plasmid). The DNA–Lipofectamine ratio should be 1:3, meaning that for 3 μg plasmid, 9 μL of Lipofectamine is used. One of the wells should be the untransfected control. Refresh cells the next day. *See* **Notes 1** and **5**.

14. 48 h after transfection, passage the cells and seed each well (10 cm²) over three 100 mm gelatin-coated dishes with 8 mL mESC culture medium per dish (60 cm² each, 180 cm² total). Start selection by adding 2.4 μL 10 mg/mL blasticidin to the 8 mL volume (3 μg/mL final concentration). From this point onward, always supplement the mESC medium with 0.5 mM IAA. Grow cells for 5–7 days until all cells on the untransfected control are dead and colonies have formed on the transfected plates. Refresh medium every 2 days. *See* **Note 5**.

15. Pick colonies as described in **step 5** and supplement medium with IAA. Blasticidin selection can be removed at this point. Grow cells until confluent, refresh every 2 days. *See* **Note 6**.

16. After cells are confluent, passage 1 in2 to two gelatin-coated flat-bottom 96-well plates. Grow cells until confluent, refresh every 2 days.

17. Freeze one of the two plates in mESC culture medium supplemented with 10% DMSO final, as described in **step 7**.

18. The other 96-well plate is used to screen the clones for genomic integration of mAID-Dam in the LaminB1 locus, by performing a PCR over the integration site. The forward primer anneals upstream of the left homology arm and the reverse on the mAID sequence. Wash once with 150 μL PBS, remove all supernatant and add 25 μL PCR-lysis mix. Incubate at 65 °C for 4 h, and transfer 5 μL to a 96-well PCR-plate containing 45 μL nuclease-free water.

19. Heat-inactivate the proteinase K by incubating the plate at 80 °C for 20 min. Transfer 2 μL of the material to a new 96-well PCR-plate already containing 8 μL Dam-PCR mix.

20. Run the following program in a thermal cycler.

94 °C, 2 min.

30× [94 °C, 30 s; 55 °C, 30 s; 72 °C, 2 min].

72 °C, 5 min.

12 °C, hold.

21. Load 8 μL of sample on a 1% agarose gel. The expected size of the band in positive clones is ~1700 bp; no band should be present in negative clones. Take 12 positive clones for subsequent analysis of methylation levels and induction efficiency using the ^{m6}A-PCR. *See* **Note 7**.

22. Take the selected mESC clones in culture from the frozen 96-well plate. Seed each clone in one well of a gelatin-coated 24-well plate. Grow cells until confluent before passaging each clone to at least 3 wells of a gelatin-coated 24-well plate. Use one well to further expand and freeze the clones and the other two for ^{m6}A-PCR.

23. The next day, induce the Dam-Lamin B1 construct for 18 h by IAA washout. This is done by washing one well three times with PBS before refreshing with mESC culture medium without IAA. The other well can be used as the uninduced control sample, by refreshing with mESC culture medium supplemented with 1 mM IAA. *See* **Note 8**.

24. Harvest the cells (~1 × 10^6 cells per well) by dissociation with TrypLE and incubating for 4 min at 37 °C. Resuspend the cells and inactivate the TrypLE by adding mESC culture medium. Transfer to an Eppendorf tube and centrifuge for 4 min at 300 × g, room temperature. Remove the supernatant and wash once with PBS. After centrifugation, remove supernatant and isolate genomic DNA using a Genomic DNA isolation kit following the manufacturer's instructions. *See* **Note 9**.

25. Measure the concentration of the purified DNA with a Nanodrop spectrophotometer.

26. For each clone, digest 200 ng of genomic DNA in a total volume of 10 μL in 1× CutSmart buffer with 2 U DpnI for 6 h at 37 °C, before heat-inactivating the enzyme at 80 °C for 20 min. It is recommended to take along a no-digest control for one or two samples. *See* **Note 10**.

27. Add 2.5 μL ^{m6}A-PCR ligase mix to each sample. Incubate overnight at 16 °C, before heat-inactivating the enzyme at 65 °C for 10 min.

28. Transfer 2 μL of each ligated sample to a PCR tube and add 18 μL ^{m6}A-PCR mix. Run the following program in a thermal cycler (*see* **Note 11**).

72 °C, 8 min.

16× [94 °C, 20 s; 58 °C, 30 s; 72 °C, 20 s].

72 °C, 2 min.

12 °C, hold.

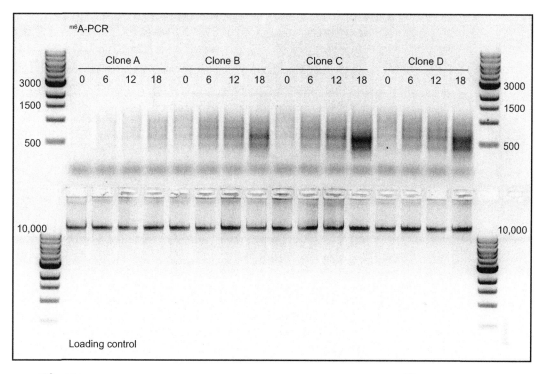

Fig. 2 ^{m6}A-PCR product loaded on a 1% agarose gel. Clones A to D with different ^{m6}-A methylation levels are shown, ranging from low to high levels of methylation. The Dam-POI fusion was induced for 0, 6, 12, or 18 h. 18 PCR cycles were performed, 200 ng genomic DNA is loaded for each sample to control for loading differences

29. Load 8 μL of sample on a 1% agarose gel and on the same gel, load 200 ng of purified genomic DNA for each sample as a loading control. If Dam methylation occurred, a product smear of around 300–1000 bp should be present (*see* Fig. 2 for a representative result). If no product is detected, repeat the PCR reaction with more cycles using the remainder of the PCR sample. Select clones based on the induction potential and product intensity. It is recommended to proceed with 4–8 clones to bulk DamID2 for evaluating specificity of the Dam-Lamin B1 fusion protein, optimizing induction times and optimal clone selection, before performing scDam&T-seq. *See* **Note 12**.

3.2 Bulk DamID2

1. Take the selected mESC clones in culture. Grow cells until confluent before passaging each clone to at least 4 wells of a gelatin-coated 24-well plate.

2. The day after seeding the cells, induce the Dam-Lamin B1 construct for 6, 12 or 18 h by IAA washout (*see* Subheading 3.1, **step 23**). One well is used as uninduced control sample, by refreshing with mESC culture medium supplemented with

1 mM IAA. Shift the induction times to be able to harvest all wells at the same moment. *See* **Notes 8** and **9**.

3. Harvest the cells (~1×10^6 cells per well) by dissociation with TrypLE and incubating for 4 min at 37 °C. Resuspend the cells and inactivate the TrypLE by adding mESC culture medium. Transfer to an Eppendorf tube and centrifuge for 4 min at $300 \times g$, room temperature. Remove the supernatant and wash once with PBS. After centrifugation, remove supernatant and isolate genomic DNA using a Genomic DNA isolation kit following the manufacturer's instructions.

4. Measure the concentration of the purified DNA with a Nanodrop spectrophotometer.

5. For each sample, digest 250 ng of genomic DNA in a total volume of 10 μL in 1× CutSmart buffer with 5 U of DpnI for 8 h at 37 °C, before heat-inactivating the enzyme at 80 °C for 20 min.

6. Add 2 μL bulk DamID2 ligase mix to each sample, then 0.5 μL of 50 μM differently barcoded DamID adapter (unique adapter for each sample). Incubate at 16 °C for 16 h, before heat-inactivating the enzyme at 65 °C for 10 min. *See* **Note 13**.

7. Pool 5 μL of each sample (with unique barcodes) together in one tube. If exceeding 16 samples, make multiple separate pools.

8. Add an equal volume of magnetic SPRI beads and purify the DNA, following manufacturer's instructions, with two washing steps with 80% ethanol and elution in 50 μL nuclease-free water.

9. Repeat **step 8** with 40 μL (0.8 volumes) SPRI beads and elution in 25 μL nuclease-free water.

10. Take 4 μL of purified sample and add 4 μL of nuclease-free water and 12 μL of assembled mastermix of the MEGAscript T7 in vitro transcription kit (2 μL of each component), to a total volume of 20 μL. Incubate samples at 37 °C for 2 h. Keep samples on ice as much as possible to prevent degradation of the RNA. *See* **Note 14**.

11. Dilute the amplified RNA (aRNA) by adding 40 μL of nuclease-free water and purify the sample using 48 μL (0.8 volumes) SPRI beads, with three washing steps in 80% ethanol and elution in 50 μL nuclease-free water.

12. Measure the RNA concentration on a NanoDrop spectrophotometer.

13. Take 1 μg RNA in 20 μL nuclease-free water and add 4 μL fragmentation buffer. Incubate samples at 94 °C for 2 min and directly place back on ice before adding 2.4 μL stop buffer. *See* **Note 15**.

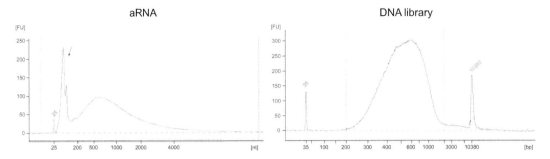

Fig. 3 Representative Agilent 2100 Bioanalyzer plots of amplified RNA (aRNA) and a DNA library. The marker peaks are indicated in green (aRNA, 25 nt) or green and purple (DNA, 35 and 10,380 bp). For the aRNA, the adapter peak is indicated with an arrow

14. Purify the sample using 21.1 μL (0.8 volumes) SPRI beads, with three washing steps and elution in 15 μL nuclease-free water.

15. Measure the RNA concentration on a NanoDrop spectrophotometer and dilute samples to 2–50 ng/μL before running 1 μL on a Pico RNA 6000 Chip (following manufacturer's instructions) on the Bioanalyzer.

16. If the RNA is of proper yield and quality (*see* Fig. 3 for a representative example), proceed with DNA library preparation and next-generation sequencing (*see* Subheading 3.5). *See* **Note 16**.

17. Process the raw data and analyze the DamID signal (*see* Subheading 3.6). Based on the specificity of the signal to the expected target sites, the induction potential and the signal-to-noise ratio, select 1–2 clones for performing scDam&T-seq. *See* **Note 12**.

3.3 Preparing Single Cell Samples for scDam&T-seq

For scDam&T-seq processing, cells need to be FACS sorted into 384-well CEL-Seq2 primer plates, which can be prepared in advance and stored until use (*see* Subheading 2.3). It is important to prevent contamination of samples as much as possible, so we advise to work in a decontaminated working environment such as a UV PCR workstation. *See* **Notes 14** and **17–19**.

1. Prepare 384-well plates with 5 μL of mineral oil per well. Thaw the source CEL-Seq2 primer plate (1500 nM, *see* Subheading 2.3) and subsequently centrifuge the plate for 2 min at 2000 × *g* to collect all liquid at the bottom of the wells. Transfer 100 nL of each well of the source plate to the same position in a primer plate using a liquid-handling robot. Avoid contamination across wells and keep plates on ice as much as possible. Cover plates with aluminum seals and centrifuge before storing at −20 °C. *See* **Note 13**.

2. Take the selected mESC clone in culture. Grow cells until confluent before passing to at least 1 well of a gelatin-coated 6-well plate. Refresh medium the next day. *See* **Note 9**.

3. Two days after seeding the cells, induce the Dam-Lamin B1 construct for 6 h by IAA washout (*see* Subheading 3.1, **step 23**; 6 h is the optimal induction time in our hands, but should be determined as described in Subheading 3.2 via ^{m6}A-PCR and bulk DamID2 for new clones). *See* **Note 8**.

4. Harvest cells after the induction is complete (~5 × 10^6 cells) by dissociation with TrypLE and incubating for 4 min at 37 °C. Resuspend the cells and inactivate the TrypLE by adding mESC culture medium. Transfer to a 15 mL Falcon tube and centrifuge for 4 min at 300 × g, room temperature. Remove the supernatant and wash once with PBS. After recentrifugation, remove supernatant and resuspend in 2 mL mESC culture medium.

5. Count cells on a hemocytometer or automated cell counter and calculate the total number of cells. Centrifuge for 4 min at 300 × g, room temperature, and resuspend in mESC culture medium to a concentration of 1 × 10^6 cells per mL.

6. Add 10 μL 10 mg/mL Hoechst per mL of cell suspension (final concentration 10 μg/mL). Incubate for 45 min at 37 °C.

7. Pass the cells through a cell-strainer cap and collect in a Falcon round-bottom tube (polypropylene or polystyrene, depending on the FACS sorter system). From this point onward, keep cells on ice as much as possible.

8. Thaw the 384-well CEL-Seq2 primer plates that were prepared in **step 1**. Keep plates on ice as much as possible.

9. Just prior to sorting, add 1 μL 1 mg/mL PI per mL of cell suspension (final concentration 1 μg/mL). Briefly resuspend by tapping the tube.

10. Sort single cells into the wells of the primer plates. Set gates to discriminate between live and dead cells, based on absence of PI signal. It is recommended to use the Hoechst levels to either set a gate to sort live cells of a preferred cell-cycle phase (such as G2/M), or to unbiasedly sort all live cells and save the Hoechst information for data analysis. Hoechst levels for cells in G2/M phase will be twice as high as for cells in G1/S phase. It is recommended to sort multiple replicate plates per condition and to include several empty wells and 10-cell samples in each plate as controls. Keep samples and plates on ice as much as possible in between sorting. *See* **Notes 18** and **20**.

11. Seal plates after sorting, centrifuge for 2 min at 2000 × g, 4 °C to collect cells at the bottom of the wells. *See* **Note 18**.

3.4 scDam&T-seq

To prevent degradation of RNA, it is very important to keep all reagents and plates on ice as much as possible, unless otherwise specified. Prevent contamination of samples by using sterilized and clean materials, by keeping plates sealed as much as possible and by working in a decontaminated area such as a UV PCR workstation. Material loss is prevented by limited sample handling, so reagents are added on top of the previous reaction. When calculating reaction mix volumes, make sure to account for the dead volume of the liquid-handling robot, pipetting loss, and the number of samples. During incubations in the thermal cycler, the heat block should already be at the starting temperature when putting the plate in and the lid temperature should be set at 100 °C, unless otherwise specified. Plates should be centrifuged for 2 min at 2000 × g, 4 °C before and after adding reagents and after incubations in the thermal cycler, to collect all liquid at the bottom of the wells. *See* **Notes 14** and **17–19**.

1. Add 100 nL lysis mix to each well with the liquid-handling robot. Incubate the plate in a thermal cycler at 65 °C for 5 min and then place on ice immediately.

2. Add 150 nL reverse transcriptase mix to each well with the liquid-handling robot. Incubate the plate in a thermal cycler at 42 °C for 2 h before holding at 4 °C. Place the plate on ice and increase the heat block temperature to 70 °C, before putting the plate back in to incubate 10 min at 70 °C min before holding at 4 °C or placing on ice.

3. Add 1850 nL second strand mix to each well with the liquid-handling robot. Incubate the plate in a thermal cycler at 16 °C for 2 h, before holding at 4 °C or placing on ice.

4. Add 500 nL proteinase K mix to each well with the liquid-handling robot. Incubate the plate in a thermal cycler at 50 °C for 10 h, then 80 °C for 20 min before holding at 4 °C or placing on ice.

5. Add 300 nL DpnI digestion mix to each well with the liquid-handling robot. Incubate the plate in a thermal cycler at 37 °C for 6 h, then 80 °C for 20 min, before holding at 4 °C or placing on ice.

6. Thaw the working-stock DamID adapter plate (1 µM, *see* Subheading 2.2) and subsequently centrifuge the plate for 2 min at 2000 × g, room temperature, to collect all liquid at the bottom of the wells. Transfer 100 nL of each well of the source plate to the same position in the sample plate using a liquid-handling robot. Avoid contamination across wells and keep plates on ice as much as possible. *See* **Notes 13** and **21**.

7. Add 900 nL scDam&T ligase mix to each well with the liquid-handling robot. Incubate the plate in a thermal cycler at 16 °C for 16 h, then 65 °C for 10 min before holding at 4 °C or placing on ice. The total accumulated volume per well is 4 μL.

8. Pool all samples with nonoverlapping barcodes into one 15 mL Falcon or 5 mL Eppendorf tube. Separate the oil from the aqueous phase by centrifuging the sample for 2 min at 2000 × *g*, room temperature, taking the bottom phase and transferring this to two clean 1.5 mL Eppendorf tubes. Repeat this step two more times to remove all the oil, since remaining oil will disturb further purification steps. *See* **Note 22**.

9. Dilute SPRI beads 1:6 to 1:10 (depending on final pooled volume) in bead binding buffer, and purify the sample with 0.8 volumes of diluted SPRI beads, with two washing steps with 80% ethanol. Elute in 8 μL nuclease-free water, but do not separate beads from eluate; instead keep the beads and resuspend them in the eluate, before transferring to a PCR tube.

10. Add 12 μL of assembled mastermix of the MEGAscript T7 in vitro transcription kit (2 μL of each component). Incubate samples at 37 °C for 14 h in a thermal cycler with the lid heated to 70 °C, before holding at 4 °C or placing on ice.

11. Separate the sample from the beads and purify the aRNA using 16 μL (0.8 volumes) SPRI beads, with three washing steps in 80% ethanol and elution in 22 μL nuclease-free water.

12. Add 4.4 μL fragmentation buffer. Incubate samples at 94 °C for 2 min and directly place back on ice before adding 2.65 μL stop buffer. *See* **Note 15**.

13. Purify the sample using 23.2 μL (0.8 volumes) SPRI beads, with three washing steps in 80% ethanol and elution in 12 μL nuclease-free water.

14. Measure the aRNA concentration on a NanoDrop spectrophotometer and dilute samples to 5–50 ng/μL before running 1 μL on a Pico RNA 6000 Chip (following manufacturer's instructions) on the Bioanalyzer.

15. If the aRNA is of proper yield and quality (*see* Fig. 3 for a representative example), proceed with DNA library preparation and next-generation sequencing. *See* **Note 16**.

3.5 Preparation of Illumina Sequencing Libraries

Library preparation of the aRNA is required before next-generation sequencing. This involves reverse transcription of the aRNA and subsequent amplification of the cDNA using PCR. During incubations in the thermal cycler, the heat block should already be at the starting temperature when putting the sample in and the lid temperature should be set at 100 °C, unless otherwise specified.

1. Take 5 μL of purified aRNA (at 5–50 ng/μL concentration) in a PCR tube and add 1 μL 20 μM random hexRT primer and 0.5 μL 10 mM dNTP mix. Incubate in a thermal cycler at 65 °C for 5 min then place directly on ice.

2. Add 2 μL 5× first-strand buffer, 1 μL 0.1 M DTT, 0.5 μL 40 U/μL RNase OUT, and 0.5 μL 200 U/μL SuperScript II reverse transcriptase. Incubate in a thermal cycler at 25 °C for 10 min, then 42 °C for 1 h (with the cycler lid at 50 °C), before placing directly on ice.

3. Add 10.5 μL nuclease-free water, 25 μL 2× NEBNext High Fidelity PCR Master Mix, and 2 μL 10 μM RP1 primer.

4. Add 2 μL 10 μM different indexed RPi primer to each sample, then run the following program on a thermal cycler.

   98 °C, 30 s.

   8–12× [98 °C, 10 s; 60 °C, 30 s; 72 °C, 30 s]; *see* **Note 23**.

   72 °C, 10 min.

   12 °C, hold.

5. Purify the sample using 40 μL (0.8 volumes) SPRI beads, with two washing steps with 80% ethanol and elution in 25 μL nuclease-free water.

6. Repeat **step 5**, but with 20 μL (0.8 volumes) SPRI beads and elution in 15 μL nuclease-free water.

7. Measure the DNA concentration using a Qubit dsDNA HS Assay (following manufacturer's instructions). A concentration of 2–5 ng/μL is optimal.

8. Dilute DNA to 1–2 ng/μL before running 1 μL on a High-Sensitivity DNA Chip (following manufacturer's instructions) on the Bioanalyzer to evaluate the fragment size distribution. *See* Fig. 3 and **Note 23**.

9. Pool libraries with different RPi indexes together and dilute to the appropriate concentration before sequencing on an Illumina sequencer. For scDam&T-seq, paired-end sequencing with 2× 75 nt output for both the forward read (R1) and reverse read (R2) is required—longer read lengths, such as 100 nt, may increase alignment efficiency. Furthermore, a sequencing depth of approximately 1.5×10^5 raw reads per high-quality single-cell sample is desirable. For bulk DamID2, sequencing single-end with 1× 75 nt output is sufficient and a sequencing depth of approximately 2×10^6 raw reads per bulk sample is adequate.

3.6 Raw Data Processing and Visualization

In brief, raw reads are first demultiplexed based on their library-specific barcode (RPi index). Then, reads are demultiplexed based on their sample-specific DamID and CEL-Seq2 barcodes, based on the forward read (i.e., R1). The reads are trimmed to remove adapter sequences and subsequently aligned to the reference genome. For the DamID reads (the trimmed R1), UMI-unique reads per GATC are counted and binned in genomic segments. For the CEL-Seq2 reads (the reverse reads, i.e., R2), arrays of UMI-unique counts per gene are made. These data can then be used for further data processing and visualization. An extensive rationalization of the data processing steps and all required codes and scripts are available in Markodimitraki et al. and at GitHub (https://github.com/KindLab/scDamAndTools) [18, 19, 28].

4   Notes

1. Generating the pUC57-Hom-BSD-P2A-mAID-Dam-Hom donor vector (Homology Chr18 59113-61113:, GRCm38/ mm10) can be achieved via standard cloning methods (e.g., PCR amplification, restriction–ligation, Gibson Assembly). The Dam sequence can be obtained from the pIND(V5)Eco-Dam vector (Addgene 59201 [29]), the BSD-P2A-mAID sequence from the pMK347 (BSD-P2A-mAID) vector (Addgene 121181, [30]), and the homology arms can be generated via PCR on isolated mouse genomic DNA. The sgRNA targeting the 5′ end of the Lamin B1 locus (5′ GCACGGGGGTCGCGGTCGCCA 3′) can be cloned into the pSpCas9(BB)-2A-GFP (PX458) vector (Addgene 48138 [25]) by following the protocol of Ann Ran et al. *Nature Protocols* [25].

2. It is highly recommended to ensure stable expression of the Dam-POI fusion at a suitable level throughout the system. In many research cases, this means that a clonal cell line has to be generated, although alternative strategies are possible in some cases [31]. The design of the construct and the delivery method strongly depend on the POI and the endogenous expression level. We have used both N-terminal and C-terminal tagging with Dam, but N-terminal tagging is most suitable for Lamin B1, Lamin A/C, and the Lamin B Receptor (LBR). Depending on the endogenous expression level of the POI, one can choose to specifically knock-in Dam into the endogenous locus or randomly integrate the entire Dam-POI fusion into the genome. Making a knock-in cell line will aid in expressing the Dam-POI at physiological levels, but creates less flexibility in tuning the expression. Especially if the POI is endogenously expressed at a very low level, random integration of (multiple copies of) the Dam-POI could be a more successful strategy to generate enough DamID signal. However, one should take care that the effective overexpression of the protein does not affect the cell line in a way that could influence the experiment.

3. During in vivo expression of the Dam-fusion, the Dam enzyme will methylate all DNA it comes into contact with, meaning that the signal will accumulate over time. It therefore represents a track-record of all cumulative protein–DNA interactions during the expression period, and it is thus important to be able to regulate this time frame. This can be done by using any induction system of choice, but a low level of leakiness and a short response time are desired. In this protocol we use the auxin-inducible degron (AID) system (Dam-POI is degraded upon addition of auxin) [27], but we have also successfully applied the ProteoTuner system (Dam-POI is stabilized upon addition of Shield1) [32, 33].

4. The Dam enzyme has a preference for methylating accessible DNA [10, 18]. It is therefore important to take along cells expressing an untethered Dam enzyme as control sample, which will represent the unspecific signal and can be used for normalization purposes during data processing. This control can either be used as side-by-side comparison to the sample, or can be used for direct normalization. The intrinsic preference of Dam for methylating open chromatin also has consequences for the suitability of the DamID method in measuring protein–DNA interactions for different proteins of interest. Dam-POI fusions targeting inaccessible chromatin will have a higher signal-to-noise ratio than fusions that target open chromatin, which will intrinsically be more similar to background signal. This makes the method very applicable for studying genome-lamina interactions, which typically represent heterochromatic domains, compared to for example transcription factors that generally bind euchromatic regions.

5. Delivery of the Dam construct into the cell is dependent on the cell line of choice. It can be achieved via transfection, or in case of random integration, also via lentiviral transduction. To select for positive clones, it is highly recommended to make use of a selectable marker, such as an antibiotic resistance cassette or a fluorescent tag on the Dam construct by which positive cells can be sorted by FACS.

6. The number of clones that should be grown for Tir1-PCR and Dam-PCR screening depends on the integration strategy and cell type. In case of random integration of the Dam-POI construct, we recommend to start with at least 96 clones for Dam-PCR screening. In case of a knock-in strategy, 48 clones should be sufficient. If the cell-type used has a low transfection or transduction efficiency, more clones should be grown.

7. It is possible that the PCR on the directly lysed cell material performs suboptimally. In that case, it is recommended to isolate genomic DNA from individual clones first and perform the PCR on 50 ng of genomic DNA material in a total reaction volume of 50 μL.

8. For initial tests in ^{m6}A-PCRs, we recommend to take long induction times such as 18–24 h, to make sure signal can be properly detected. Subsequently, bulk DamID2 can be performed to determine optimal induction times to gain the highest signal-to-noise levels. We usually perform time series of 6, 12, and 18 h. For both ^{m6}A-PCRs and bulk DamID2, it is recommended to take along an uninduced control sample (0 h). One should consider that scDam&T-seq experiments generally require a higher methylation level than bulk DamID2, to obtain enough signal for single-cell measurements. In our experience, induction for 6–12 h is suitable for Dam-Lamin B1 fusion constructs. Increasing the IAA concentration from 0.5 mM (used during regular culturing) to 1 mM during the 48 h before induction reduces the background methylation levels.

9. Cells should be ~90% confluent at the end of induction for an optimal result—adjust the seeding density before inductions accordingly, depending on the growth rate of the cell line. Passaging rates depend on the cell line, in our hands a passage of 1 in 6 allows the mES cells to reach confluency in 48 h.

10. Take along a no-digest control without DpnI for the ^{m6}A-PCR. This control should not show any product smear. In case product is present, this is most likely caused by fragmentation of the genomic DNA, which is then no longer suitable for bulk DamID2 experiments. It is important not to vortex the genomic DNA sample to prevent fragmentation, resuspend by pipetting instead.

11. In the ^{m6}A-PCR, the number of cycles that is required depends on the level of methylation present in the cells. Start with a low number of cycles, such as 14–16 rounds of amplification, and check part of the PCR reaction on gel. In case no product smears are detected, one can take the remaining of the PCR reaction and add a chosen number of cycles. Usually, between 14 and 22 cycles should be enough to detect product smears. If more cycles are needed, the methylation levels are probably too low for successful scDam&T-seq experiments.

12. Performing bulk DamID2 is optional, but highly advised before performing the scDam&T-seq experiment. It is a fast and low-cost approach to confirm that the Dam-POI fusion is marking the appropriate genomic regions. Furthermore, a first selection can be made from the tested clones and induction times. However, it is difficult to identify the best clones and induction times on bulk DamID2 alone, since the cleanest bulk data is often obtained at expression levels that are too low for single-cell experiments. It is therefore recommended to select multiple promising clones for further single-cell testing.

13. The DamID adapters and CEL-Seq2 primers contain unique barcode sequences. These are required to link the DamID signal in silico to the corresponding transcriptomic signal for each individual cell and allow pooling of multiple samples in one library. Pooling as many samples as possible is highly recommended to limit batch effects and ensure high in vitro transcription efficiency. Using 96–384 unique barcodes for both the DamID adapters and CEL-Seq2 primers is therefore advised for scDam&T-seq. For bulk DamID2, individual samples are also ligated with unique DamID adapters and pooled before in vitro transcription. Sequences for a set of 384 unique DamID adapters and compatible CEL-Seq2 primers are available in Markodimitraki et al. [19].

14. To prevent degradation of mRNA and contaminations of the samples, it is very important to work in a clean and decontaminated environment. Use a designated set of pipettes for single-cell work. Make sure all materials such as Eppendorf tubes, pipette-tips and PCR tubes/strips are DNase- and RNase-free, either as guaranteed by the supplier or by autoclaving. UV light or DNAZap and RNase ZAP cleaning reagents (Invitrogen) can be used to decontaminate the working area and materials such as pipettes, tube racks etc. Always use sterile and nuclease-free water.

15. RNA fragmentation is an important step to create fragments of an optimal length. The correct size distribution is about 200–2000 nt, peaking at 600 nt. In case fragment sizes are too short, the fragmentation time at 94 °C can be slightly reduced. In case the fragment size is too large and the yield is high, the amount of input RNA into the fragmentation can be decreased to increase efficiency.

16. Purified aRNA usually consists of the desired product (ranging from 200 to 2000 nt) as well as "adapter" RNA that was amplified from free DamID adapters that were still present in the sample during in vitro transcription. These fragments have a size below 200 nt and appear as a peak on the Pico RNA 6000 Chip (*see* Fig. 3). If the ratio between the maximum value of product–adapter is smaller than 1:4 (in Fig. 3, the ratio is 1: 2.5), we recommend repeating the cleanup step with 0.8 volumes of SPRI beads. Purified aRNA can be stored at −80 °C.

17. The scDam&T-seq protocol uses liquid-handling robots to pipette small volumes, from 100 nL to 1.85 μL. We strongly recommend using these robots, since it increases the sample throughput and reduces the costs and processing time. In case this is unavailable, it is theoretically possible to upscale all the reagents to pipettable volume ranges, though we have never

attempted this ourselves. We routinely use the Nanodrop II robot (BioNex) for pipetting all reaction mixes. It is important to thoroughly clean the robot using the "Daily clean" program with a 2% Micro-90 cleaning solution before dispensing a reaction mix. This cleaning step is also recommended after dispensing restriction enzymes and proteases, such as DpnI and proteinase K, to prevent cross-contaminations. For pipetting DamID adapters or CEL-Seq2 primers, we use the Mosquito HTS robot (TTP Labtech), which is more suitable for the 384-well plate format. To prevent contaminations between wells, we program the robot to change tips after each DamID adapter dispension. For CEL-Seq2 primer plates, we program the robot to change tips when moving to a new column in the mother plate, this is not required when moving between the same columns in primer plates.

18. Always collect all liquid at the bottom of the wells in the 384-well plate before removing the plate seal. This is done before and after adding reaction mixes and preferably also after incubations in the thermal cycler. Collect liquid at the bottom of the plate by centrifuging at $2000 \times g$ for 2 min, 4 °C. In case multiple samples are sorted into the same 384-well plate during FACS sorting and the plate is removed from the FACS in between, it is also recommended to centrifuge the plate. Make sure the plate is always properly sealed, to prevent cross-contaminations between wells. 384-well plates containing sorted cells can be stored for several months at −80 °C before starting the scDam&T-seq protocol.

19. We have successfully applied the scDam&T-seq protocol using half of the indicated volumes. This applies to the volumes used before pooling of the samples, meaning CEL-Seq2 primers, reaction mixes and DamID adapters. This greatly reduces the costs of the experiment and facilitates sample handling, but since this could result in lower product yield, we recommend performing initial testing and optimization of the protocol with the indicated volumes.

20. The ^{m6}A mark becomes hemimethylated after DNA replication in S phase. Since DpnI does not recognize hemimethylated DNA, the mark will no longer be detected. This means that the accumulation of signal is reset every cell cycle, which prevents the saturation of the genome over time. Dam is able to restore the fully methylated ^{m6}A mark from the hemimethylated ^{m6}A when it comes into renewed contact with the DNA during the G2 phase. We therefore recommend to harvest cells at the G2/M transition, to allow a maximum ^{m6}A recovery time within the same cell cycle. It is optional to harvest cells at the G1/S transition, but one has to consider that protein–DNA contacts that were present in the G2 phase in the mother cell

will also contribute to the signal. Hoechst staining is used to evaluate DNA content during FACS sorting, acting as an indicator for cell cycle phase. Alternatively, the FUCCI reporter system can be used [34].

21. In the given scDam&T-seq protocol, the working concentration of DamID adapter is 25 nM. The concentration of DamID adapter required during the ligation step is largely determined by the expected level of ^{m6}A methylation in the cells. In case a large amount of methylation is present, a high number of G^{m6}ATC sites will be digested by DpnI, and more adapters are thus required to ligate all fragments. DamID adapters need to be present in some excess to increase ligation efficiency. However, the final amount of CEL-Seq2 derived sequencing reads decreases with increasing DamID adapter concentrations, as DamID derived fragments will take over the in vitro transcription reaction [18]. Furthermore, if the concentration of DamID adapters is too high, there will be a large amount of free DamID adapter present during the in vitro transcription step, which decreases the efficiency and can result in adapter contaminations of the amplified RNA and DNA sequencing library. It is therefore important to carefully titrate the amount of DamID adapters and to test multiple concentrations during initial scDam&T-seq experiments. We have successfully used concentrations from 5 to 50 nM. Usually, 25 or 50 nM is most suitable.

22. To limit batch effects (especially in the transcriptomic readout), we recommend to pool as many samples together as possible before in vitro transcription. In general, maintain best practices for single-cell experiments and ensure that different conditions are not distributed over different batches. We usually pool an entire 384-well plate with single cell samples together. It is important to make sure that no overlapping barcodes are present in the pooled sample. In case a large amount of 10 cell samples (>5) is present in the 384-well plate, or in case >10 cell samples were sorted, it is recommended to pool these separately from the single-cell samples.

23. The number of PCR cycles during the preparation of the Illumina sequencing libraries depends on the aRNA concentration and yield. We usually perform 8–12 cycles. In case of the example in Fig. 3, with an aRNA concentration of ~25 ng/μL, 8 cycles are recommended. Increasing cycles will increase the DNA concentration in the library, but will also reduce the complexity by increasing the amount of PCR duplicate reads. In case more than 12 cycles are required, the amount of aRNA is too low to ensure proper data quality. A final DNA library concentration of 2–5 ng/μL is considered optimal. The correct size distribution for the DNA library is from 200 to 2000 bp. Store the DNA library at −20 °C.

Acknowledgments

We would like to thank the members of the Kind lab for their critical reading of the manuscript and their helpful suggestions and comments. Work in the Kind lab on the scDam&T-seq technique was funded by the European Research Council Starting grant (no. ERC-StG 678423-EpiID) and a Dutch Research Council (NWO) Open grant (no. 824.15.019). J.K. and S.J.A.L. are funded by an NWO ALW/VIDI grant (no. 161.339). The Oncode Institute is supported by the KWF Dutch Cancer Society.

References

1. Sexton T, Schober H, Fraser P et al (2007) Gene regulation through nuclear organization. Nat Struct Mol Biol 14(11):1049–1055. https://doi.org/10.1038/nsmb1324

2. Bickmore WA (2013) The spatial organization of the human genome. Annu Rev Genomics Hum Genet 14:67–84

3. Cremer T, Cremer C (2001) Chromosome territories, nuclear architecture and gene regulation in mammalian cells. Nat Rev Genet 2(4): 292–301

4. Wendt KS, Yoshida K, Itoh T et al (2008) Cohesin mediates transcriptional insulation by CCCTC-binding factor. Nature 451(7180): 796–801. https://doi.org/10.1038/nature06634

5. van Steensel B, Furlong EE (2019) The role of transcription in shaping the spatial organization of the genome. Nat Rev Mol Cell Biol 20:327

6. Ibrahim DM, Mundlos S (2020) The role of 3D chromatin domains in gene regulation: a multi-facetted view on genome organization. Curr Opin Genet Dev 61:1–8

7. Lieberman-Aiden E, Van Berkum NL, Williams L et al (2009) Comprehensive mapping of long-range interactions reveals folding principles of the human genome. Science 326(5950):289–293

8. Dekker J, Rippe K, Dekker M et al (2002) Capturing chromosome conformation. Science 295(5558):1306–1311

9. de Wit E, De Laat W (2012) A decade of 3C technologies: insights into nuclear organization. Genes Dev 26(1):11–24

10. Vogel MJ, Peric-Hupkes D, Van Steensel B (2007) Detection of in vivo protein–DNA interactions using DamID in mammalian cells. Nat Protoc 2(6):1467

11. Beagrie RA, Scialdone A, Schueler M et al (2017) Complex multi-enhancer contacts captured by genome architecture mapping. Nature 543(7646):519–524. https://doi.org/10.1038/nature21411

12. Boettiger A, Murphy S (2020) Advances in chromatin imaging at kilobase-scale resolution. Trends Genet 36(4):273–287. https://doi.org/10.1016/j.tig.2019.12.010

13. Bintu B, Mateo LJ, Su J-H et al (2018) Super-resolution chromatin tracing reveals domains and cooperative interactions in single cells. Science 362(6413):eaau1783. https://doi.org/10.1126/science.aau1783

14. Nagano T, Lubling Y, Stevens TJ et al (2013) Single-cell Hi-C reveals cell-to-cell variability in chromosome structure. Nature 502(7469): 59–64

15. Kind J, Pagie L, de Vries SS et al (2015) Genome-wide maps of nuclear lamina interactions in single human cells. Cell 163(1): 134–147

16. Skene PJ, Henikoff S (2017) An efficient targeted nuclease strategy for high-resolution mapping of DNA binding sites. elife 6:e21856

17. Flyamer IM, Gassler J, Imakaev M et al (2017) Single-nucleus Hi-C reveals unique chromatin reorganization at oocyte-to-zygote transition. Nature 544(7648):110–114

18. Rooijers K, Markodimitraki CM, Rang FJ et al (2019) Simultaneous quantification of protein–DNA contacts and transcriptomes in single cells. Nat Biotechnol 37(7):766–772

19. Markodimitraki CM, Rang FJ, Rooijers K et al (2020) Simultaneous quantification of protein–DNA interactions and transcriptomes in single cells with scDam&T-seq. Nat Protoc 15(6):1922–1953

20. Hashimshony T, Senderovich N, Avital G et al (2016) CEL-Seq2: sensitive highly-

multiplexed single-cell RNA-Seq. Genome Biol 17(1):77

21. Hashimshony T, Wagner F, Sher N et al (2012) CEL-Seq: single-cell RNA-Seq by multiplexed linear amplification. Cell Rep 2(3):666–673

22. Guelen L, Pagie L, Brasset E et al (2008) Domain organization of human chromosomes revealed by mapping of nuclear lamina interactions. Nature 453(7197):948

23. Satijn DP, Gunster MJ, van der Vlag J et al (1997) RING1 is associated with the polycomb group protein complex and acts as a transcriptional repressor. Mol Cell Biol 17(7): 4105–4113. https://doi.org/10.1128/mcb. 17.7.4105

24. Nora EP, Goloborodko A, Valton A-L et al (2017) Targeted degradation of CTCF decouples local insulation of chromosome domains from genomic compartmentalization. Cell 169(5):930–944.e922

25. Ran FA, Hsu PD, Wright J et al (2013) Genome engineering using the CRISPR-Cas9 system. Nat Protoc 8(11):2281–2308

26. Illumina (2020). https://www.illumina.com

27. Nishimura K, Fukagawa T, Takisawa H et al (2009) An auxin-based degron system for the rapid depletion of proteins in nonplant cells. Nat Methods 6(12):917–922

28. Rang FJ (2019) scDamAndTools. https:// github.com/KindLab/scDamAndTools

29. Vogel MJ, Guelen L, de Wit E et al (2006) Human heterochromatin proteins form large domains containing KRAB-ZNF genes. Genome Res 16(12):1493–1504

30. Yesbolatova A, Natsume T, Hayashi K-I et al (2019) Generation of conditional auxin-inducible degron (AID) cells and tight control of degron-fused proteins using the degradation inhibitor auxinole. Methods 164:73–80

31. Borsos M, Perricone SM, Schauer T et al (2019) Genome–lamina interactions are established de novo in the early mouse embryo. Nature 569(7758):729–733

32. Banaszynski LA, Chen L-C, Maynard-Smith LA et al (2006) A rapid, reversible, and tunable method to regulate protein function in living cells using synthetic small molecules. Cell 126(5):995–1004

33. Haugwitz M, Garachtchenko T, Nourzaie O et al (2008) Rapid, on-demand protein stabilization and destabilization using the ProteoTuner™ systems. Nat Methods 5(10):iii–iv

34. Sakaue-Sawano A, Kurokawa H, Morimura T et al (2008) Visualizing spatiotemporal dynamics of multicellular cell-cycle progression. Cell 132(3):487–498. https://doi.org/10.1016/j. cell.2007.12.033

Part V

Visualizing Spatial Genome Organization

Chapter 12

High-Throughput DNA FISH (hiFISH)

Elizabeth Finn, Tom Misteli, and Gianluca Pegoraro

Abstract

High-throughput DNA fluorescence in situ hybridization (hiFISH) combines multicolor combinatorial DNA FISH staining with automated image acquisition and analysis to visualize and localize tens to hundreds of genomic loci in up to millions of cells. hiFISH can be used to measure physical distances between pairs of genomic loci, radial distances from genomic loci to the nuclear edge or center, and distances between genomic loci and nuclear structures defined by protein or RNA markers. The resulting large datasets of 3D spatial distances can be used to study cellular heterogeneity in genome architecture and the molecular mechanisms underlying this phenomenon in a variety of cellular systems. In this chapter we provide detailed protocols for hiFISH to measure distances between genomic loci, including all steps involved in DNA FISH probe design and preparation, cell culture, DNA FISH staining in 384-well imaging plates, automated image acquisition and analysis, and, finally, statistical analysis.

Key words Nuclear architecture, 3D genome, DNA FISH, High-throughput imaging

1 Introduction

Chromosome conformation capture (3C) techniques that take advantage of next generation sequencing (NGS), such as high-throughput chromosome conformation capture (Hi-C), Hi-C paired with chromatin immunoprecipitation (Hi-ChIP), chromatin interaction analysis with paired-end tags (ChIA-PET), and others, have enabled the study of the 3D genome organization in a systematic, genome-wide fashion [1]. These methods rely on fixation, shearing, and subsequent ligation of DNA fragments as a proxy measurement for spatial colocalization, since fixatives will bind together genomic regions in close spatial proximity. As a result, physically interacting regions will be enriched in the pool of ligation products when compared to noninteracting ones. These biochemical techniques allow easy comparisons between most pairs of loci in the genome, rather than being limited to a few targets of interest. For this reason, they provide an unprecedented genome-wide view of genome organization at multiple length scales.

Tom Sexton (ed.), *Spatial Genome Organization: Methods and Protocols*,
Methods in Molecular Biology, vol. 2532, https://doi.org/10.1007/978-1-0716-2497-5_12,

Unfortunately, though, it is still challenging to routinely produce single cell [2] or single allele [3] data with these techniques. In addition, Hi-C techniques only provide information on pairs of genomic loci that are physically closer than a spatial threshold determined by the fixative used, and genomically far enough apart to be discriminated by restriction enzyme digestion [4]. As a result, these methods are limited in their ability to generate information about very short promoter-enhancer loops or long-distance interactions between intra- or inter-chromosomal regions that can be mediated by nuclear bodies and/or nuclear condensates [5].

Beside 3C biochemical techniques, DNA fluorescence in situ hybridization (DNA FISH) has historically been the other major class of technique used to study genome architecture. This is because DNA FISH provides spatially resolved information for genomic regions at the single-allele and single-cell level, thus complementing many of the technical limitations of 3C techniques mentioned above. In addition, DNA FISH, as any other microscopy technique, can be adapted to visualize DNA in combination with RNA or proteins, thus providing additional context for the localization of genomic regions relative to functional gene expression, to specific cellular compartments, or to cellular subpopulations defined by RNA FISH or antibody-based immunofluorescence (IF). Up until recently, though, DNA FISH had fairly limited throughput in terms of how many distinct genomic regions could be detected in a single experiment, given that fluorescence microscopes are generally limited to 4 or 5 spectral channels. Furthermore, traditional FISH experiments generated data for a limited number of cells per experiment, and in a fairly limited number of experimental conditions. This was mostly due to the lack of automation in the DNA FISH staining protocols, in the image acquisition, in the image analysis, and finally in the statistical analysis of the data. These weaknesses hindered the generation and availability of imaging-based estimations of physical proximity for hundreds or thousands of genomic loci in millions of cells per study [6–8].

More recently, however, several different approaches relying on automated microscopy and image analysis have partially overcome these limitations. Among these, high-throughput imaging (HTI) was used to study cell-to-cell variability for hundreds of genomic loci [9], to identify extremely rare biological events such as chromosomal DNA translocations [10, 11], and to perform functional genomics screens with DNA FISH as a readout [12, 13].

In this protocol, we describe in detail an end-to-end HTI protocol to image and analyze up to hundreds of genomic loci in millions of cells using DNA FISH, automated microscopy, high-content image analysis, and custom statistical analysis, a technique we refer to as high-throughput DNA FISH (hiFISH). In particular, we will focus on important experimental design and execution steps, provide suggestions on how to quality control the data, and finally provide practical solutions to common experimental issues.

2 Materials

2.1 Nick Translation for BAC and Fosmid Probes

1. 40 ng/μL fosmid or BAC probe DNA template, in ddH$_2$O (*see* **Note 1**).

2. 10× nick translation buffer: 0.5 M Tris-HCl pH 8.0, 50 mM MgCl$_2$, 0.5 mg/mL BSA.

3. Nucleotide mixture stock: 20 mM Tris-HCl pH 7.4, 166 μM dATP, 166 μM dGTP, 166 μM dCTP, 166 μM fluorophore-conjugated dUTP (*see* **Note 2**).

4. 0.01 M β-mercaptoethanol.

5. 100 μg/mL DNase I, freshly made each time by diluting a 1 mg/mL stock solution (in 50% glycerol) in ice-cold ddH$_2$O (*see* **Note 3**).

6. 10 U/μL *E. coli* DNA Polymerase I.

7. 0.5 M EDTA.

8. 3 M sodium acetate pH 5.2.

9. 1 mg/mL human Cot-1 DNA (*see* **Note 4**).

10. 10 mg/mL yeast tRNA.

11. 100% and 70% ethanol.

12. Hybridization buffer: 15% (w/v) dextran sulfate, 50% (v/v) formamide, 2× SSC (saline sodium citrate), 1% (v/v) Tween 20.

2.2 Amplification and Preparation of OligoFISH Libraries (Adapted from [14])

1. Oligonucleotides pool library (Twist Bioscience, CustomArray, or Agilent).

2. TE buffer: 10 mM Tris-HCl pH 8.0, 0.1 mM EDTA.

3. PCR master mix 1: 0.2 mM dNTP mix, 1× KAPA Fidelity buffer, 1 μM Forward F′ PCR primer (specific to library, includes T7 sequence), 1 μM Reverse R′ PCR primer (specific to library), 0.01 U/μL KAPA High Fidelity DNA polymerase.

4. 20 mg/mL glycogen.

5. 5 M ammonium acetate (unbuffered).

6. NanoDrop spectrophotometer.

7. PCR master mix 2: 0.2 mM dNTP mix, 1× KAPA Fidelity buffer, 1 μM Forward F′ PCR primer (specific to library, includes T7 sequence), 1 μM Reverse R′ PCR primer (specific to library), 0.005 U/μL KAPA High Fidelity DNA polymerase.

8. DNA clean and concentrate-5 cleanup kit (Zymo).

9. 50 U/μL T7 RNA polymerase and 10× T7 buffer.

10. 100 mM ATP.

11. 100 mM GTP.

12. 100 mM UTP.

13. 100 mM CTP.

14. 1 U/μL RNasin Plus.

15. RT master mix: 2 mM dNTPs, 11.4 U/μL Maxima reverse transcriptase, 2× Maxima RT buffer, 0.1 U/μL RNasin Plus.

16. Alkaline solution: 0.25 M EDTA, 0.5 M NaOH. Make fresh each time from 0.5 M EDTA and 1 M NaOH stocks.

17. Oligo Clean and Concentrate-5 cleanup kit (Zymo).

18. Readout probes: 20 nt custom-designed oligos, synthesized as ssDNA, HPLC-purified, and labeled at the 3′ or 5′ end with the appropriate fluorophore (e.g., Alexa 488, Alexa 568, or Alexa 647). Resuspended at 5 μM in TE. Sample sequences for the readout probes can be found in **Note 5**.

19. Vacuum concentrator (e.g., Speedvac).

2.3 Cell Culture in 384-Well Imaging Plate Format

1. 384-well imaging plates (PerkinElmer CellCarrier-384 Ultra), coated in poly-D-lysine if required for the cell type cultured.

2. Liquid Dispenser (Multidrop Combi, ThermoFisher, or similar), or multichannel pipette.

3. 1× PBS (phosphate-buffered saline).

4. 8% PFA (paraformaldehyde) in PBS, made from a 16% PFA stock.

5. Appropriate cell dissociation reagent (for adherent cells). We routinely use Trypsin EDTA (0.25%) for epithelial and adherent cells, and ReLeSR for stem cells. We have also successfully used Accutase for fragile cell lines.

6. Appropriate cell growth medium. Our human telomerase (hTERT) immortalized fibroblasts are grown in DMEM with 20% FBS, supplemented with sodium pyruvate and glutamate. Our stem cells are grown in complete mTeSR media.

7. Sponge for changing media on spun-down cells: We have had success with utility sponges cut in half.

8. Matrigel.

2.4 DNA FISH in 384-Well Imaging Plate Format

1. Permeabilization solution: 0.5% (v/v) Triton X-100, 0.5% (w/v) saponin in 1× PBS.

2. 0.1N HCl.

3. SSC (saline sodium citrate) at 2×, 1× and 0.1× concentrations.

4. Equilibration buffer: 50% (v/v) formamide in 2× SSC.

5. 500 nm Tetraspeck fluorescent beads (ThermoFisher).

6. Aluminum adhesive seal.

7. Slide moat or heat block at 85 °C.

8. Humidified chamber or a water bath at 37 °C for probe hybridization.

9. Three utility sponges (*see* **Note 6**).

10. 5 mg/mL 4′,6-diamidino-2-phenylindole (DAPI) stock solution. Dilute 1:1000 to 1:5000 in PBS to generate a working staining solution.

2.5 Automated Image Acquisition and Analysis

1. High-throughput spinning disk confocal microscope (Such as Yokogawa CV7000/CV8000 or PerkinElmer Opera Phenix) equipped, at a minimum, with the following:

   – Automated stage in *x*, *y*, and *z*.

   – Autofocus mechanism.

   – Excitation light sources in four fluorescence channels.

   – Matched dichroic and emission mirrors.

   – High NA (numerical aperture) 40× or 60× water objectives.

   – One or more sCMOS cameras.

   – Microscope-controlling software.

2. Image analysis software (can be either commercial or open-source software, e.g., Python, ImageJ, KNIME, CellProfiler).

3. Computing infrastructure. Large image datasets are usually analyzed on dedicated servers with tens of CPUs (16 or more) and large amounts of RAM (Ideally 256 GB or more), or on scientific high-performance computing (HPC) clusters.

4. Statistical analysis software. We generally use R (https://cran.r-project.org/) and RStudio Desktop (https://rstudio.com) for statistical analyses, as they are open source and provide a balance between powerful scripting for bulk analysis, and more user-friendly, interactive exploratory analyses. However, all the statistical analysis steps described in the following sections can be performed with any other scientific programming language or platform, such as Python, KNIME, Julia, or Matlab.

3 Methods

3.1 Selection of Fluorescent Probe Types for hiFISH

The first step of any DNA FISH experiment is the selection of appropriate probes to hybridize and visualize the genomic regions of interest. This is a critical step because much of the technical variability of DNA FISH comes down to differences in

250   Elizabeth Finn et al.

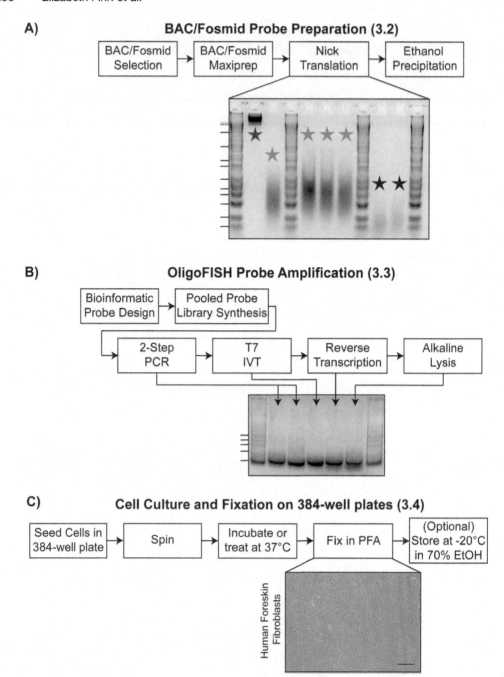

Fig. 1 FISH probe preparation and cell culture workflows. (**a**) Flowchart for the preparation of BAC or fosmid FISH probes. The image of the agarose gel represents the quality control step to ensure the success of the nick translation labeling reaction. A successful reaction should appear as a smear between 100 and 1000 bp on the gel (Green star symbols). An unsuccessful reaction due to failed digestion will run as the original full-size BAC or fosmid (Red star symbol). An unsuccessful reaction due to excessive digestion will run as a small smear at very short lengths (Blue star symbols). (**b**) Flow chart for the oligoFISH probe amplification reactions. The picture of the agarose gel represents the quality control/yield step obtained by loading a small sample volume (2–5 μL) of the completed reaction. A successful pooled library of oligos should run as a single band at approximately 150–200 bp. (**c**) Flow chart for cell culture in 384-well imaging and fixation. As shown in the image, users should use a brightfield microscope to check that the confluency at the time of fixation is approximately 70–80%

hybridization efficiency between different probes. Unfortunately, though, there are very few rules that can successfully predict DNA FISH probe hybridization efficiency. As a consequence, some trial and error in the probe selection process will likely be necessary. For hiFISH protocols, fluorescent probes can be prepared either by nick translation or by OligoFISH. Either of these two approaches has its own pros and cons (*see* **Note** 7) and the user should choose between one of them as the first step in the design of a hiFISH experiment.

Nick-translated probes have long been used for DNA FISH [15]. Their preparation starts with the selection of a fragment of genomic DNA cloned into a vector: plasmids for short sequences (up to 20 kb), fosmids for moderate length sequences (around 50 kb), or bacterial artificial chromosomes (BACs) for long sequences (around 200 kb). These bacterial vectors and their eukaryotic genomic cargo are then used as substrates in an in vitro enzymatic reaction called nick translation (Fig. 1a), which includes modified oligonucleotides (chemically labeled either with a fluorophore or with a hapten such as biotin or digoxigenin), DNase I, and DNA Pol I. In the reaction, DNase first nicks the dsDNA of the BAC or of the fosmid vector, and DNA Polymerase I then uses the newly generated 3′-OH ends at the nick as priming sites for the synthesis and amplification of new dsDNA molecules that include the modified oligonucleotides. Nick translation produces large quantities of fluorescently labeled dsDNA probes, optimally between 100 and 1000 bp, and visible as a smear in a QC gel (Fig. 1a), which are subsequently precipitated in ethanol to concentrate and purify them and resuspended in hybridization solution (*see* **Note** 7). Please refer to Subheading 3.2 for a detailed nick translation protocol.

As an alternative to nick-translated BAC and fosmid probes, and to address disadvantages such as nonuniformity of FISH efficiency, preparation steps that must be done on a per-probe basis, and the inability to decide with base-pair resolution the placement of a probed region, several laboratories have optimized and implemented libraries of tens or hundreds of thousands of small, chemically synthesized single stranded DNA (ssDNA) oligos for DNA FISH, an approach originally named Oligopaint [16, 17], which was further adapted into other applications including MERFISH [14, 18, 19], and oligoFISSEQ [20]. Here, we will refer to this family of approaches as OligoFISH and cover their most basic variant.

ssDNA oligo libraries used in OligoFISH are synthesized and delivered by commercial vendors as pools of up to millions of oligonucleotides, where the effective concentration of each nucleotide is in the low femtomolar range [21] (*see* **Note 8**). To generate sufficient quantities of the oligo libraries to be used in OligoFISH protocols and to regenerate them indefinitely, they need to first be amplified by two-step PCR (Fig. 1b). An additional step converting the library to RNA via in vitro transcription and subsequent reverse transcription amplifies the library up to 1000-fold with a comparatively low rate of errors or imbalances, and a final alkaline treatment degrades the RNA strand and results in a ssDNA probe that will bind genomic DNA very efficiently (*see* **Note 9**) (Fig. 1b). Refer to Subheading 3.3 for a detailed OligoFISH probe set preparation protocol.

3.2 Nick Translation for BAC and Fosmid Probes

What follows is a fairly typical protocol, adapted from [15, 22], to amplify and fluorescently label BAC and fosmid probes via nick translation, followed by probe concentration via precipitation. In our hands, nick translation reactions can be successfully scaled up from a total volume of 50 μL as described here to a maximum of 500 μL. We do not recommend scaling up an individual reaction farther than this. If a volume larger than 500 μL of probe is required, set up several 500 μL reactions in parallel, and then pool the products.

1. Add the following reagents in a microcentrifuge tube in the following order, on ice: 50 μL probe DNA, 5 μL 10× nick translation buffer, 5 μL nucleotide mixture stock, 5 μL 0.01 M β-mercaptoethanol, 2 μL 100 μg/mL DNase I, 1 μL 10 U/μL DNA polymerase I. Tap tubes to remove bubbles and spin briefly in a desktop centrifuge to collect all liquid.

2. Incubate at 16 °C in a water bath in a cold room for 80 min.

3. To stop the reaction, add 2 μL 0.5 M EDTA, and incubate at 72 °C for 10 min.

4. As a quality control step, run 5 μL of the reaction volume on a 2% agarose gel at 60 V. Verify that the nick-translated BAC or fosmid probes run as a smear between 100 and 1000 bp (Fig. 1a) (*see* **Note 10** *for storage information*).

5. Precipitate the probes by mixing the following, where "1 part" is defined as 1/20th the total volume of nick-translated probe. As an example, for precipitation of three probes labeled with three different fluorophores, at 20 μL for each probe: 60 μL total probe mix/20 = 3 μL. The volumes for this example (three probes) are included below in parentheses.

   – 3 M sodium acetate: 5 parts (15 μL).

   – 1 mg/mL human Cot-1 DNA: 3 parts (9 μL).

   – Yeast tRNA: 2 parts (6 μL).

   – Each nick-translated probe: 20 μL.

   – 100% ethanol: 100 parts (300 μL).

6. Vortex to mix and briefly spin down in a desktop centrifuge, then chill at −20 °C for 30 min.

7. Centrifuge at 4 °C, 20,000 × *g* for 30 min.

8. Gently decant supernatant and air-dry tubes by inverting for 10 min at room temperature.

9. Resuspend pellet in 45 µL hybridization buffer by heating to 75 °C and vortexing, or using a thermomixer at 75 °C and 750+ rpm for at least 60 min (*see* **Note 10** *for storage information*).

3.3 Amplification and Preparation of OligoFISH Libraries (Adapted from [14])

3.3.1 Primary PCR Amplification and Aliquoting

1. Dilute the oligonucleotide pool library stock to 50 ng/µL total concentration in TE buffer (*see* **Note 11**).

2. Perform primary amplification to generate library stocks: mix 0.1 µL oligonucleotide library plus 49.9 µL PCR master mix 1 in PCR tubes on ice. We recommend starting with at least 20 reactions at this step (*see* **Note 11**).

3. Run the following program on a thermocycler.

   95 °C, 5 min.

   25× [95 °C, 30 s; 59 °C, 30 s; 72 °C, 15 s].

   72 °C, 5 min.

   4 °C, hold.

4. Run 10 µL of the reaction on a 2% agarose gel at 60 V to check product size and purity (Fig. 1b).

3.3.2 Precipitation and Determination of Concentration of PCR Products

1. In a 15 mL conical tube, mix 990 µL PCR reaction, 3.5 µL 20 mg/mL glycogen, 108 µL 5 M ammonium acetate and 2.25 mL 100% ethanol. Divide the total volume into three 1.5 mL centrifuge tubes (1.1 mL per tube). Chill at −80 °C for 30 min.

2. Centrifuge at $14,000 \times g$ for 30 min at 4 °C.

3. Carefully aspirate the supernatant and wash the pellet with 70% ethanol. Dry at room temperature for 15 min.

4. Dissolve the pellets in 20 µL of PCR-quality water per tube (combined final volume: 60 µL).

5. Check concentration using a Nanodrop spectrophotometer; the desired concentration is 1.5–2 µg/µL.

6. Optionally, check product size and purity by running 2 µL of the reaction on a 2% agarose gel at 60 V (Fig. 1b).

3.3.3 Secondary PCR Amplification

1. Mix the PCR product with PCR master mix 2 on ice, so that at least ten 50 µL aliquots of 0.125 µL PCR product plus 49.875 µL master mix are made in PCR tubes. Run the following program on a thermocycler.

   95 °C, 5 min.

   40× [95 °C, 30 s; 59 °C, 30 s; 72 °C, 15 s].

   72 °C, 5 min.

   4 °C, hold.

2. Combine 1.25 mL of Zymo DNA binding buffer with 250 µL of the PCR reaction (pool five PCR reaction tubes in two columns) (*see* **Note 12**).

3. Add 0.75 mL of the above mixture to the Zymo DNA spin column, then spin for 30 s at >10,000 × g in a benchtop microcentrifuge.

4. Discard the flowthrough and repeat **step 3** with the remaining mixture.

5. Add 200 µL of DNA Wash buffer to each column and spin for 30 s at >10,000 × g.

6. Discard the flowthrough and repeat **step 5** with an additional 200 µL of DNA Wash buffer.

7. Discard the flowthrough, then spin for 1 min at >10,000 × g to remove residual DNA Wash buffer.

8. Transfer each column to a clean 1.5 mL Eppendorf tube and add 11 µL of nuclease-free water per column. Incubate for 1 min, then spin for 1 min at >10,000 × g.

9. For quality control (QC) purposes, save 0.5 µL of the eluate to be run on a 2% agarose gel at 90 V (Fig. 1b).

3.3.4   T7 In Vitro Transcription (IVT), Reverse Transcription (RT) and Alkaline Degradation of the RNA Strand

1. Add the following reagents to 40 µL cleaned PCR product on ice: 8 µL 10× T7 buffer, 8 µL 100 mM ATP, 8 µL 100 mM UTP, 8 µL 100 mM GTP, 8 µL 100 mM CTP, 4 µL 1 U/µL RNasin Plus, 8 µL 50 U/µL T7 RNA polymerase. Incubate at 37 °C for 4–16 h (the yield is slightly better after longer incubation) (*see* **Note 13**).

2. Add 88 µL RT master mix and split to four 50 µL PCR tubes (45 µL per tube). Incubate in a thermocycler at 50 °C for 50 min, then at 85 °C for 5 min.

3. Add 45 µL of alkaline solution to each tube, and heat to 95 °C for 10 min.

3.3.5   *Final Cleanup*

We use Zymo Oligo Clean and Concentrator kits for this step. This optimizes the pull-down of single stranded short DNA oligos which might not be as efficiently purified with the DNA Clean and Concentrator kits.

1. In a 15 mL conical tube, combine all tubes of the final ssDNA library (360 µL total) obtained at the end of Subheading 3.3.4 with 720 µL of Oligo Binding Buffer. Add 2.88 mL 100% ethanol and mix.

2. Transfer 750 µL of the mix to each of two Oligo-CC columns, then spin for 30 s at >10,000 × g to load the columns.

3. Discard the flowthrough and repeat **step 2** at least twice, until all of the mixture is loaded (for a total of 3 loading steps).

4. Discard the flowthrough and add 750 μL of DNA Wash buffer to each column. Spin for 30 s at >10,000 × g.

5. Discard the flowthrough and spin at >10,000 × g to remove any excess of DNA Wash buffer.

6. Transfer the columns to clean 1.5 mL Eppendorf tubes. Add 20 μL of nuclease-free water to each column. Incubate for 1 min, then spin the column for 1 min at >10,000 × g.

7. Repeat **step 6**, eluting again in an additional volume of 20 μL of nuclease-free water.

8. Measure the final probe DNA concentration using a Nanodrop spectrophotometer. The desired target concentration should be around 1.4 μg/μL.

3.3.6 Dilution in Hybridization Buffer for FISH

1. Mix the following reagents, fresh each time, in a 1.5 mL tube (volumes given for 20 wells): 20 μL Cot-1 DNA, 3 μL probe library, 20 μL 2 pmol/μL green readout probe, 20 μL 2 pmol/μL red readout probe, 20 μL 2 pmol/μL far-red readout probe.

2. Optional: concentrate/dry in a vacuum concentrator on low setting for 10 min. The residual volume of the mix should be ~2 μL per well (*see* **Note 14**).

3. Add 12 μL of hybridization solution (Subheading 2.1) to the mix, and mix thoroughly by pipetting. The final volume of probes in the hybridization solution should be 14–15 μL per well.

3.4 Cell Culture in 384-Well Imaging Plate Format

In order to minimize the DNA FISH hybridization volumes, and to maximize the number of DNA FISH probe combinations used in a single experiment, our hiFISH protocol was optimized in 384-well imaging plates (Fig. 1c). Cell culture procedures in this format are not much different from lower-throughput formats (6-wells, 12-wells, etc.). For this reason, we will not describe generic mammalian cell culture reagents and procedures in detail, but rather focus on the steps that are specific to the 384-well format, and to automated image acquisition and analysis.

3.4.1 Growing Adherent Cells in 384-Well Imaging Plates

This procedure works for fibroblasts, human bronchial epithelial cells, and other immortalized and/or transformed cell lines that do not form clumps or grow in colonies.

1. Plate cells in the 384-well imaging plate using a multichannel pipette or an automated liquid dispenser. The appropriate number of cells should be plated in a volume of 50 μL of growth media per well to obtain ~80% cell confluency at the time of fixation (*see* **Note 15**) (Fig. 1c).

2. Centrifuge at 250 × g for 1 min, room temperature, to collect cell suspension at the bottom of the well.

3. Leave the plate at room temperature for 30 min. This has been shown to help solve issues with uneven cell growth patterns in the wells [23].

4. Incubate the plate at 37 °C, 5% CO_2, 80% humidity for the necessary amount of time required by the experiment (*see* **Note 16**).

5. Using an automated liquid handler or a multichannel pipette, add 50 µL of 8% PFA in PBS directly to the cells and incubate for 10 min (*see* **Notes 17** and **18**).

6. Wash the imaging plate three times in PBS using a multichannel pipette, an automated liquid handler, or a plate washer. The liquid from the plates can be discarded using a dedicated plate washer, a couple of firm manual "taps," a glass needle connected to a vacuum line (taking care not to dislodge the fixed cells), and/or a sponge (*see* **Notes 18** and **19**).

7. Store the fixed and washed imaging plates in PBS at 4 °C for up to 1 week, or in 70% ethanol at −20 °C for up to 2 months.

3.4.2 Spinning Suspension Cells in 384-Well Imaging Plates

This procedure has been used on patient derived and purified peripheral blood mononucleated cells (PBMCs), dissociated colonies of human stem cells, and cell lines growing in suspension, such as Jurkat. For detailed instructions on how to perform delicate washes and not dislodge these cell types from 384-well plates, please *see* **Note 19**.

1. Warm a poly-D-lysine–coated 384-well imaging plate for 30 min at 37 °C.

2. Dissociate the cells and resuspend them in growth medium at a concentration of 1×10^7 cells/mL (50,000 cells per 50 µL of media per well). Add 50 µL cell suspension per well.

3. Centrifuge at $250 \times g$ for 5 min, room temperature. Immediately add 50 µL 8% PFA in PBS. Incubate at room temperature for 15 min.

4. Rinse once in PBS using a sponge to remove liquid.

5. Store in PBS at 4 °C up to 1 week or transfer to 70% ethanol and store at −20 °C for up to 2 months.

3.4.3 Growing Colony-Forming Cells on Plates

This procedure has been used for colonies of human embryonic stem cells (hESCs). A similar technique could be used for other colony-forming cells or organoids.

1. Coat wells of 384-well plates with matrigel (*see* **Note 20**).

2. Lift colonies from one well of a 6-well dish and resuspend them into 45 mL of complete medium.

3. Plate 50 µL of the cell suspension, inverting the plate or gently rocking it back and forth to mix between each well. This results in an approximately 1/10 dilution by surface area. Our best results were obtained using a reservoir and a multichannel

pipette for consistency between wells. Shake briefly side to side and front to back to spread. Incubate overnight at 37 °C, 5% CO_2, 80% humidity.

4. Replace the medium with 50 µL of fresh complete medium. Repeat daily medium changes until desired confluency is reached (*see* **Notes 19** and **21**).

5. Replace medium with 50 µL 4% PFA in PBS and incubate at room temperature for 10 min.

6. Wash three times in PBS, flicking the plate to empty it and using a multichannel pipette to add 100 µL per well of PBS each time (*see* **Notes 18** and **19**).

7. Store in PBS at 4 °C up to 1 week or transfer to 70% ethanol and store at −20 °C for up to 2 months.

3.5 DNA FISH in 384-Well Plate Format (Fig. 2a)

1. If stored at −20 °C, thaw the fixed 384-well plate first at 4 °C overnight, then equilibrate at room temperature for at least 2 h.

2. Wash the plate three times with 100 µL/well of PBS.

3. Incubate in 100 µL/well of permeabilization solution for 20 min at room temperature.

4. Rinse twice in PBS.

5. Add 100 µL/well of 0.1N HCl to the plate and incubate for 15 min at room temperature.

6. Remove the HCl and neutralize the acid by first adding 100 µL/well of 2× SSC, and then incubating for 5 min at room temperature.

7. Incubate in 100 µL/well of equilibration buffer for at least 30 min at room temperature (*see* **Note 22**).

8. Remove the equilibration buffer and add 13.5 µL of hybridization solution stock containing the FISH probes (final products of Subheading 3.2 or 3.3) to each well (*see* **Notes 23** and **24**).

9. We use fiducial beads within control wells not containing cells to check and, if necessary, adjust the alignment of the microscope in every experiment (*see* **Note 25**). Our best results were obtained with 500 nm Tetraspeck beads. Vortex the vial containing the beads for 30 s to bring them into suspension and reduce clumps.

10. Dilute the bead suspension 1:10 in nuclease-free water (*see* **Note 26**). Vortex the diluted beads for 30 s.

11. Immediately add 15 µL of diluted bead suspension to each of two empty wells in the 384-well imaging plate.

12. Seal the plate with adhesive aluminum foil to block light and to ensure an airtight seal. Tap the plate firmly against a benchtop to remove bubbles. Leave the wells with Tetraspeck beads unsealed to allow the water to evaporate.

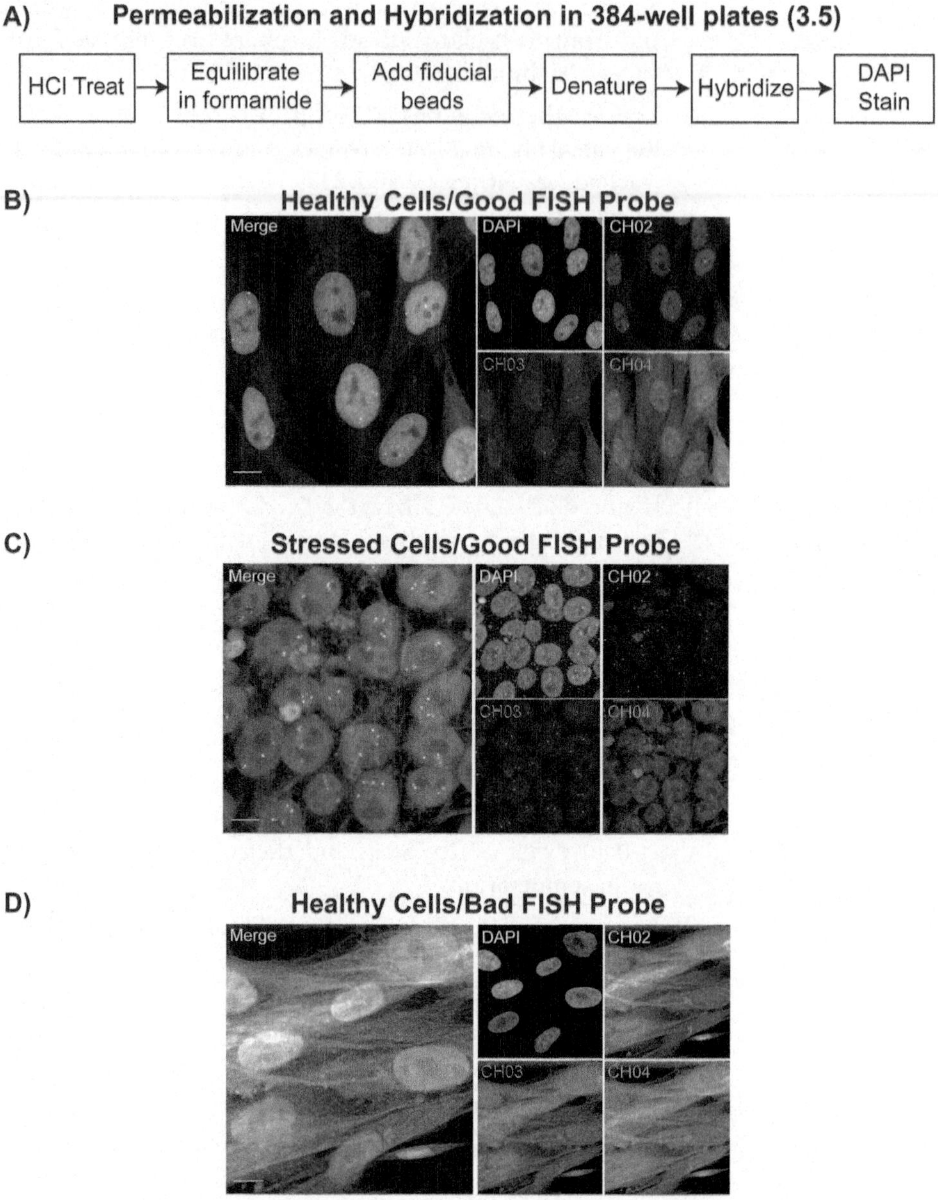

A)   **Permeabilization and Hybridization in 384-well plates (3.5)**

B)   **Healthy Cells/Good FISH Probe**

C)   **Stressed Cells/Good FISH Probe**

D)   **Healthy Cells/Bad FISH Probe**

Fig. 2 FISH staining workflow. (**a**) Flowchart for FISH staining of PFA-fixed 384-well plates. (**b**) 3D maximally projected images of ~70–80% confluent healthy human immortalized fibroblasts successfully stained with the FISH protocol. CH02, CH03 and CH04 Images represent the channels assigned to the FISH probes. As expected, since they are diploid, most cells contain two FISH signals in each channel. Scale bar: 10 μm. (**c**) Same as (**b**), except that in this case the cells are either too confluent or clumped while they were seeded, making it difficult or impossible to automatically segment nuclei with automated image analysis pipelines (*see* **Note 15**). This clumping also causes cell stress, resulting in the enlarged/fused nucleoli visible in DAPI staining and cellular debris visible in all channels. (**d**) Same as (**b**), except that in this case the FISH probes used in the experiment did not efficiently bind to their genomic targets, and/or the FISH staining conditions were not properly optimized (*see* **Notes 14, 28** and **29**)

13. Spin for 1 min in a plate centrifuge at room temperature. The centrifugation speed does not matter much at this step, since this step is mostly to collect liquid at the bottom of the well and remove bubbles.

14. Denature DNA on a slide moat, or on a hot block, for 7 min and 30 s at 85 °C (*see* **Note 27**).

15. Immediately move the plate to the hybridization chamber. Hybridize overnight or longer at 37 °C (*see* **Note 28**).

16. Before starting the washes (**steps 17–19**), preheat $1\times$ SSC and $0.1\times$ SSC wash buffers to 45 °C or 60 °C, as needed.

17. Rinse the plate once in $2\times$ SSC at room temperature.

18. If using nick-translated probes, wash three times for 5 min in $1\times$ SSC at 45 °C (*see* **Note 29**). For OligoFISH probes, wash three times for 5 min in $2\times$ SSC at 42 °C.

19. If using nick-translated probes, wash three times for 5 min in $0.1\times$ SSC at 45 °C (*see* **Note 29**). For OligoFISH probes, wash three times for 5 min in $2\times$ SSC at 60 °C.

20. Stain DNA with 50 µL/well of a working solution of DAPI (1–5 µg/mL) in PBS for 10 min at room temperature.

21. Wash once in PBS, add 50 µL/well of PBS to the plate, then seal with aluminum adhesive foil.

22. Store the plate at 4 °C for up to 2 months (*see* **Note 30**).

3.6 Automated Image Acquisition Setup

1. Equilibrate the DNA FISH stained 384-well imaging plate to room temperature (*see* **Note 31**).

2. Clean the bottom of the plate with a Kimwipes tissue and 70% ethanol.

3. Load the 384-well plate onto the high-throughput microscope stage.

4. Select the appropriate objective. We routinely use a $60\times$ water objective (NA (numerical aperture) 1.2), or a $40\times$ water objective (NA 1.15).

5. Select camera binning 1×1.

6. Program the four acquisition channels. We routinely use these combinations of excitation light sources and emission bandpass filters.

    Ex: 405 nm, Em: 445/45 nm, DAPI.

    Ex: 488 nm, Em: 525/50 nm, Dyomics DY488.

    Ex: 561 nm, Em: 600/37 nm, Dyomics DY549P1.

    Ex: 640 nm, Em: 676/29 nm, Dyomics DY647P1.

7. Program the microscope to acquire a 3D z-stack in four channels in a random well and a random field of view (FOV) (*see* **Note 32**). Use the DAPI channel for this process, and then verify that the z-stack settings apply to all other channels. The

z-stack should include the whole volume of most nuclei, but not empty space below or above them. This is performed in snapshot mode (i.e., the images are not saved).

8. Verify the signal to background ratio in the DAPI channel. This should have a value of at least 5 at the focal plane (Fig. 2b).

9. Verify that the fluorescence signal is not saturated (i.e., the FOV does not include pixels with the maximum grayscale value for the camera used).

10. If conditions in **steps 8** and **9** are not met, change the camera exposure time and/or the excitation light power.

11. Repeat **steps 7–10** for all three FISH channels (Fig. 2b).

12. Move to other wells and acquire a few z-stacks in snapshot mode to verify that the image acquisition settings apply to the rest of the plate. If it looks like the cells were stressed or over-crowded when cultured (Fig. 2c), repeat the experiment by adjusting the initial number of cells seeded per well (*see* **Note 15**). If a large fraction of the FISH probe sets on the plate does not show the expected spot-like FISH signals and/or high fluorescence background (Fig. 2d), repeat the experiment by modifying the DNA FISH conditions (*see* **Notes 14, 28** and **29**).

13. Program the microscope to acquire a sufficient number of FOVs per well to acquire ~1000 cells per condition.

14. Launch the image acquisition routine.

3.7 Image Analysis

3.7.1 Image Analysis Pipeline Setup

Analyze the hiFISH image datasets acquired with the high-throughput microscope using an automated pipeline for automated image analysis, which is custom-designed by the user and implemented using commercial or open source software (e.g., Python, ImageJ, KNIME, CellProfiler) (Fig. 3a).

1. Apply a shading correction to compensate for illumination field artifacts. This is an absolutely necessary step for images acquired on spinning disk confocal microscopes, but it is also good practice if widefield microscopes are used [24].

2. Register the DAPI and DNA FISH channels using images from the wells containing fluorescent beads as a reference to compensate for camera misalignment and for chromatic aberrations.

3. Maximally project the 3D z-stacks for all channels (*see* **Note 32**).

4. Segment nuclei using the maximally projected DAPI image (Fig. 3a). Interactively adjust the nuclear segmentation parameters so that ~90–95% of the nuclei in random test images are properly segmented. Iterative cycles of parameter alteration and visual inspection might be needed.

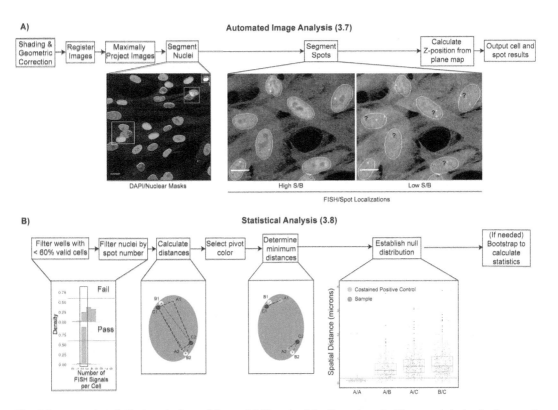

Fig. 3 Image and statistical analysis workflows. (**a**) Flowchart for the automated image analysis pipeline used to identify single cells and localize FISH signals in *X*, *Y*, and *Z* (Optional). The pseudocolored nucleus mask image shows a representative output of the automated nuclear segmentation step. The yellow box indicates an oversegmentation error. The cyan box represents a merge error. The teal box represents segmentation of a nucleus in mitosis. All these segmentation events need to be filtered out from the final analysis by applying a combination of filters set based on morphology, fluorescence intensity, and/or fluorescence texture features calculated from the DAPI channel. Scale bar: 20 μm. The FISH/Spot localization images show detection of FISH signals (Pseudocolored circles), as output by the image analysis software. The white masks represent the border of the nucleus masks where the spots are detected. The question mark represents questionable or missed FISH signal detections, which are particularly common in situations when the FISH channel images show low fluorescence signal to background ratio (S/B). Scale bar: 15 μm. (**b**) Flow chart for the automated statistical analysis pipeline. The histogram represents the distribution of the number of detected FISH spots signals per cell in one channel for diploid human immortalized fibroblasts for a successful FISH experiment (bottom), or a failed FISH experiment (top), respectively. The two pictograms represent the nucleus in gray, the detected FISH spot centers as pseudocolored circles (A, B, and C for FISH probes in different colors), and all the Euclidean distances between FISH spots in different channels as dashed colored lines, for the whole or minimum sets of distances per nucleus, respectively. The boxplot represents the single chromosome level distribution of distances between different FISH probes labeled with two different colors (Pink), or the positive colocalization control where cells are FISH stained with a mix of the same probe labeled with two different colors (Teal)

5. Find and segment DNA FISH signals in the nucleus ROI (region of interest) in each of the maximally projected FISH channels (Fig. 3a). Interactively adjust the spot segmentation parameters so that ~90–95% of the FISH spots in each channel for random test images are properly segmented (Fig. 3a). This step will output the *X-Y* coordinates of the FISH signals.

6. Optional: if 3D distance analysis needs to be performed, calculate a plane map image for each FISH channel where every pixel contains the coordinate of the Z-plane containing the maximum greyscale value at that X-Y position. This step will output the Z coordinates of the FISH signals (*see* **Note 32**).

7. Optional: if radial position needs to be determined, calculate the distance transform for each nucleus to convert X-Y position within the nucleus to radial position relative to the nuclear edge. This step will output the radial positions of the FISH signals.

8. Launch the optimized image analysis in batch on the entire dataset.

9. Output the results of the analysis as tabular text files (One file per population, per well). Two types of results are generated, cell level results and spot level results for each FISH channel. The cell level results should contain, for each cell, an index for the plate, for the well, and for the FOV it belongs to. The spot level results should contain, for each spot, an index for the plate, for the well, for the FOV, and for the cell it belongs to. In addition, it should also contain the spot X, Y, and Z (if calculated) coordinates.

*3.7.2 Visual Inspection of Image Analysis Results as a Quality Control Step (See **Note 33**)*

1. Randomly select a pool of ~10 random images and their corresponding image analysis results to visually examine the results of automated nuclear segmentation. This step can alert the user to the need to go back to Subheading 3.7.1, **step 4** and optimize the parameters of nuclear segmentation to correct for variability in the DAPI staining, cell shape, and cell density (Fig. 3a).

2. Once the results of the nuclear segmentation are confirmed, examine a random pool of ~20 cells per probe combination to visually validate the accurate detection of the FISH spots. This allows optimization for variability in the intensity, background, and size of the FISH spots (Fig. 3a).

3.8   Statistical Analysis

3.8.1   Pairwise FISH Signal Distance Calculations

1. In the statistical analysis pipeline, analyze only nuclei that contain the expected number of FISH signals (based on ploidy, i.e., two spots for a diploid cell in G1 phase) in every channel (*see* **Note 34**).

2. Include in the analysis only wells where at least 60% of the nuclei have the expected number of FISH signals in each channel (*see* **Note 35**) (Fig. 3b).

3. For each cell, calculate all the possible Euclidean distances between FISH signals in different channels (i.e., Red/Green, Red/FarRed, and Green/FarRed) in the same cell (Fig. 3b). The output of the calculation should include an index for each FISH signal for which the distances are calculated.

4. If the experiment design employs FISH probes against regions on the same chromosome, select one "pivot" color, and calculate the set of minimum distances based on the indexes of the pivot color (Fig. 3b). In the case of diploid cells, each cell should contain sets of minimum distances with unique FISH signal indexes for each color combination (Red/Green, etc.).

3.8.2 Determine Your Null Hypothesis: Experimentally Define What "Colocalization" Means

All microscopes have technical limitations (resolution, optical aberrations, mechanical drift, etc.), and as such there is a certain amount of built-in drift and technical noise in the acquisition that can be difficult to correct post-hoc without selecting fields with well-placed fiducial markers in them, which automated imaging makes impossible. To ensure the validity of the results despite this technical noise, it is imperative to check microscope alignment using fluorescence beads (as described in the section above) and determine colocalization thresholds experimentally (Fig. 3b).

1. Determine technical noise by imaging beads in every experiment.

2. Find bead spot centers for each color used in the FISH experiment and determine distances between channels: distance to nearest bead of a different color.

3. Option 1: use bead distances as a quality control metric, ruling out experiments with "significant" misalignment. In our experimental conditions, median offsets between channels of around 90 nm and observing 95% of beads within 180 nm is considered normal.

4. Option 2: use bead images to calculate an isometric transformation to computationally align the channels. Most high throughput microscopes have built-in functionality to do this, but generating a transformation on the basis of bead positions or images can sometimes improve channel alignment. In our hands, this provided at best a marginal improvement, as spot centers were rarely adjusted by more than 40 nm, but for some experiments this may be important.

5. Determine an experimental threshold for "colocalization" by including a costained control (one FISH probe labeled in all three colors) in each experiment.

6. Determine spot-to-spot distances as normal in the costained control (*see* **Note 36**) (Fig. 3b).

7. Determine an experimental distance cutoff for defining colocalization. We generally use the 95th percentile as cutoff: the distance within which 95% of the distribution of costained spots is found. In our hands, this has generally been comparable to our threshold for beads, or around 180 nm.

One complication of working with large datasets is that even small effects may nonetheless be statistically significant in a very large dataset. As such, in reporting on the results of high-throughput imaging analyses, we consider it absolutely imperative to report the effect size as well as the statistical significance. Furthermore, for more complicated analyses it may be worth bootstrapping the dataset to reduce this effect (Fig. 3b).

1. After calculating parameters for all spots/cells, take a random subsample of the same size as a low-throughput experiment (for FISH datasets, this is often around 100–200 spots).

2. Calculate relevant statistics (in particular, mean, median and standard deviation) on this subsample.

3. Repeat the process multiple times, each time starting from the entire dataset (i.e., subsamples are generated *with replacement*). In practice, this is often done hundreds of thousands of times to generate a sampling distribution of each statistic.

4. Perform subsequent tests directly on these sampling distributions. Calculate a 90% confidence interval for the mean by obtaining fifth and 95th percentiles on the sampling distribution of means. To determine whether the median is greater than 200 nm with a confidence threshold of 0.05, calculate whether the fifth percentile of the sampling distribution of medians is greater than 200 nm.

4 Notes

1. We routinely use a large-scale plasmid preparation kit for preparation of BAC and fosmid DNA. We find that the Nucleobond BAC 100 Maxiprep kit from Takara works particularly well, but any alkaline lysis column preparation method optimized for low-copy plasmids should provide similar results. Start by streaking the bacteria from a frozen glycerol stock onto an agar plate containing the appropriate selection antibiotic to obtain single bacterial colonies. Pick a single bacterial colony and use it to inoculate 4 mL of bacterial growth media plus the appropriate selection antibiotic and grow the culture overnight at 37 °C. Use the whole volume of this starter bacterial culture to inoculate 400 mL of fresh bacterial growth media plus the appropriate selection antibiotic and grow the culture overnight at 37 °C. Use this culture as the input for the plasmid preparation kit. Spinning the lysate to pellet the supernatant and warming the elution buffer have both increased our yields and purities and improved the subsequent nick translation steps. In general, a good DNA prep will have a concentration above 80 ng/μL, A_{260}/A_{280} ratio between 1.7 and 2, and A_{260}/A_{230} ratio between 1.9 and 2.2. Furthermore, when

run on a gel, it should run as a single very long band (~50 kb for a fosmid and up to 200 kb for a BAC).

2. We have had success with this protocol using a variety of fluorescently conjugated nucleotides including Alexa dyes, Cy-dyes, and Dyomics dyes, as well as two-step labeling processes with incorporation of hapten-conjugated nucleotides (Biotin, Digoxygenin), followed by visualization with fluorescently tagged conjugates (such as Avidin) or antibodies. Our current preference is for Dyomics dyes, which we find to be very bright and stable, but we have reason to believe that many if not most synthetic fluorophores will be amenable to this technique. Importantly, repeated freeze-thaw cycles will degrade most fluorophores, and as such freeze-thaw cycles should be kept to a minimum (two times max) both before and after nick translation.

3. As DNase I activity determines the frequency of nicks and hence the efficiency of the nick translation, and as lyophilized DNase I is frequently variable in its activity, we recommend titrating the appropriate DNase concentration on a well-known probe whenever one makes a fresh 1 mg/mL DNase stock. To do this, perform the nick translation reaction using a probe that is known to reliably work as the substrate, and with several DNase dilutions between 1:100 and 1:2000. Examine the results on a 2% agarose gel to see the size of the products, with a uniform smear between 100 and 1000 bp as the desired outcome (Fig. 1a).

4. Use of Cot-1 DNA specific to the organism of origin of the cells used in the FISH experiment is critical.

5. 20 nt barcode/readout sequences should have no homology to genomic regions, be orthogonal relative to other barcode sequences, be relatively uniform in GC content, and be unlikely to form hairpins or other secondary structures. Here are sequences for ten of the readout probes we have used successfully in the past:

5'-CTCAGTGGCGGCCAATGTCAACTCTGCATCTT
5'-GTGCAAAGGAGTACCGGGCACCTACCATGGAA
5'-CCTTCCCGACGAGTTGCCGCGATCCGATTTGT
5'-TTTTGGTTGGCAGCTGAACCGGGCGGGCAACA
5'-TCCATGGCTAGCCACATACTGCGCGACTATCT
5'-TGTGGGTGATCATGACTGTTGACCGGGTGGCA
5'-CTTAGAACCGTCCCGAGCGAAGCGAGGATTGA
5'-GCGGCGAGTTCCGAAGCCGACCGGCGTCCAAC
5'-CTACACGGGCTGTAGCTTGGTCGAATCAACGA
5'-GTCGAATTGAGCTTTGGCCGACTCCCCAACTT

Also *see* **Note 8**.

6. Because some residual liquid will remain in the sponge even after a thorough wringing, it is best to dedicate specific sponges for use with hazardous chemicals that need specific handling. We dedicate one sponge to be used with PFA for fixation, one to be used with the formamide equilibration buffer, and one to be used with nonhazardous buffers. This limits cross-contamination between waste categories and ensures that the sponge itself, after several uses, can be disposed of according to chemical waste disposal regulations.

7. One of the principal advantages of nick translation is that the long (up to 1 kb) randomly dispersed segments of DNA that are created as probes have high melting temperatures, and therefore bind stably to their genomic targets. Another advantage is that probes generated by nick translation have a high density of labeled oligonucleotides (up to 25% of the oligonucleotides in the target sequence), thus making them very bright and relatively easy to detect under the microscope. On the other hand, while it is possible to select the appropriate genomic region of interest from large libraries of BAC or fosmids, this approach does not allow the programmatic, in silico design of short probes with bioinformatics tools. Another disadvantage of probes generated by nick translation is that the large contiguous genomic regions included in BAC and fosmid vectors can include short repeated sequences and secondary structures that can affect the specificity and efficiency of the hybridization of the probe. Furthermore, it is difficult to precisely determine the boundaries of genomic regions stained with BAC or fosmid probes.

8. To generate an OligoFISH library, search the genome with bioinformatic algorithms for hundreds of thousands of short (~30–40 bp) DNA sequences that are genomically unique, that are nonrepetitive, that lack secondary structures, and that have uniform GC content and melting temperatures [16, 17, 25, 26]. Each genome-binding oligonucleotide is then concatenated in silico with at least two target-specific barcode sequences (~20–30 bp), which allow the primer-specific PCR amplification of the library. OligoFISH libraries are chemically synthesized as oligonucleotide pools, and they generally contain thousands of oligonucleotides per genomic region, for up to a few hundred genomic regions per library. The subset of oligonucleotides targeting a single genomic region all have different genome-binding regions, but they all share the same synthetic barcode sequences. After hybridization of the library to the genome in an "encoding" step, and since the oligonucleotide pool library is not fluorescently labeled, each genomic region can be visualized by a secondary hybridization with fluorescently labeled "decoding/readout" oligonucleotide probes that are specific for each library barcode. As an alternative to short "decoding/readout" oligonucleotides,

hybridization chain reaction (HCR) [27, 28] or signal amplification by exchange reaction (SABER) [29] can be used to amplify the signal of OligoFISH.

9. Bioinformatics algorithms allow the programmatic design of oligoFISH libraries where the oligonucleotide probes have uniform biophysical properties and low nonspecific hybridization background. In addition, this also enables users to precisely control where the probe sets of probe sequences start and end for each genomic region targeted. OligoFISH also allows for the design of complex combinatorial barcode schemes, which together with multiple rounds of hybridization, stripping, and imaging of the shorter "readout," drastically increases the number of regions that can be visualized in a single cell [14, 18, 19]. However, as opposed to more traditional BAC or fosmid DNA FISH approaches, in our hands OligoFISH results in lower signal and higher noise in a multi-well plate format, and a greater sensitivity to degradation of DNA within the fixed cells. This might be due to the relatively shorter stretches of genomic homology in oligo probes or the much lower fluorophore density of OligoFISH as opposed to BAC or fosmid DNA FISH, resulting in lower fluorescence signal at imaging time.

10. Nick translated probes can be frozen immediately after the reaction has been stopped and stored at -20 °C for several weeks. The same is true for probe mixes resuspended in hybridization buffer. In either case, since the fluorophores are still sensitive to freeze-thaw cycles, we recommend that you limit these to one at most.

11. The concentrated library can be stored for long times at -80 °C. We strongly recommend aliquoting it into 5–10 µL aliquots and doing primary amplification steps with 40–100 wells, as the reactions of the primary amplification are also stable at -20 °C for several weeks to use in multiple experiments. In every case, aliquot and thaw only what you need to prevent freeze-thaw cycles which can cause shearing and degradation of the DNA.

12. We use the Zymo DNA Clean and Concentrator kits to purify the PCR product (Zymo DNA Clean and Concentrator-5 for low volume). We expect similar kits, or other purification methods, would yield similar results, but these performed the best in our hands.

13. Because the RNA intermediate is quickly reverse-transcribed and degraded, and because it is always incubated with RNase inhibitors, we have not found RNA degradation to be a common or significant risk. That said, we do take care to ensure an RNase-free environment, by wiping down tube racks, ice buckets, and benches with spray RNase-inactivating cleansers before the RNA transcription step.

14. The final concentration of reagents in the hybridization buffer is important for the efficiency of the FISH reaction itself, and as probe mixes in this technique are not being precipitated to concentrate them in the hybridization buffer, we generally recommend that total volumes of encoding probe, readout probes, and C_0t-1 not exceed ~3 μL per well. Using a vacuum concentrator to dry and concentrate the aqueous solutions containing OligoFISH probes can be a good, fast alternative to ethanol precipitation since the reagents in question have already been purified.

15. The appropriate number of cells plated per well should be optimized as part of the assay development. Plating too few cells will force acquisition of a much larger number of images, whereas plating too many cells will likely cause issues with the automated nuclear segmentation algorithms.

16. Edge effects, where cells in wells located at the edge of the plate behave differently than in other wells on the plate, are a notorious issue with 384-well plates and long incubation times (i.e., ≥72 h). This is largely due to increased media evaporation rates at the outer edge of the plate. To mitigate these issues, ensure that the humidity in the incubator is kept constant, by not opening and closing the doors of the incubator multiple times during the incubation, and by making sure the incubator has enough water in the water pan.

17. In our experience, cells can sometimes detach if the growth medium is discarded from the 384-well plates before fixation. For this reason, direct fixation of cells in growth media is recommended. This step can sometimes result in fluorescent background, but in our hands this has not been a major drawback.

18. DNA FISH wash steps with nonviscous and nonhazardous buffers (e.g., PBS, SSC, permeabilization, and acid treatment) can be greatly sped up by using plate washers and/or automated liquid handlers (Such as the BlueCatBio BlueWasher or the Biotek EL406). Such devices will also work to dispense hazardous but nonviscous solutions such as fixative and equilibration mix, although steps must be taken to ensure safe handling disposal of formaldehyde and formamide, such as ensuring that they are operated in a chemical fume hood, and that the liquid waste lines are connected to sealed chemical waste containers using appropriate air-tight fittings and filtered air relief valves.

19. Our preferred method for changing wash buffers is tapping the plate manually to remove the old buffer and immediately replacing it using a multichannel pipette or plate washer. Care must be taken not to dry out the wells. However, this sometimes results in cell loss, in particular for delicate cells grown in

suspension and not well adhered to the plate bottom, such as patient derived and purified PBMCs, dissociated colonies of human stem cells, and suspension cell lines. In these cases, the use of a sponge to very gently remove the old buffer from the plate is recommended. First, soak a sponge in 70% ethanol and squeeze it to remove residual liquid before starting. Then, use the sponge to remove the wash buffer from the wells: fill the plate with wash buffer to the top, invert the plate face down on the sponge, and gently press down. Press down again over a specific well, or move the plate, to get better liquid removal if some wells still show residual liquid. After each use, squeeze out the liquid from the sponge for proper waste disposal. Then thoroughly wash the sponge with water, return it to a 70% ethanol bath, and squeeze to permeate. Store the sponge at room temperature in 70% ethanol.

20. We have had reasonable success in coating individual wells of a 384-well plate with matrigel in entirely the same way one would coat a larger dish with matrigel. To do this, start by diluting matrigel to an appropriate concentration with DMEM-F12 while on ice, then add 25 µL diluted matrigel per well onto an ice-cold 384-well plate using ice-cold pipette tips. Let the matrigel solidify at room temperature for at least 30 min. For longer-term storage, add 25 µL of additional cell growth media to each well, seal with parafilm, and store at 4 °C for up to 5 days.

21. We do not recommend fixing cells just 1 day after seeding, as colonies will not have flattened and the nuclear segmentation step in the image analysis will be much more challenging.

22. The required timing for equilibration depends on the DNA FISH probes used. Nick-translated probes work best with an overnight equilibration at 4 °C, and also work well with plates that have been stored in equilibration buffer at 4 °C for up to a week. OligoFISH probes need freshly fixed cells, work best with a 30 min equilibration, and can tolerate an overnight storage at 4 °C, but not longer.

23. DNA FISH hybridization mixes tend to be very viscous due to the presence of high concentrations of dextran sulfate. Furthermore, some applications will require many different combinations of probes on the same plate. For these reasons, plate washers and automated liquid handlers with a "one-to-many dispensing" setup will be insufficient. We have had some success using tips-based automated liquid handlers to dispense hybridization mix for both DNA FISH and RNA FISH. As an alternative, first preparing a global source plate of hybridization mix in a 96-well format and then transferring the mix to a 384-well imaging plate using a multichannel pipette is a lower-throughput approach that nonetheless saves time compared to individually pipetting each well.

24. Because cells are not hybridized with DNA FISH probes between a coverslip (or piece of parafilm) and a slide, as is done in conventional FISH methods, and despite the small cross section of the wells, one will need to use more hybridization buffer to prevent wells from drying out during hybridization. We typically use 12–14 μL of hybridization buffer per well for a 384-well plate. Note that this is twice as much as needed for a standard coverslip, despite containing comparatively fewer cells. It can be tempting therefore to reduce the probe concentration. We have had success doing this with probes to small regions (10 kb or less), or with oligo libraries, but BAC probes and fosmid probes generally perform better when provided in excess.

25. While Tetraspeck beads stick to cells, they will float in PBS or water and will not provide good localization unless they are dried to the bottom of the well. We generally do not add Tetraspeck beads in wells containing cells, as we have observed that channel alignment is sufficient without image-specific fiducial markers to individually align images. In addition, Tetraspeck beads tend to be much brighter than DNA FISH spots and can occlude them or can saturate images. Since the fields-of-view (FOVs) in each well is predetermined but randomized as to what is present in each FOV, it is not possible to select FOVs to ensure that beads and FISH signals are resolvable.

26. This dilution can be stored at 4 °C indefinitely, but vortex well before using.

27. While cyclic olephin bottom plates are cheaper and easier to handle than glass bottom plates, they have worse optical properties. On the other hand, differences in thermal expansion coefficients between glass and plastic and reductions in adhesive strength at high temperatures mean that some glass plates cannot withstand high denaturation temperatures without warping or without the glass bottom detaching from the frame of the plate.

28. Optimal hybridization time depends strongly on the probes and fluorophores used. We have observed that brighter fluorophores in the green channel tend to perform better with shorter hybridization times, while fainter fluorophores such as Cy-5 or other far-red dyes tend to prefer longer hybridization times. Similarly, shorter probes such as oligonucleotide-based probes tend to perform better with shorter hybridization times while longer probes such as nick translated probes tend to do better with longer hybridization times. In our hands, an incubation anywhere from overnight to 5 days has yielded good results.

29. The small cross section and the geometry of wells in a 384-well plate make it more difficult to remove buffers when compared to traditional DNA FISH formats such as slides or coverslips. For this reason, the fluorescence background will generally be higher than in traditional DNA FISH protocols performed on slides. Careful design and testing of probes can partially mitigate these issues but altering wash conditions has also been useful. For gentler washes, consider using only $2\times$ SSC or warming the buffers and doing the washes at room temperature. For more stringent washes, consider lengthening the time, switching to $0.1\times$ SSC earlier, or raising the temperature to 60 °C.

30. Our best results were obtained by imaging immediately after DAPI staining. However, we have reimaged old plates with good FISH signals months later and signal strength was only modestly reduced.

31. This is especially relevant if the plate was stored at 4 °C after washing and before imaging. A plate that is colder than the ambient temperature can generate condensation on its bottom, which will interfere with the ability of the microscope to autofocus and can even damage the microscope if using air objectives.

32. Image acquisition is performed in 3D to capture entire nuclear volumes, even if the analysis is done on 2D maximal projections. We have previously observed some discrepancies in distance measurements between DNA FISH signals calculated from maximally projected 2D and distances calculated using 3D image stacks [30]. In short, 3D measurements of distance tend to be less precise but more accurate than 2D distances (as in the measured distance will be closer to the real value, but the variance in the distribution around the measured value will be larger) and imaging for 3D measurements requires oversampling in the z-stack according to the Nyquist criterion (which defines the optimal measurement sampling value in any dimension as two- to threefold smaller than the resolution of the measurement in that dimension) and post-processing alignment steps to register channels in z. In our hands, a $\sim 3\times$ oversampling rate worked well: for objectives with a depth-of-field of 1 μm, a z-stack height of 300 nm provided good results. It is worth noting that in optimal acquisition conditions, the Nyquist criterion should also be applied in the X and Y dimensions. As an example, using a $60\times$ water objective (NA 1.2) and a camera with a single sensor physical size of 6.5 μm, the pixel size is 108 nm, which is ~ 2.5-fold to threefold less than the lateral resolution in these conditions with a light source at 488 nm. However, 2D deviations are not usually systematic and we have consequently observed that trends observable in 3D data are resolvable in 2D data as well.

33. In traditional microscopy pipelines, images are often examined individually, and this provides a quality control step as researchers can exclude cells that did not stain well. But when generating thousands to millions of images per experiment and analyzing them in a fully automated fashion (i.e., in batch), it is not feasible to individually examine millions of nuclei. We employ visual "spot-checking" on a few random images for quality of the nuclear segmentation, as well as setting thresholds for spot detection efficiency to ensure that experiments that failed are properly flagged and the data we use is of high quality.

34. When working with nontransformed, diploid human cells, the number of detected FISH signals per cell in each channel is 2 or 4, depending on whether interphase cells are in G1 or in S/G2 (mitotic cells are not considered at the nuclear segmentation step). This assumption generally does not apply to cancer cell lines, and for this reason it is critical to karyotype them before starting any FISH experiment. Filtering by ploidy ensures that nuclei and spots within them have been properly segmented. Furthermore, this step allows relative confidence that each genomic region can properly be matched with its pairs on the same chromosome without applying an additional spatial distance-based threshold. Note however, that the likelihood of staining and segmenting all copies of a locus within a cell degrades as ploidy gets higher: an experiment with 80% overall detection efficiency will generally have 64% of diploid cells being properly segmented, but only 51% of triploid cells being properly called. We recommend adjusting your expectations accordingly.

35. We use this filter because in DNA FISH experiments that have failed the FISH signal-to-noise ratio is often low and the spot detection algorithms tend to pick up a high proportion of false positives. This is undesirable, because in these conditions a certain number of nuclei will randomly have the expected number of FISH signals, which are nonetheless likely to be background. This represents a "Garbage in, garbage out" situation, and we instead focus on troubleshooting FISH probe design, probe generation, permeabilization conditions, hybridization conditions, and washes to get better staining, as opposed to attempting to analyze low quality data that have a high potential to lead to incorrect measurements.

36. The triple-stained locus is never stained as well as experimental probes, due to competition between the three colors. For this reason, we always select robust probes for this control and we sometimes do not apply our well-based QC filtering step to it. It is plausible that there are background spots that have been segmented and analyzed in this sample, and therefore the thresholds determined here are likely to be upper bounds.

Acknowledgments

Research in the Misteli lab and at HiTIF was supported by funding from the Intramural Research Program of the National Institutes of Health, National Cancer Institute and Center for Cancer Research, projects 1-ZIA-BC010309 and 1-ZIC-BC011567-01, respectively. The authors declare no conflict of interest.

References

1. Sati S, Cavalli G (2017) Chromosome conformation capture technologies and their impact in understanding genome function. Chromosoma 126:33–44

2. Tan L, Xing D, Chang C-H et al (2018) Three-dimensional genome structures of single diploid human cells. Science 361:924–928

3. Collombet S, Ranisavljevic N, Nagano T et al (2020) Parental-to-embryo switch of chromosome organization in early embryogenesis. Nature 580:142–146

4. Lajoie BR, Dekker J, Kaplan N (2015) The Hitchhiker's guide to Hi-C analysis: practical guidelines. Methods 72:65–75

5. Williamson I, Berlivet S, Eskeland R et al (2015) Spatial genome organization: contrasting views from chromosome conformation capture and fluorescence in situ hybridization. Genes Dev 28:2778–2791

6. Lanctôt C, Kaspar C, Cremer T (2007) Positioning of the mouse Hox gene clusters in the nuclei of developing embryos and differentiating embryoid bodies. Exp Cell Res 313: 1449–1459

7. Brickner DG, Sood V, Tutucci E et al (2016) Subnuclear positioning and interchromosomal clustering of the GAL1-10 locus are controlled by separable, interdependent mechanisms. Mol Biol Cell 27:2980–2993

8. Harada A, Mallappa C, Okada S et al (2015) Spatial re-organization of myogenic regulatory sequences temporally controls gene expression. Nucleic Acids Res 43:2008–2021

9. Finn EH, Pegoraro G, Brandão HB et al (2019) Extensive heterogeneity and intrinsic variation in spatial genome organization. Cell 176:1502–1515.e10

10. Roukos V, Voss TC, Schmidt CK et al (2013) Spatial dynamics of chromosome translocations in living cells. Science 341:660–664

11. Burman B, Misteli T, Pegoraro G (2015) Quantitative detection of rare interphase chromosome breaks and translocations by high-throughput imaging. Genome Biol 16: 146–146

12. Joyce EF, Williams BR, Xie T et al (2012) Identification of genes that promote or antagonize somatic homolog pairing using a high-throughput FISH-based screen. PLoS Genet 8: e1002667

13. Shachar S, Voss TC, Pegoraro G et al (2015) Identification of gene positioning factors using high-throughput imaging mapping. Cell 162: 911–923

14. Chen KH, Boettiger AN, Moffitt JR et al (2015) RNA imaging. Spatially resolved, highly multiplexed RNA profiling in single cells. Science 348:aaa6090

15. Langer PR, Waldrop AA, Ward DC (1981) Enzymatic synthesis of biotin-labeled polynucleotides: novel nucleic acid affinity probes. Proc Natl Acad Sci U S A 78:6633–6637

16. Beliveau BJ, Joyce EF, Apostolopoulos N et al (2012) Versatile design and synthesis platform for visualizing genomes with Oligopaint FISH probes. Proc Natl Acad Sci U S A 109: 21301–21306

17. Beliveau BJ, Boettiger AN, Avendaño MS et al (2015) Single-molecule super-resolution imaging of chromosomes and in situ haplotype visualization using Oligopaint FISH probes. Nat Commun 6:7147

18. Bintu B, Mateo LJ, Su J-H et al (2018) Super-resolution chromatin tracing reveals domains and cooperative interactions in single cells. Science 362:eaau1783

19. Su J-H, Zheng P, Kinrot SS et al (2020) Genome-scale imaging of the 3D organization and transcriptional activity of chromatin. Cell 182:1641

20. Nguyen HQ, Chattoraj S, Castillo D et al (2020) 3D mapping and accelerated super-resolution imaging of the human genome using in situ sequencing. Nat Methods 17: 822–832

21. Medina-Cucurella AV, Steiner PJ, Faber MS et al (2019) User-defined single pot mutagenesis using unamplified oligo pools. Protein Eng Des Sel 32:41–45

22. Meaburn KJ (2010) Fluorescence in situ hybridization on 3D cultures of tumor cells. In: Bridger JM, Volpi EV (eds) Fluorescence in situ hybridization (FISH). Humana Press, Totowa, NJ, pp 323–336

23. Lundholt BK, Scudder KM, Pagliaro L (2003) A simple technique for reducing edge effect in cell-based assays. J Biomol Screen 8:566–570

24. Singh S, Bray M-A, Jones TR et al (2014) Pipeline for illumination correction of images for high-throughput microscopy. J Microsc 256:231–236

25. Beliveau BJ, Apostolopoulos N, Wu C-T (2014) Visualizing genomes with Oligopaint FISH probes. Curr Protoc Mol Biol 105:Unit 14.23

26. Hershberg EA, Close JL, Camplisson CK et al (2020) PaintSHOP enables the interactive design of transcriptome- and genome-scale oligonucleotide FISH experiments. Nat Methods

18:937. https://doi.org/10.1038/s41592-021-01187-3

27. Choi HMT, Beck VA, Pierce NA (2014) Next-generation in situ hybridization chain reaction: higher gain, lower cost, greater durability. ACS Nano 8:4284–4294

28. Choi HMT, Schwarzkopf M, Fornace ME et al (2018) Third-generation in situ hybridization chain reaction: multiplexed, quantitative, sensitive, versatile, robust. Development 145: dev165753

29. Kishi JY, Lapan SW, Beliveau BJ et al (2019) SABER amplifies FISH: enhanced multiplexed imaging of RNA and DNA in cells and tissues. Nat Methods 16:533–544

30. Finn EH, Pegoraro G, Shachar S et al (2017) Comparative analysis of 2D and 3D distance measurements to study spatial genome organization. Methods 123:47

Chapter 13

Versatile CRISPR-Based Method for Site-Specific Insertion of Repeat Arrays to Visualize Chromatin Loci in Living Cells

Thomas Sabaté, Christophe Zimmer, and Edouard Bertrand

Abstract

Hi-C and related sequencing-based techniques have brought a detailed understanding of the 3D genome architecture and the discovery of novel structures such as topologically associating domains (TADs) and chromatin loops, which emerge from cohesin-mediated DNA extrusion. However, these techniques require cell fixation, which precludes assessment of chromatin structure dynamics, and are generally restricted to population averages, thus masking cell-to-cell heterogeneity. By contrast, live-cell imaging allows to characterize and quantify the temporal dynamics of chromatin, potentially including TADs and loops in single cells. Specific chromatin loci can be visualized at high temporal and spatial resolution by inserting a repeat array from bacterial operator sequences bound by fluorescent tags. Using two different types of repeats allows to tag both anchors of a loop in different colors, thus enabling to track them separately even when they are in close vicinity. Here, we describe a versatile cloning method for generating many repeat array repair cassettes in parallel and inserting them by CRISPR-Cas9 into the human genome. This method should be instrumental to studying chromatin loop dynamics in single human cells.

Key words Live-cell imaging, Repeat array, Chromatin 3D architecture, TADs, Loop extrusion

1 Introduction

The combination of Hi-C [1] or related genomic techniques and imaging methods such as DNA FISH [2–4] led to a growing understanding of the 3D architecture of the genome and to the discovery of chromatin structures including topologically associating domains (TADs) and chromatin loops [5, 6]. It has been shown that these structures emerge, at least partly, from a DNA loop extrusion process driven by the cohesin complex [7, 8]. However, most studies have been restricted to fixed cells and/or to cell population averaged snapshots of the genome architecture. The

The original version of this chapter was revised. The correction to this chapter is available at https://doi.org/10.1007/978-1-0716-2497-5_16

Tom Sexton (ed.), *Spatial Genome Organization: Methods and Protocols*,
Methods in Molecular Biology, vol. 2532, https://doi.org/10.1007/978-1-0716-2497-5_13,
© The Author(s), under exclusive license to Springer Science+Business Media, LLC, part of Springer Nature 2022,
Corrected Publication 2022

temporal dynamics of loops and TADs in single cells, although of major importance to understanding their role in gene regulation and chromosome organization, has been insufficiently character- ized [9–11]. An important requirement to study the dynamics of chromatin regions by microscopy is to visualize specific genomic loci with live-cell compatible fluorescent labels. This can be achieved in various ways: using sgRNAs and catalytically dead Cas9 (dCas9) [12], using fluorescent reporter operator systems (FROS), such as lacO, TetO, or CuO [13, 14] arrays, or using the parS/ParB system [15]. Here, we describe a general cloning method to generate repair cassettes of CuO repeat arrays and to insert them in the genome of human cells by CRISPR-Cas9. This method can be extended to all types of repeat arrays for multicolor imaging.

Inserting the repeat array at a nontranscribed region of the genome will lead to more off-targets than tagging a coding sequence, because for nontranscribed regions the expression of the antibiotic resistance gene is under its own promoter while for a coding region it can be transcribed from the endogenous promoter via the use of an Internal Ribosome Entry Site (IRES) or 2A self-cleaving peptides. Therefore, we use a Cas9 D10A nickase that limits the off-target rate compared to a WT Cas9 [16]. We also use an improved scaffold of guide RNA that vastly increases their efficiency [12].

CRISPR-Cas9–mediated insertion of repeat arrays in the genome of human cells requires addition of homology arms on both sides of the repeat array. This can be achieved by PCR amplifi- cation of the long repeat array with primers already containing the homology arms [17]. However, this technique can only produce short homology arms (<200 bp), which reduces the efficiency of homology-directed repair insertion by Cas9, depends on the PCR yield and limits the choice of insertion sites. Therefore, we instead clone the homology arms into the repeat array plasmid. This pro- cedure is longer than with PCR amplified repair cassettes but allows generation of longer homology arms (>800 bp). Hence, it improves the integration rate compared to PCR-amplified products and increases the choice of insertion sites. The combination of the Cas9 nickase, optimized guide scaffold, and long homology arms allows for large operator repeats to be integrated relatively easily and with a good efficiency (10–65% of antibiotic-resistant clones contain the repair cassette inserted at the right location depending on the targeted locus).

The cloning method consists of PCR amplification of the homology arms by genomic PCR, which are then integrated by Gibson assembly at each side of the repeat array. This method allows to generate quickly and in parallel multiple repair cassettes for specific tagging of multiple chromatin loci, increasing the throughput of such experiments. Using two different types of repeats enables two-color imaging, which allows localization and tracking of two loci with subpixel and subdiffraction precision, even when the distance between them falls below the optical resolution.

Thus, it is possible to analyze and quantify the movements and distances between anchors of TADs or chromatin loops in living cells. We propose that this protocol is a useful tool to study the dynamics of chromatin structures.

2 Materials

1. A U6-based expression plasmid containing an improved scaffold for guide RNAs (12; available upon request). The plasmid bears two mutations compared to the original sequence. First, a run of four Ts after the guide sequence is modified to prevent Pol-III termination. Second, a stem-loop in the guide scaffold is stabilized to favor its formation at 37 °C (Fig. 1a). The guides are designed such that they do not cleave the repair cassette (Fig. 1b).

2. 10 U/μL BpiI enzyme (provided with 10× buffer G, ThermoFisher).

3. Nuclease-free water.

4. Gel extraction and PCR purification kits (e.g., NucleoSpin Gel and PCR Clean-up; Macherey-Nagel).

5. NanoDrop spectrophotometer or equivalent.

6. Separately synthesized forward and reverse oligos containing the sgRNA sequences resuspended at 1 μg/μL in TE buffer (10 mM Tris–HCl pH 8, 1 mM EDTA). *See* Subheading 3.1 for design strategy.

7. 10 mM ATP.

8. 100 mM DTT.

9. 400 U/μL T4 DNA ligase (NEB).

10. Electrocompetent, recombinant-deficient *E. coli* (e.g., XL1-Blue).

11. Bacterial electroporation system.

12. SOB medium.

13. LB medium.

14. 100 μg/mL ampicillin LB agar plates.

15. 100 μg/mL ampicillin.

16. Plasmid miniprep kit (e.g., QIAprep miniprep kit; Qiagen).

17. 20 U/μL BbsI-HF, provided with 10× CutSmart buffer (NEB).

18. 20 U/μL ScaI-HF (NEB).

19. Plasmid midiprep kit (e.g., Nucleobond Xtra Midi kit; Macherey-Nagel).

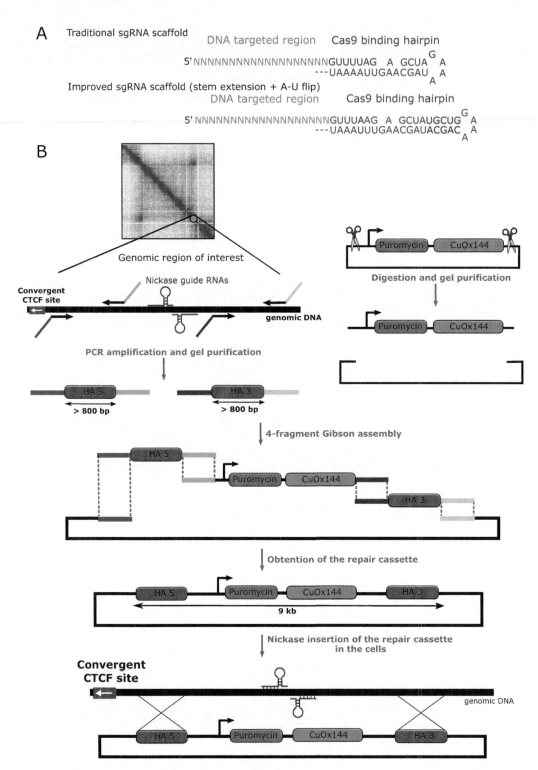

Fig. 1 Cloning method to generate repeat array repair cassettes at loop anchors. (**a**) Sequence of the traditional and optimized guide RNA scaffold designed in [12]. Modified nucleotides in the optimized scaffold are shown in purple. The dashes represent the *S. pyogenes* terminator. (**b**) Schematic representation of the cloning steps to generate the repeat array repair cassette. Top: Hi-C contact frequency map of a targeted

20. 10 μM forward and reverse primers to amplify homology arms (*see* Subheading 3.2 for design strategy).

21. HCT116 cells (American Type Culture Collection) expressing a fluorescently tagged CymR protein in the nucleus that will bind the CuO repeats (*see* **Note 1**).

22. Genomic DNA extraction kit (e.g., GenElute™ Mammalian Genomic DNA Miniprep Kit Protocol; Sigma-Aldrich).

23. 2 U/μL Phusion High-Fidelity DNA polymerase, supplied with 5X Phusion HF Buffer (ThermoFisher).

24. 10 mM dNTPs.

25. pDEST-CuOx144_Bxb1attB_loxP-PGKpuro-loxP containing the CuOx144 repeats and a puromycin-resistant gene, available on Addgene (#119903).

26. 20 U/μL NheI-HF (NEB).

27. 20 U/μL HindIII-HF (NEB).

28. Gibson assembly mix (NEBuilder HiFi DNA Assembly Master Mix; NEB).

29. Other restriction enzymes to verify Gibson assembly, will depend on the sequences of the cloned homology arms (*see* Subheading 3.3, **step 15**).

30. Cell culture medium: McCoy's medium supplemented with penicillin–streptomycin and 10% (v/v) fetal bovine serum (FBS).

31. pX335-U6-Chimeric_BB-CBh-hSpCas9-nickase_D10A which encodes the nickase D10A Cas9, available on Addgene (#42335).

32. JetPRIME buffer and reagent (Polyplus transfection).

33. 1× PBS (phosphate-buffered saline).

34. 0.5% trypsin–EDTA (diluted from 10× stock solution, Gibco).

35. Selection medium: Cell culture medium supplemented with 0.8 μg/mL puromycin.

36. Sterile cloning disks (diameter 3.2 mm) (Sigma-Aldrich).

37. 10 μM forward and reverse primers to screen array insertion into genome (*see* Subheading 3.5 for design strategy).

38. QuickExtract™ DNA Extraction Solution (Lucigen).

Fig. 1 (continued) chromatin loop. Long homology arms (>800 bp) are amplified by PCR on genomic DNA at the location of the loop anchor. The primers for homology arm amplification contain the overhangs (colored) for a Gibson assembly with the digested plasmid. Dotted gray lines show the overlap between the overhangs. The obtained repair cassette is then inserted in the cell genome by transfection of a nickase Cas9 together with two guide RNAs. Abbreviations: HA, homology arm

39. 5 U/μL GoTaq™ G2 Flexi DNA Polymerase (Promega), provided with 5× Green GoTaq Flexi Buffer and 25 mM MgCl₂ solution, or equivalent.

40. 4% (w/v) formaldehyde, made by diluting 16% formaldehyde (Thermofisher) in 1× PBS.

41. Vectashield antifade mounting medium with DAPI (LifeSpan Biosciences).

42. Widefield fluorescent microscope.

43. Glass-bottom petri dish suited for the chosen live-cell microscope (e.g., Petri fluorodish; Dutscher).

44. Live-imaging medium: antibleaching live cell visualization medium DMEM^{gfp}-2 (Evrogen), supplemented with 10% FBS, penicillin–streptomycin, and rutin (added fresh from 100× stock; Evrogen).

45. Fluorescent microscope setup for live imaging (see below).

3 Methods

3.1 sgRNA Design and Cloning

1. Use ChopChop (https://chopchop.cbu.uib.no/) to design the nickase sgRNAs targeting the region of interest (using the "CRISPR/Cas9 nickase" and "knock-in" options) and choose the highest ranked pair of guides which satisfies requirements (*see* **Note 2**).

2. Make the following modifications, if required, to the sgRNA oligonucleotide sequence design (*see* **Note 3** for an example).

   Do not include the NGG PAM sequence in the ordered oligonucleotides; On the side opposite to the removed NGG, add a G at the 5′ end if the oligonucleotide starts with a different nucleotide (and add a C at the 3′ end of the complementary oligo) for U6 transcription.

   Add cohesive ends identical to those of the digested vector containing the improved sgRNA scaffold: (a) before the 5′ end, on the side opposite to the NGG, add 5′ CACC on the oligonucleotide bearing the sgRNA sequence; (b) before the 5′ end, on the NGG side, add 5′ AAAC to the complementary oligonucleotide.

3. In a microtube, mix 5 μg improved sgRNA scaffold plasmid, 3 μL 10× BpiI Buffer G, 1 μL 10 U/μL BpiI and make the volume up to 30 μL with nuclease-free water. Incubate digestion reaction overnight at 37 °C.

4. Purify DNA with a PCR cleanup kit, following the manufacturer's instructions.

5. Measure DNA concentration with a NanoDrop spectrophotometer.

6. In a PCR tube, mix 1 μL 1 μg/μL forward sgRNA oligonucleotide, 1 μL 1 μg/μL reverse sgRNA oligonucleotide, 7 μL nuclease-free water, and 1 μL 10X BpiI Buffer G. Anneal the oligonucleotides with the following program in a thermal cycler.

95 °C, 2 min;

65 °C, 3 min;

Cool to room temperature slowly (over 1 h).

7. In a microtube, mix 400 ng digested plasmid (from **step 4**), 2 μL 10 mM ATP, 2 μL 100 mM DTT, 2 μL 10× BpiI Buffer G, 1 μL 400 U/μL T4 DNA ligase, 1 μL 10 U/μL BpiI, and make the volume up to 20 μL with nuclease-free water. Add the annealed oligonucleotides (from **step 6**) to make a 30 μL ligation reaction volume and incubate overnight at 37 °C (*see* **Note 4**).

8. Add 0.5 μL 10 U/μL BpiI, 0.5 μL 10× BpiI Buffer G, and 4 μL nuclease-free water and incubate for 1 h at 37 °C.

9. Mix 2 μL of the ligation reaction with 50 μL of electrocompetent XL1-Blue bacteria and leave on ice for 10 min.

10. Electroporate the bacteria with the electroporator settings: 2.5 kV; 25 μF.

11. Add the electroporated bacteria to 700 μL SOB medium and incubate for 1 h at 37 °C.

12. Centrifuge for 3 min at 10,000 × *g*, room temperature, and discard the supernatant. Resuspend the bacteria in 100 μL LB and plate on ampicillin agar plate. Incubate overnight at 37 °C.

13. Pick four clones for each sgRNA with a sterile pipette tip. Make streaks of each clone on one ampicillin agar plate before putting the pipette tip in 5 mL of LB supplemented with 100 μg/mL ampicillin. Incubate the liquid culture overnight at 37 °C, 180 rpm, and the plate at 37 °C without shaking.

14. Extract plasmid DNA with a miniprep kit, following manufacturer's instructions.

15. In a microtube, mix 1 μg plasmid DNA, 3 μL 10× CutSmart buffer, 1 μL 20 U/μL BbsI-HF, and 1 μL 20 U/μL ScaI-HF, and make volume up to 30 μL with nuclease-free water. Incubate at 37 °C for 2 h.

16. Analyze digest on a 1% agarose gel. A unique 3196 bp linear band should be seen if the sgRNA sequence is incorporated; two bands are seen if the incorporation failed. Confirm sequence of digestion-verified clone by sequencing using the M13 forward primer.

17. Inoculate a colony of the verified clone (from the plate made at the same time as the miniprep culture in **step 13**) in 50 mL LB culture and incubate overnight at 37 °C, 180 rpm.

18. Purify plasmid DNA with a midiprep kit, following manufacturer's instructions.

3.2 Primer Design for Genomic PCR Amplification of the Homology Arms

1. Select a genomic region encompassing at least 800 bp starting just at the end of each of the two sgRNA sites (*see* **Note 5**). Note the homology arms' restriction maps for strategies to verify final Gibson assembly (*see* Subheading 3.3, **step 15**).

2. Design primers using the NCBI primer designing tool to specifically amplify the homology arms. Add the following 5′ overhangs for Gibson assembly (*see* **Note 6** and Fig. 1b).

   Forward primer for 5′ homology arm ("Red" in Fig. 1b): 5′-gagct cggtacgagtttacgtccagccaag.

   Reverse primer for 5′ homology arm ("Green"): 5′-ccggtag aatttcgacgacctgcagccaag.

   Forward primer for 3′ homology arm ("Blue"): 5′-aattcgagctcg aggtcgacggtatcgata.

   Reverse primer for 3′ homology arm ("Yellow"): 5′-ccctgccc gggctgcaggaattcgatatca.

3.3 Repair Cassette Cloning

1. Extract genomic DNA (gDNA) from 5×10^6 cells with the genomic DNA extraction kit, following the manufacturer's instructions. Use the exact same cell line that will undergo repeat array insertion.

2. In a PCR tube, mix 10 μL 5× Phusion HF buffer, 1 μL 10 mM dNTPs, 2.5 μL 10 μM forward primer, 2.5 μL 10 μM reverse primer, 60 ng genomic DNA, 0.5 μL 2 U/μL Phusion High Fidelity DNA polymerase, and make up the volume to 50 μL with nuclease-free water.

3. Run the following program on a thermal cycler (*see* **Note 7**).

   98 °C, 30 s;

   35×: [98 °C, 10 s; annealing temperature, 30 s; 72 °C, 30 s/kb product size];

   72 °C, 5 min;

   4 °C, hold.

4. Load the whole PCR reaction on a 0.8% agarose gel. Gel-purify the two homology arms using a gel purification kit, following the manufacturer's instructions (Fig. 1b).

5. In a microtube, mix 5 μg CuOx144 plasmid, 3 μL 10× CutSmart buffer, 0.5 μL 20 U/μL NheI-HF, and 0.5 μL HindIII-HF, and make up volume to 30 μL with nuclease-free water. Incubate overnight at 37 °C.

6. Load the whole digestion reaction on a 0.8% agarose gel. Gel-purify the two digestion fragments separately (6926 bp and 3199 bp) using a gel purification kit.

7. Assess DNA concentration of the PCR products and the digestion fragments with a NanoDrop spectrophotometer (*see* **Note 8**).

8. Perform the Gibson assembly with the two PCR amplified homology arms and the two digested plasmid fragments (4 fragments in total) (Fig. 1b). Use 0.2 pmol of total DNA fragments at a 1:1 vector–insert DNA molar ratio with 10 μL of Hifi Gibson assembly Master Mix in a 20 μL total reaction volume. Incubate the Gibson assembly reaction for 2 h at 50 °C in a thermocycler (*see* **Note 9**).

9. Mix 2 μL of the Gibson reaction with 50 μL of XL1-Blue bacteria and leave in ice for 10 min.

10. Electroporate the bacteria with the electroporator settings: 2.5 kV; 25 μF.

11. Transfer immediately the bacteria to 700 μL of SOB medium and allow for recovery 1 h at 30 °C, 180 rpm (*see* **Note 10**).

12. Centrifuge for 3 min at $10,000 \times g$, room temperature, discard supernatant, and resuspend bacteria in 100 μL LB.

13. Spread the bacteria on an ampicillin agar plate and incubate for 24 h at 30 °C (*see* **Note 10**).

14. Pick clones, grow them for 24 h in 5 mL LB medium at 30 °C, 180 rpm, and purify the plasmid with a miniprep kit, following the manufacturer's instructions (*see* **Note 10**).

15. Verify each plasmid by digesting 1 μg of DNA in a 30 μL reaction volume and analyze digested bands on a 1% agarose gel. Check for presence and orientation of homology arms and integrity of the full repeat array (*see* **Note 11**).

16. Make a 50 mL liquid LB culture of the verified clone, grow bacteria for 24 h at 30 °C, 180 rpm and perform the midiprep for high plasmid yield, following the manufacturer's instructions (*see* **Note 10**).

17. Verify the length of repeat array and correct insertion of the homology arms by restriction enzyme digestion (*see* **Note 11**).

3.4 Nickase-Mediated Knock-in

1. Seed 300,000 cells expressing a nuclear fluorescent CymR protein in a 6-well plate with 2 mL of complete McCoy's medium. Grow cells for 24 h in a humidified 37 °C incubator, 5% CO_2.

2. Mix 0.4 μg of nickase-expressing plasmid, 0.5 μg each of the two sgRNA plasmids and 0.6 μg of the repair cassette plasmid in 200 μL of jetPRIME buffer in a 1.5 mL tube and vortex.

3. Add 4 µL of jetPrime reagent, vortex and incubate for 10 min at room temperature.

4. Remove medium and add 2 mL of fresh medium to the cells. Add the transfection mix from **step 3** dropwise onto cells. Grow cells for 24 h.

5. Rinse cells with 2 mL of 1× PBS and add 300 µL 0.5% trypsin–EDTA to the cells. Incubate the plate for 3 min at 37 °C.

6. Prepare four different 10 cm plates with selection medium.

7. Dilute the detached cells in 10 mL of selection medium, and seed 200 µL, 500 µL, 1 mL, and the rest of the suspension into the different 10 cm plates (*see* **Note 12**).

8. Rinse cells with 5 mL of 1× PBS and add fresh selection medium twice a week until clones are clearly seen by eye (about 2 weeks).

9. Pick single colonies using cloning disks, put each of them in 24-well plates with selection medium and shake the plate vigorously to detach cells from the cloning disk (*see* **Note 13**).

10. Change the selection medium twice a week until cells reach 70% confluency.

3.5 Primer Design for Genomic DNA PCR Screening

1. For each side of the repair cassette (it is very important to check both sides to identify proper recombination events; *see* **Note 14**), select a genomic region outside of the homology arm (in the genome, as close as possible to the homology arm) and a region in the repair cassette that is close to the homology arm (Fig. 2a). The primers used inside the repair cassette we have used are.

   5′ side: 5′-ctaaagcgcatgctccagac

   3′ side: 5′-ttgtacaaagtggttgatgggg

2. Design PCR screening primers with the NCBI primer designing tool (*see* **Note 15**).

3.6 Clone Screening

1. Rinse cells with PBS and add 100 µL of 0.5% trypsin–EDTA. Incubate for 3 min at 37 °C.

2. Once the cells are detached, resuspend them in 600 µL of selection medium. Transfer 300 µL of the cells to a 1.5 mL tube and seed the other half in 1 mL of McCoy's medium in a 12-well plate for clone expansion.

3. Centrifuge the tubes for 5 min at 300 × *g*, room temperature. Discard supernatant and add 1 mL of PBS. Centrifuge for 5 min at 300 × *g*, room temperature and remove PBS.

4. Extract the DNA using the QuickExtract solution, following the manufacturer's instructions.

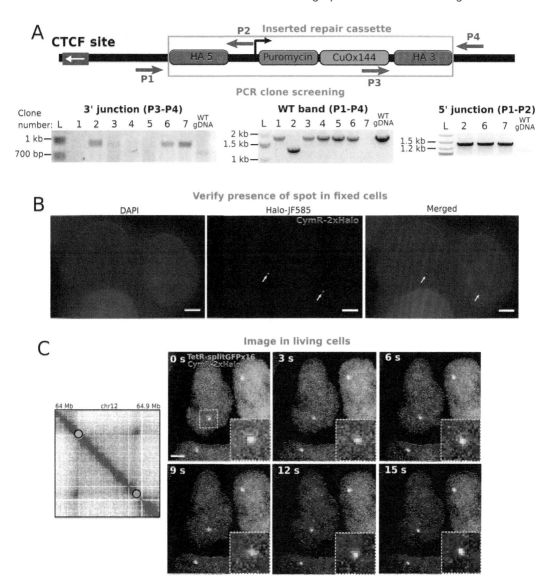

Fig. 2 Clone screening and imaging. (**a**) PCR screening to verify the insertion of the repair cassette. Top: Schematic of the inserted repair cassette in the genome. Pink arrows represent the location of screening primers at both sides of the repair cassette. Bottom: Agarose gel of 7 different clones after antibiotic selection. Left: Junction-PCR on the 3′ side (908 bp band is expected if the repair cassette is inserted). Middle: WT allele PCR (a band at 1.8 kb is expected if at least one allele is WT). Clone 7 has a homozygous insertion since no WT band is seen. Right: Junction-PCR on the 5′ side for clones that were positive on the 3′ side (a band at 1.4 kb is expected if the repair cassette is inserted). L: DNA ladder. (**b**) Fixed cell images of a PCR-screened positive clone. White arrows show the fluorescently tagged locus. Scale bar: 3 μm. (**c**) Live-imaging of loop anchors in HCT116 cells. Left: Hi-C contact frequency map of a 2 Mb region on chromosome 12 containing the two inserted operator arrays (circles) at the anchors of a 400 kb-long chromatin loop. Right: Spinning disk images of the loop anchors. Left anchor (green) was labelled by TetOx96 (inserted with the same cloning method described here for CuOx144) bound by TetR-splitGFPx16 and right anchor (red) was labeled by CuOx144 bound by a CymR-2xHalo (imaged with JF585). The inset shows a magnification of the region outlined by a white dotted line. The images are maximum intensity projections of 11 z-stacks each separated by 1 μm. One full z-stack was imaged every 3 s. Scale bar: 3 μm

5. In a PCR tube for each clonal gDNA preparation, mix 5 μL 5× Green GoTaq Flexi reaction buffer, 2.5 μL 25 mM MgCl$_2$, 0.5 μL 10 mM dNTPs, 0.5 μL 10 μM forward primer, 0.5 μL 10 μM reverse primer, 3 μL clonal gDNA, 12.88 μL nuclease-free water, 0.125 μL 5 U/μL GoTaq G2 Flexi DNA polymerase (*see* **Note 16**).

6. Run the following program on a thermocycler, with the annealing temperature determined using the Promega Tm calculator.

94 °C, 2 min;

35×: [94 °C, 30 s; annealing temperature, 30 s; 72 °C, 1 min/ kb product size];

72 °C, 10 min.

7. Analyze the PCR products on a 1% agarose gel (Fig. 2a).

3.7 Fixed-Cell Imaging for Verification of PCR-Positive Clones

1. From the 12-well plates containing PCR-positive clones, seed half of the cells in a 6-well plate for clone expansion and the other half on a glass coverslip inside a 6-well plate with 2 mL of selection medium.

2. 48 h after seeding the cells on the glass slide, rinse them once with 1× PBS and add 2 mL of 4% formaldehyde in 1× PBS and incubate for 15 min at room temperature.

3. Wash cells three times (1 min each) with 2 mL of 1× PBS at room temperature.

4. Put a drop of Vectashield + DAPI on a glass coverslip and transfer the slide with the cells onto the drop. Incubate for at least 2 h at 4 °C and protect from light.

5. Image the clones with a classic widefield microscope (acquire the whole cell height in z with each step separated by 0.3 μm) to verify the presence of one or two spots (depending on the zygosity of the insertion) in the nucleus (Fig. 2b).

3.8 Live-Cell Imaging of the Chromatin Loop Anchors

1. Seed 250,000 cells of a verified clone on a glass-bottom culture dish (suitable for the microscope to be used). Grow cells for 48 h.

2. Rinse the cells once with 1× PBS and change the medium to live-imaging medium.

3. Incubate for 2 h at 37 °C in the microscope chamber before imaging the living cells (*see* **Note 17**; Fig. 2c).

4   Notes

1. To increase the signal-to-noise ratio (SNR), and hence localization precision, of the visualized spot, it is preferable to sort the low fluorescence cells by FACS to limit background

intensity. We used an ePiggyBac CymR-NLS-2xHalo plasmid available on Addgene (#119907) to generate stable cell lines with low expression of the CymR protein for visualization of the integrated CuOx144 repeats. Diploid cell lines are preferred over polyploid ones.

2. Detailed and clear information on how to choose regions of interest are provided in [18]. Pick sgRNA sequences that are at least 50 bp away from repeated sequences so that the homology arm sequences closest to the DNA cut site are unique. The repeat array insertion sites should be located downstream of the last convergent CTCF site in order to not interfere with extruding cohesin but should be as close as possible to the loop anchor for precise quantification of the anchor-anchor distance. Insert the repeat array plasmid so that the antibiotic resistance transcription is directed outward of the TAD or chromatin loop to avoid interference with the loop extrusion process. Homology arms in euchromatin regions [17] and low percentage of repeated sequences will increase the insertion rates.

3. Sequence targeted by the sgRNA pair (in red the PAM sequence, in bold and underlined: sgRNA binding region).

AGTCCA**AGTTATGGAAGATCTCGATA**TCGAAGCCGTAGTCGTAGCAGAAGC

CGTAGCAGAAGCAACAGCA**GGAGTCGCAGTTACTCCCCA**GGAG

    sgRNA1: **TATCGAGATCTTCCATAACT**TGG

Oligonucleotides to order for sgRNA1 (in red: overhang to clone in U6 plasmid, in blue: on sgRNA side, in green: nucleotide added for U6 transcription).

    5′ CACCGTATCGAGATCTTCCATAACT

    5′ AAACAGTTATGGAAGATCTCGATAC

    sgRNA2: **GGAGTCGCAGTTACTCCCCA**GG

Oligonucleotides to order for sgRNA2 (in red: overhang to clone in U6 plasmid, in blue: on sgRNA side).

    5′ CACCGGAGTCGCAGTTACTCCCCA

    5′ AAACTGGGGAGTAACTGCGACTCC

4. Since sticky ends from BpiI sites are destroyed upon cloning the sgRNA sequence, only plasmids that incorporated the sgRNA sequence will not be digested by BpiI.

5. The beginning of the homology arm should be as close as possible to the NGG cut site (which should not be included in the homology arm) and no more than 100 bp away. The small sequence between the two sgRNAs will be deleted from the genome.

6. The overhangs should begin at the 3′ end of the cut NheI and HindIII sites on the CuOx144 plasmid. The enzymatic restriction site is destroyed after successful Gibson assembly since the full restriction site is not included in the overhangs. It is recommended to use 20 to 30 bp of Gibson overhang.

7. Use a high-fidelity DNA polymerase for the amplification of the homology arms. Use the ThermoFisher Tm Calculator to define the annealing temperature (using the "Phusion or Phire DNA polymerase" option). If unspecific bands are seen (higher annealing temperature is needed) or PCR amplification is not successful, run a gradient PCR with annealing temperatures separated by 3 °C each and choose the annealing temperature where the expected PCR product and no other bands are seen.

8. Ensure that plasmid and PCR fragment DNA concentrations are not lower than 25 ng/μL and that the A_{260}/A_{230} ratio is not lower than 1.5 for best Gibson assembly efficiency.

9. The Gibson assembly reaction should stay at 50 °C for at least 1 h. Increasing the incubation time leads to higher efficiency.

10. Repeat arrays can recombine in bacteria. It is therefore mandatory to grow them at 30 °C in recombinant-deficient bacteria to avoid losing a fraction of the repeats. Verify the length of the repeat array (5217 bp) at each step by restriction enzyme digestion (e.g., BamHI/ScaI digestion) to ensure full integrity of the size of the repeat array.

11. For restriction enzyme digestion verification of the plasmid, try to obtain one fragment containing the full repeat array to verify that no repeat was lost during bacteria growth. It is highly recommended to use restriction enzyme sites present inside the homology arms to verify their integration and orientation.

12. Using different dilutions of the cell solution ensures well-separated clonal colonies to pick after antibiotic selection.

13. The percentage of right insertions among resistant clones varies between 10–65% depending on the targeted locus. A few clones will not recover from the splitting. Picking 40 clones should ensure at least a few positive clones in low-efficiency conditions.

14. If using degenerated repeat cassettes, amplification of the whole array of repeats can be done to identify clones that did not lose some repeats during the recombination process. Optimization of the PCR conditions might be necessary.

15. Similar melting temperature and a GC clamp are preferable for the PCR screening primers.

16. Always add a PCR negative control with WT gDNA (no band is expected) to visualize potential unspecific bands. Once positive clones have been identified, perform a PCR to identify the WT allele and determine whether the insertion is heterozygous (presence of WT band) or homozygous (absence of WT band) before imaging the cells. Once a clone has been completely genotyped and verified by imaging, sequence the PCR products to verify correct recombination.

17. Detailed discussions of suitable live-cell imaging conditions can be found in [18, 19]. Briefly, organic dyes such as Janelia Fluor dyes (ligands of the Halo and Snap tags) [20] are preferable to fluorescent proteins due to their higher photostability, brightness, and versatility for multicolor imaging. Parameters to consider include the speed of acquisition or temporal resolution (especially for z-sectioning to avoid motion blur which decreases the localization accuracy), signal-to-noise ratio (SNR) and hence localization precision, total duration of imaging before photobleaching, and the ability of detecting two colors simultaneously if two separate structures are visualized.

References

1. Rao SSP, Huntley MH, Durand NC et al (2014) A 3D map of the human genome at kilobase resolution reveals principles of chromatin looping. Cell 159:1665–1680

2. Cardozo Gizzi AM, Cattoni DI, Fiche J-B et al (2019) Microscopy-based chromosome conformation capture enables simultaneous visualization of genome organization and transcription in intact organisms. Mol Cell 74(1):212–222.e5

3. Bintu B, Mateo LJ, Su J-H et al (2018) Super-resolution chromatin tracing reveals domains and cooperative interactions in single cells. Science 362:eaau1783

4. Boettiger A, Murphy S (2020) Advances in chromatin imaging at kilobase-scale resolution. Trends Genet 36(4):273–287

5. Rao SSP, Huang S-C, Glenn St Hilaire B et al (2017) Cohesin loss eliminates all loop domains. Cell 171:305–320.e24

6. Parmar JJ, Woringer M, Zimmer C (2019) How the genome folds: the biophysics of four-dimensional chromatin organization. Annu Rev Biophys 48:231–253

7. Fudenberg G, Abdennur N, Imakaev M et al (2017) Emerging evidence of chromosome folding by loop extrusion. Cold Spring Harb Symp Quant Biol 82:45–55

8. Davidson IF, Bauer B, Goetz D et al (2019) DNA loop extrusion by human cohesin. Science 366(6471):1338–1345

9. Sikorska N, Sexton T (2019) Defining functionally relevant spatial chromatin domains: it's a TAD complicated. J Mol Biol 432(3):653–664

10. Hansen AS, Cattoglio C, Darzacq X et al (2018) Recent evidence that TADs and chromatin loops are dynamic structures. Nucleus 9:20–32

11. Beagan JA, Phillips-Cremins JE (2020) On the existence and functionality of topologically associating domains. Nat Genet 52:8–16

12. Chen B, Gilbert LA, Cimini BA et al (2013) Dynamic imaging of genomic loci in living human cells by an optimized CRISPR/Cas system. Cell 155:1479–1491

13. Alexander JM, Guan J, Huang B et al (2018) Live-cell imaging reveals enhancer-dependent Sox2 transcription in the absence of enhancer proximity. Elife 8:e41769

14. Mullick A, Xu Y, Warren R et al (2006) The cumate gene-switch: a system for regulated expression in mammalian cells. BMC Biotechnol 6:43

15. Germier T, Audibert S, Kocanova S et al (2018) Real-time imaging of specific genomic

loci in eukaryotic cells using the ANCHOR DNA labelling system. Methods 142:16–23

16. Cong L, Ran FA, Cox D et al (2013) Multiplex genome engineering using CRISPR/Cas systems. Science 339:819–823

17. Tasan I, Sustackova G, Zhang L et al (2018) CRISPR/Cas9-mediated knock-in of an optimized TetO repeat for live cell imaging of endogenous loci. Nucleic Acids Res 46:e100

18. Brandão HB, Gabriele M, Hansen AS (2021) Tracking and interpreting long-range chromatin interactions with super-resolution live-cell imaging. Curr Opin Cell Biol 70: 18–26

19. Schermelleh L, Ferrand A, Huser T et al (2019) Super-resolution microscopy demystified. Nat Cell Biol 21:72–84

20. Grimm JB, Muthusamy AK, Liang Y et al (2017) A general method to fine-tune fluorophores for live-cell and in vivo imaging. Nat Methods 14:987–994

Part VI

Functionally Dissecting Chromatin Architecture

Chapter 14

CLOuD9: CRISPR-Cas9-Mediated Technique for Reversible Manipulation of Chromatin Architecture

Wei Qiang Seow, Poonam Agarwal, and Kevin C. Wang

Abstract

The spatial organization of the genome plays a critical role in cell-specific biological functions such as gene expression. Existing genome-wide technologies reveal a dynamic interplay between chromatin looping and gene regulation, but the mechanisms by which regulatory interactions between genetic elements are established or maintained remain unclear. Here, we present CLOuD9, a CRISPR-based technology that can create de novo, pairwise chromatin interactions in cells. This technique for chromatin loop reorganization employs dCas9-targeting and ABI1-PYL heterodimerization. It is reversible, but can also establish epigenetic memory under certain conditions, which provides a way to dissect gene regulation mechanisms.

Key words Chromatin architecture, Genome organization, CRISPR, Gene regulation, Epigenetics

1 Introduction

Eukaryotic nuclear DNA is packaged into chromatin, which provides the foundation for transcription and gene regulation. Emerging evidence suggests that there is a dynamic interplay between three-dimensional (3D) chromatin architecture and gene expression machinery. The fine-tuning of this relationship is required for the normal functioning of cells. Indeed, failure in establishing proper 3D chromatin contacts can lead to epigenetic disorders and disease progression. However, the mechanisms by which 3D chromatin contacts are established between regulatory elements, such as enhancers and promoters, are unclear due to constraints in the availability of tools to study them. Methods to manipulate 3D chromatin architecture without altering the linear DNA sequence are needed to interrogate higher order genome organization.

Wei Qiang Seow and Poonam Agarwal contributed equally with all other contributors.

Tom Sexton (ed.), *Spatial Genome Organization: Methods and Protocols*,
Methods in Molecular Biology, vol. 2532, https://doi.org/10.1007/978-1-0716-2497-5_14,
© The Author(s), under exclusive license to Springer Science+Business Media, LLC, part of Springer Nature 2022

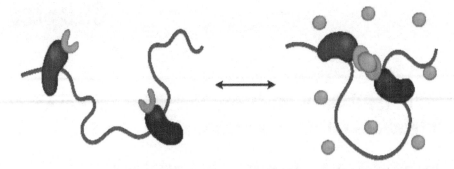

Fig. 1 Schematic of CLOuD9. Addition of dimerizer abscisic acid (ABA, green) brings the two complementary CLOuD9 constructs, CSP-red (CLOuD9 *S. pyogenes*) and CSA-blue (CLOuD9 *S. aureus*) into proximity, remodeling chromatin structure. ABA removal restores endogenous chromatin conformation

We developed a technique for targeted and reversible chromatin loop reorganization using nuclease-deficient Cas9 (dCas9) that enables the forced juxtaposition between any two genomic loci [1]. Our tool, chromatin loop reorganization using CRISPR-dCas9 (CLOuD9), utilizes chemically induced proximity to mediate chromatin contacts between two DNA-bound dCas9 proteins. Specifically, two orthologous dCas9 proteins from *Streptococcus pyogenes* and *Staphylococcus aureus* are each fused to different components of the *Arabidopsis thaliana* abscisic acid (ABA) signaling pathway—PYL1 and ABI1, which undergo heterodimerization in the presence of ABA. The dimerization physically localizes two disparate genomic loci close together in nuclear space (Fig. 1). Additionally, removal of ABA restores the endogenous chromatin conformation. This approach to induce chromatin interactions to regulate gene expression could also be applied to understand the role of 3D chromatin architecture in cell state transitions during development and disease.

2   Materials

2.1   sgRNA Vector Construction

1. 100 μM sgRNA cloning oligonucleotides (*see* Subheading 3.1 for design strategy):

   Oligonucleotide 1: 5′-CACCG[sgRNA sequence]-3′.

   Oligonucleotide 2: 5′-AAAC[reverse complement of sgRNA sequence]C-3′.

2. pC9-sadCas9-ABI-puro plasmid (System Biosciences CASCL9-100A-KIT) (*see* **Note 1**).

3. pC9-spdCas9-PLY1-hyg plasmid (System Biosciences CASCL9-100A-KIT) (*see* **Note 1**).

4. FastDigest Esp3I (Thermo Scientific), provided with 10× FastDigest buffer.

5. 1 U/μL FastAP alkaline phosphatase (Thermo Scientific).

6. 100 mM DTT.

7. Nuclease-free water.

8. QiaQuick gel extraction kit (Qiagen), or equivalent.

9. 400 U/μLT4 DNA ligase (NEB), provided with 10× T4 DNA ligase buffer.

10. 10 U/μLT4 polynucleotide kinase (NEB).

11. Quick Ligase (NEB), provided with 2× Quick Ligase buffer.

12. Stbl3 chemically competent *E. coli* (Thermo Scientific) (*see* **Note 2**).

13. SOC outgrowth medium.

14. LB medium and agar plates with 100 μg/mL carbenicillin.

15. Plasmid mini-prep kit.

16. ZymoPURE II Plasmid Maxiprep Kit, or equivalent.

2.2 Lentiviral Transduction

1. Lenti-X 293T cells (Takara Bio).

2. DMEM culture medium: DMEM with GlutaMAX, 10% (v/v) fetal bovine serum (FBS).

3. Lipofectamine 2000 transfection reagent (Invitrogen).

4. Opti-MEM I reduced serum medium (Gibco).

5. psPAX2 plasmid (Addgene 12260).

6. pMD2.G plasmid (Addgene 12259).

7. 0.45 μm polyethersulfone (PES) filters.

8. Lenti-X concentrator (Takara Bio).

9. 1× PBS (phosphate-buffered saline) pH 7.4, calcium- and magnesium-free.

10. Lenti-X GoStix Plus (Takara Bio).

11. RPMI culture medium: RPMI 1640 with GlutaMAX, 10% (v/v) FBS.

12. 10 mg/mL polybrene.

13. 100× penicillin/streptomycin (Gibco).

14. 10 mg/mL puromycin dihydrochloride.

15. 50 mg/mL hygromycin B.

2.3 Western Blot

1. RIPA buffer: 50 mM Tris-HCl pH 8, 150 mM NaCl, 1% (v/v) Igepal CA-630, 0.5% (w/v) sodium deoxycholate, 0.1% (v/v), 1× EDTA-free protease inhibitor cocktail (PIC).

2. NanoDrop spectrophotometer.

3. 2× Laemmli buffer: 125 mM Tris-HCl pH 6.8, 4% (w/v) SDS, 10% (v/v) 2-mercaptoethanol, 20% (v/v) glycerol, 0.004% (w/v) bromophenol blue.

4. 4–20% Mini-PROTEAN TGX precast protein gel and Mini-PROTEAN Tetra cell (Bio-Rad), or equivalent.

5. SDS gel running buffer: 25 mM Tris base, 190 mM glycine, 0.1% (v/v) SDS, pH 8.3.

6. 0.45 μm nitrocellulose or PVDF membrane.

7. Transfer buffer: 25 mM Tris base, 190 mM glycine, 20% (v/v) methanol, pH 8.3.

8. TBST buffer: 20 mM Tris-HCl pH 7.5, 150 mM NaCl, 0.1% (v/v) Tween 20.

9. Block buffer: 3% (w/v) BSA in TBST.

10. Primary antibody solution: 1:1000 dilution of either monoclonal anti-FLAG Tag M2 mouse antibody (Sigma) or monoclonal anti-HA-Tag rabbit antibody (Cell Signalling) in 5% (w/v) BSA/TBST.

11. HRP-conjugated secondary antibody solution: 1:1000 dilution of either anti-mouse HRP-conjugated secondary antibody (Cell Signaling) or anti-rabbit HRP-conjugated secondary antibody (Santa Cruz) in TBST.

12. ECL Western blotting reagents.

13. Bio-Rad Gel Doc XR, or equivalent, with CCD camera.

2.4 CLOuD9 Induced Dimerization and ChIP

1. 1 mM (S)-(+)-abscisic acid (ABA) (Thermo Scientific), made up in DMSO and kept protected from light (*see* **Note 3**).

2. Induction culture medium: DMEM or RPMI culture medium with 1× penicillin/streptomycin and 100 μM ABA, added fresh before each addition of medium to cells.

3. 2% and 1% (w/v) formaldehyde solutions in 1× PBS, made fresh from 16% stocks in glass ampoules before use.

4. 2 M glycine.

5. EDTA-free protease inhibitor cocktail (PIC). Tablet dissolved in 1 mL nuclease-free water to make a 100× stock fresh before use.

6. 0.1 M PMSF in isopropanol.

7. 1× PIC and 1 mM PMSF in 1× PBS.

8. Zymo-Spin ChIP kit (Zymo Research), containing Nuclei Prep Buffer, Chromatin Shearing Buffer, Chromatin Dilution Buffer, ZymoMag Protein A beads, Chromatin Wash Buffer I, Chromatin Wash Buffer II, Chromatin Wash Buffer III, Chromatin Elution Buffer, 5 M NaCl, Proteinase K, ChIP

DNA Binding Buffer, Zymo-Spin IC columns, DNA Wash Buffer and DNA Elution Buffer.

9. Diagenode Bioruptor Pico with 1.5 mL Bioruptor microtubes, or equivalent.

10. Magnetic stand.

2.5   CLOuD9 3C

1. Lysis buffer: 10 mM Tris–HCl pH 7.5, 10 mM NaCl, 0.2% Igepal CA-630, 1× PIC.

2. 0.5% (w/v) SDS.

3. 10% (v/v) Triton X-100.

4. Restriction enzyme and manufacturer-supplied 10× buffer (*see* **Note 4**).

5. Ligation master mix: 1.33× T4 DNA ligase buffer, 1.11% (v/v) Triton X-100, 0.13 mg/mL BSA, 2.22 U/μL T4 DNA ligase.

6. 20 mg/mL proteinase K.

7. 10 mg/mL RNase A.

8. Phenol–chloroform–isoamyl alcohol (25:24:1).

9. 3 M sodium acetate pH 5.2.

10. 100% and 70% ethanol.

11. 10 mM Tris-HCl pH 7.5.

12. Qubit fluorimeter and dsDNA BR assay, or equivalent.

13. LightCycler 480 SYBR Green I Mastermix (Roche), or other PCR reaction kit.

14. 3C PCR primers (*see* **Note 5**).

3   Methods

3.1   sgRNA Vector Construction

The sgRNAs to target the CLOuD9 to the desired loop anchor regions are designed and engineered into the CLOuD9 vectors.

1. Identify the regions of the genome to be targeted. Generally, active chromatin is more amenable to dCas9 binding, so where possible, use available chromatin accessibility (e.g., DNase I hypersensitivity or ATAC-seq) and/or histone modification ChIP-seq data (e.g., H3K27ac to identify active chromatin; H3K27me3 to identify repressive chromatin) to assess the chromatin landscape. sgRNA sites are best situated away from the tail ends of DNase I hypersensitive peaks to reduce interference with endogenous protein binding. For inducing promoter-enhancer contacts, position one sgRNA upstream of the promoter to avoid encroaching on transcription initiation and/or elongation, and position the other sgRNA

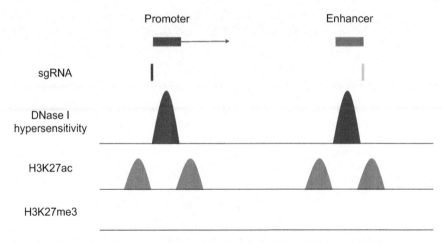

Fig. 2 Schematic for sgRNA positioning. sgRNAs should be located away from DNase I hypersensitive sites and be upstream of promoters. sgRNA from the partner dCas9 should be situated downstream of the element of interest wherever possible to avoid interference with potential loop extrusion activity

downstream of the enhancer to recapitulate CTCF-mediated chromatin looping (Fig. 2). For promoter-promoter contacts, position both sgRNA sites upstream of the direction of transcription.

2. Design gRNAs for each targeting site with CRISPOR [2], ensuring that one site uses the protospacer adjacent motif (PAM) for *S. aureus* dCas9, and the other uses the PAM for *S. pyogenes* dCas9 (*see* **Note 6**).

The following steps are performed to clone gRNAs into pC9-sadCas9-ABI-puro and pC9-spdCas9-PYL1-hyg plasmids.

3. In separate microtubes, mix 5 µg each CLOuD9 plasmid with 6 µL 10× FastDigest buffer, 3 µL FastDigest Esp3I, 3 µL 1 U/µL FastAP alkaline phosphatase, and 0.6 µL 100 mM DTT. Make volume up to 60 µL with nuclease-free water. Incubate for 30 min at 37 °C.

4. Perform gel electrophoresis of the digested product on a 1% agarose gel. Excise the larger band (14 kb for the S. pyogenes plasmid and 13 kb for the S. aureus plasmid) and purify the DNA with a QiaQuick gel extraction kit, following the manufacturer's instructions.

5. Separately, mix 1 µL 100 µM oligonucleotide 1, 1 µL 100 µM oligonucleotide 2, 6.5 µL nuclease-free water, 1 µL 10× T4 DNA ligase buffer (*see* **Note 7**), and 0.5 µL 10 U/µL T4 PNK. Run the following program in a thermal cycler to phosphorylate and anneal the oligonucleotides.

37 °C, 30 min.

95 °C, 5 min.

Cool to 25 °C at a rate of 5 °C/min.

Dilute annealed oligonucleotides at a 1:200 dilution.

6. Mix 4 μL 12.5 ng/μL digested plasmid (from **step 4**), 1 μL 500 nM annealed oligonucleotides, 5 μL 2× Quick Ligase buffer, and 1 μL Quick Ligase. Incubate for 10 min at room temperature.

7. Add 1 μL ligation product to an aliquot of Stbl3 E. coli, thawed on ice (*see* **Note 2**).

8. Heat shock at 42 °C for 45 s and incubate on ice for 2 min.

9. Add 100 μL of SOC outgrowth medium and plate 50 μL of transformants on LB agar plates with 100 μg/mL carbenicillin overnight at 37 °C.

10. Pick a colony from the plate and transfer to 5 mL LB with 100 μg/mL carbenicillin and incubate overnight at 37 °C, 950 rpm.

11. Take 4.5 mL of culture, purify plasmid with a miniprep kit, following manufacturer's instructions, and confirm sgRNA incorporation by Sanger sequencing.

12. Inoculate the remaining 500 μL of culture (from **step 10**) into 200 mL LB with 100 μg/mL carbenicillin and incubate overnight at 37 °C, 950 rpm.

13. Purify the two plasmids (pC9-sadCas9-ABI-puro-sgRNA and pC9-spdCas9-PLY1-hyg-sgRNA, respectively, with their incorporated sgRNAs) with the Zymo Pure II Plasmid Maxiprep kit, following manufacturer's instructions.

3.2 Lentiviral Transduction

The CLOuD9 vectors are placed into lentiviruses and transduced into the target cells.

1. Seed 750,000 Lenti-X 293T cells per well in a 6-well plate in DMEM culture medium. Culture to ~80% confluency at 37 °C, 5% CO_2 (~20 h).

2. Aspirate medium and replace with 2 mL fresh DMEM culture medium.

3. In a 1.5 mL microtube, mix 10 μL lipofectamine 2000 and 150 μL Opti-MEM I reduced serum medium. In a separate tube, mix 2 μg of either pC9-sadCas9-ABI-puro-sgRNA or pC9-spdCas9-PLY1-hyg-sgRNA plasmid (i.e., perform a separate reaction for each CLOuD9 plasmid), 1 μg psPAX2, and 1 μg pMD2.G plasmids into 150 μL opti-MEM I reduced serum medium. Add the lipofectamine-containing solution to the plasmid-containing mix and incubate for 5 min at room temperature.

4. Gently transfer the mixture dropwise to the cells, without dislodging them, and incubate overnight at 37 °C, 5% CO_2. The next morning, gently aspirate medium and replace with 2 mL fresh DMEM culture medium.

5. 48 h, 72 h, and 96 h posttransfection, transfer the medium (containing lentiviruses) to the same 15 mL tube, storing the medium at 4 °C between collections. Add 2 mL fresh DMEM culture medium per well for each collection and culture at 37 °C, 5% CO_2.

6. Filter collected medium through a 0.45 μm PES filter into a sterile 15 mL tube. Add 1 volume of Lenti-X-concentrator to 3 volumes of clarified supernatant. Mix by gentle inversion and incubate overnight at 4 °C.

7. Centrifuge for 45 min at 1500 × g, 4 °C. Carefully remove supernatant without disturbing the pellet. Briefly recentrifuge at 1500 × g to remove residual supernatant and resuspend in 1 mL ice-cold 1× PBS.

8. Estimate viral titer with Lenti-X GoStix, following the manufacturer's instructions (*see* **Note 8**).

The transduction protocol differs for suspension cells (Subheading 3.2.1) and adherent cells (Subheading 3.2.2).

3.2.1 Suspension Cells

1. Transfer 80,000–100,000 cells to a 1.5 mL tube, centrifuge for 5 min at 200 × g, room temperature, and remove supernatant.

2. Resuspend cells in 25 μL each viral construct (i.e., add both CLOuD9 lentiviruses in the same 50 μL transduction). Add 500 μL RPMI culture medium and 0.4 μL 10 mg/mL polybrene.

3. Centrifuge for 30 min at 800 × g, room temperature, resuspend cells briefly without removing supernatant, and transfer to cell culture plates. Incubate for 24 h at 37 °C, 5% CO_2.

4. Centrifuge for 5 min at 200 × g, room temperature and aspirate supernatant. Resuspend cells in 500 μL RPMI culture medium and 5 μL 100× penicillin/streptomycin. Incubate overnight at 37 °C, 5% CO_2.

5. Culture the cells in a 6-well plate with RPMI culture medium supplemented with 0.25–10 μg/mL puromycin and 10–500 μg/mL hygromycin B final concentrations (*see* **Notes 9–12**).

3.2.2 Adherent Cells

1. Seed cells in a 6-well plate and culture to 80% confluency.

2. Mix 25 μL each viral construct with 1.5 mL DMEM culture medium and 1.2 μL 10 mg/mL polybrene.

3. Aspirate medium from cells and replace with the mixture containing viral constructs. Incubate overnight at 37 °C, 5% CO_2.

4. Aspirate medium containing viruses and replace with 2 mL DMEM culture medium and 20 μL 100× penicillin–streptomycin. Incubate overnight at 37 °C, 5% CO_2.

5. Culture the cells with DMEM culture medium supplemented with 0.25–10 μg/mL puromycin and 10–500 μg hygromycin B final concentrations (*see* **Notes 9–12**).

3.3 Western Blot

The expression of both CLOuD9 constructs is verified by western blot for presence of their FLAG and HA tags.

1. Place the cell culture dish on ice, aspirate culture media, and wash the cells once with ice cold 1× PBS.

2. Add 150 μL RIPA buffer per well to prepare cell lysates. Scrape adherent cells with a cell scraper.

3. Transfer lysed cell suspension to a prechilled microcentrifuge tube and incubate on ice for 5 min.

4. Centrifuge for 20 min at 15,000 × g, 4 °C. Carefully transfer the supernatant to a fresh, prechilled microfuge tube.

5. Determine the protein concentration of each cell lysate using a Nanodrop with 10–20 μL of the lysate (*see* **Note 13**).

6. Take 10–20 μg protein and add an equal volume of 2× Laemmli buffer. Boil each cell lysate at 95 °C for 5–10 min.

7. Load each sample lysates on a 4%–20% Mini-Protean TGX gel along with molecular weight markers standard and run the gel for 1 h at 150 V in SDS gel running buffer.

8. Transfer the proteins from the gel onto a 0.45 μm nitrocellulose membrane at 4 °C using the wet transfer method by running electrophoresis in transfer buffer at 400 mA for 2 h (*see* **Note 14**).

9. Block the blot in block buffer for 30 min at room temperature (*see* **Note 15**).

10. Incubate the blot overnight in the primary antibody solution at 4 °C.

11. Rinse the blot three times with 1× TBST for 10 min each.

12. Incubate the blot in HRP-conjugated secondary antibody solution for 1 h at room temperature.

13. Rinse the blot three times with 1× TBST for 10 min each.

14. Develop the blot using the ECL substrate following the manufacturer's instructions.

15. Image the blot on a Bio-Rad Gel Doc XR system with a CCD camera to capture the chemiluminescent signals.

3.4 Induced Dimerization and Washout

Culturing cells with ABA induces dimerization of CLOuD9 constructs and thus looping interactions. Within a certain time window, loop induction is reversible by washing out the ABA from the culture medium.

1. Culture cells at a density of up to two million cells/mL with induction culture medium (*see* **Notes 3** and **16**). For adherent cells, aspirate medium and replace daily with fresh induction culture medium. For suspension cells, each day transfer to 15 mL tube, centrifuge for 5 min at 200 × g, room temperature and aspirate supernatant. Resuspend cells in fresh induction culture medium.

2. Wash out ABA to assess reversibility of chromatin looping (*see* **Note 17**). For adherent cells, aspirate supernatant, wash cells once with 1× PBS, then replace with DMEM culture medium with 1× penicillin/streptomycin. Incorporate DMSO if necessary as a control when comparing to ABA treatment. For suspension cells, transfer to a 15 mL tube, centrifuge for 5 min at 200 × g, room temperature and aspirate supernatant. Wash cells in 5 mL 1× PBS, centrifuge for 5 min at 200 × g, room temperature and aspirate supernatant. Resuspend cells in fresh RPMI culture medium with 1× penicillin/streptomycin. Incorporate DMSO if necessary as a control when comparing to ABA treatment.

3.5 CLOuD9 ChIP

ChIP-qPCR or ChIP-seq with antibodies against HA and FLAG tags confirm recruitment of the CLOuD9 fusion proteins to their expected sites. Use of other antibodies can assess whether induced looping affects recruitment of other proteins (e.g., RNA PolII) or histone modifications.

1. Wash five million cells twice in 1× PBS and fix in 1 mL 1% formaldehyde in 1× PBS for 10 min at room temperature with gentle shaking.

2. Quench the cross-linking reaction by adding 62 μL 2 M glycine and incubating for 5 min at room temperature with gentle shaking.

3. Centrifuge for 1 min at 3000 × g, 4 °C. Remove supernatant and resuspend cell pellet in 1 mL 1× PIC and 1 mM PMSF in 1× PBS. Repeat once more for a total of two washes.

4. Centrifuge for 1 min at 3000 × g, 4 °C. Remove supernatant and resuspend cell pellet in 500 μL chilled Nuclei Prep Buffer supplemented with 5 μL 0.1 M PMSF and 5 μL 100× PIC.

5. Resuspend completely by vortexing briefly and incubate on ice for 5 min.

6. Centrifuge for 1 min at 3000 × g, 4 °C. Remove supernatant and resuspend cell pellet in 500 μL chilled Chromatin Shearing Buffer supplemented with 5 μL 0.1 M PMSF and 5 μL 100× PIC.. Resuspend cells by gentle grinding and incubate on ice for 5 min.

7. Transfer 250 µL aliquots of cells into 1.5 mL Bioruptor micro-tubes and sonicate with a Bioruptor Pico at 4 °C for 12 cycles (30 s ON, 30 s OFF) (*see* **Note 18**).

8. Centrifuge for 5 min at 12,000 × *g*, room temperature, and pool supernatants (500 µL total) to a prechilled microcentrifuge tube.

9. In microtubes, prepare ChIP reaction mixes for each target antibody and a negative control (e.g., IgG): 100 µL sheared chromatin, 10 µg monoclonal anti-FLAG Tag M2 mouse anti-body or monoclonal anti-HA-Tag rabbit antibody (*see* **Note 19**), 10 µL 100× PIC, 10 µL 0.1 M PMSF, Chromatin Dilu-tion Buffer to a total volume of 880 µL. Keep 10 µL sheared chromatin as a DNA input control.

10. Incubate ChIP reactions for 3 h to overnight at 4 °C with rotation.

11. Add 15 µL ZymoMag Protein A beads and incubate for 1 h at 4 °C with rotation.

12. Place tubes on a magnetic stand, allow the beads to cluster, and remove supernatant when the solution clears.

13. Resuspend beads in 1 mL of prechilled Chromatin Wash Buffer I. Incubate for 4 min with rotation, followed by supernatant removal on magnet.

14. Repeat **step 13** with Chromatin Wash Buffer II and Chromatin Wash Buffer III sequentially.

15. Resuspend beads in 500 µL DNA Elution Buffer and transfer to a new 1.5 mL microcentrifuge tube.

16. Place tubes on a magnetic stand, allow the beads to cluster, and remove supernatant when the solution clears.

17. Resuspend beads in 150 µL of 1× Chromatin Elution Buffer and mix gently. For DNA input control, add 140 µL of 1× Chromatin Elution Buffer to 10 µL of sheared chromatin.

18. Add 6 µL 5 M NaCl and mix by flicking the tube.

19. Incubate at 75 °C for 5 min and centrifuge at >10,000 × *g* for 30 s.

20. Place tubes on a magnetic stand, allow the beads to cluster, and transfer supernatant containing ChIP DNA to a new 1.5 mL microcentrifuge tube.

21. Incubate at 65 °C for 30 min.

22. Add 1 µL Proteinase K and incubate at 65 °C for 90 min.

23. Add 785 µL (5 volumes) of ChIP DNA Binding Buffer to each ChIP DNA sample and mix briefly.

24. Transfer mixture to a Zymo-Spin IC Column in a collection tube.

25. Centrifuge at >10,000 × g for 30 s. Discard the flow-through.

26. Add 200 μL DNA Wash Buffer to the column. Centrifuge at >10,000 × g for 30 s. Repeat once more for a total of two washes.

27. Transfer column to a new 1.5 mL microcentrifuge tube and add 8 μL DNA Elution Buffer directly onto the column matrix. Incubate for 1 min at room temperature and centrifuge at >10,000 × g for 30 s to elute the ChIP DNA.

28. Repeat **step 27** for a total elution volume of 16 μL.

29. Used the ChIP DNA for qPCR at selected regions, or make into libraries for next-generation sequencing.

3.6 CLOuD9 3C

Induction of chromatin interactions is assessed by 3C. Here is described the basic protocol to qualitatively confirm interactions by PCR. Refer to Chap. 1 for an in-depth discussion of the controls and setup required for quantitative 3C.

1. Fix five million cells in 9.5 mL 2% formaldehyde/PBS and incubate at room temperature for 10 min with gentle rocking (*see* **Note 20**).

2. Quench cross-linking reaction by adding 712 μL 2 M ice-cold glycine.

3. Centrifuge for 5 min at 300 × g, 4 °C and remove supernatant. Resuspend cells in 1 mL cold 1× PBS.

4. Centrifuge for 5 min at 300 × g, 4 °C and remove supernatant. Resuspend the cell pellet in 300 μL of ice-cold lysis buffer and incubate on ice for 15 min.

5. Centrifuge for 5 min at 1000 × g, 4 °C and remove supernatant. Resuspend the cell pellet in 500 μL of ice-cold lysis buffer.

6. Centrifuge for 5 min at 1000 × g, 4 °C and remove supernatant. Resuspend cell pellet in 50 μL 0.5% SDS and incubate at 62 °C for 5 min. Add 145 μL nuclease-free water and 25 μL 10% Triton X-100. Mix while avoiding foaming then incubate at 37 °C for 15 min.

7. Take a 5 μL aliquot to be used as the "genomic DNA" and store at −20 °C until **step 11**.

8. Add 25 μL of 10× restriction enzyme buffer (depending on the restriction enzyme used; *see* **Note 4**) to the remaining sample and 100 U of restriction enzyme and digest overnight at 37 °C with shaking.

9. Incubate at 65 °C for 20 min to inactivate the restriction enzyme, then cool to room temperature. Take a 25 μL aliquot as the "digested" control and store at 4 °C until **step 11**.

10. Centrifuge for 5 min at $1000 \times g$, 4 °C and remove supernatant. Add 900 μL ligation master mix. Incubate for 4 h at 16 °C, followed by 30 min at room temperature.

11. Add 50 μL 20 mg/mL proteinase K to the ligation sample. Add 40 μL 10 mM Tris-HCl pH 7.5 and 2.5 μL 20 mg/mL proteinase K to the "genomic DNA" control; add 25 μL 10 mM Tris-HCl pH 7.5 and 2.5 μL 20 mg/mL proteinase K to the "digestion" control. Incubate overnight at 65 °C to decrosslink samples.

12. Add 30 μL 10 mg/mL RNase A (1.5 μL to the controls). Incubate for 30 min at 37 °C.

13. Add 1.4 mL phenol/chloroform/isoamyl alcohol and mix vigorously (60 μL to the controls). Centrifuge for 5 min at $13,000 \times g$, room temperature. Transfer upper aqueous phase to a new 15 mL tube, or to new microtubes for the controls.

14. Add 100 μL (5 μL to the controls) 3 M sodium acetate pH 5.2, followed by 2.5 mL (125 μL to the controls) ice-cold 100% ethanol. Mix thoroughly and leave at −80 °C for 1 h.

15. Centrifuge for 45 min at $2220 \times g$, 4 °C. Remove supernatant and wash pellet in 5 mL 70% ethanol. Centrifuge for 15 min at $2220 \times g$, 4 °C. For the controls, centrifuge for 45 min at $13,000 \times g$, 4 °C, wash pellet in 100 μL 70% ethanol, then recentrifuge for 15 min at $13,000 \times g$, 4 °C.

16. Remove supernatant and air-dry pellets at room temperature, then dissolve DNA pellet in 150 μL (20 μL for controls) 10 mM Tris-HCl pH 7.5. Label as "3C library."

17. Quantify 3C library and controls with a Qubit fluorimeter and dsDNA BR assay.

18. Perform gel electrophoresis using 1% agarose with ~100 ng each of "genomic DNA," "digestion," and "3C library." "Genomic DNA" should be present as a tight, high molecular weight band. "Digestion" should reflect a smear. "3C library" should show an upward band shift and a tighter band (Fig. 3).

19. Set up 30-cycle PCR reactions using 100 ng of input DNA for both "3C library" and "genomic DNA," LightCycler 480 SYBR Green I Mastermix (or another PCR reaction setup) with buffer and dNTPs according to manufacturer's recommendations, and 500 nM each 3C primer (*see* **Note 5**).

20. Perform gel electrophoresis on PCR products. "Genomic DNA" should not have any visible product, and "3C library" is expected to give single bands with the same size as that predicted of the cognate 3C product. Bands can be excised, gel-purified, and sent for Sanger sequencing to confirm sequence.

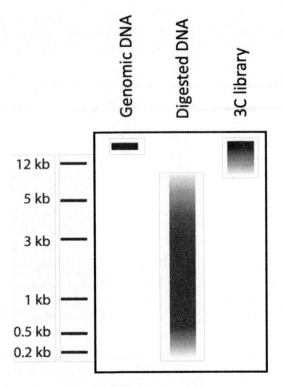

Fig. 3 Quality control for 3C (adapted from [5]). Sample schematic for a series of 3C quality controls after gel electrophoresis. Exact banding and smearing patterns may differ based on cell-type used

4   Notes

1. Plasmids pC9-sadCas9-ABA-puro (13 kb) and pC9-spdCas9-PLY1-hyg (14 kb) are third generation lentiviral vectors with self-inactivating (SIN) 3′ LTR and hybrid LTR (CMV-5′ LTR) features. They express dCas-fusion proteins that can target a pair of genomic loci and subsequently bring them into physical proximity via heterodimerization. Both plasmids have a gRNA cloning site and corresponding gRNA scaffold that allows for cloning of a single gRNA sequence. The lentivector plasmid pC9-sadCas9-ABA-puro expresses a nuclease-deficient *S. aureus* Cas9 protein fused to the ABI protein from the plant abscisic acid (ABA) signaling pathway which is utilized as one of the protein dimerization partners in the CLOuD9 system. The second lentivector plasmid pC9-spdCas9-PYL1-hyg expresses a nuclease-deficient *S. pyrogenes* Cas9 protein fused to the complementary dimerization partner PYL1 protein from the plant ABA signaling pathway (Fig. 4). The CLOuD9 plasmids were adapted from LentiCRISPR plasmids from the Zhang lab [3].

Fig. 4 Schematic for CLOuD9 lentivector plasmids. CLOuD9 system consists of two plasmids that are integrated into the host genome via lentiviral transduction. Gray circles represent linkers between the fusion proteins

2. Lentiviral plasmids contain long terminal repeats and must be transformed into recombinase-deficient bacteria such as *Stbl3* or equivalent.

3. ABA is commercially available as a racemic mixture or as (+) or (−) stereoisomers. We specifically use the (S)-(+) Abscisic Acid with >98.0% purity. Protect stock solution from light and keep for up to 6 months at −80 °C.

4. Select a 4-base cutting restriction enzyme that produces cohesive ends according to cut size availability around the target locus. Validated four-cutters include DpnII, NlaIII, Csp6I, and MboI. Restriction enzymes sensitive to DNA methylation should not be used.

5. Restriction fragments used should be <1 kb from the genetic element of interest. Forward primers need to be within 150 bp of the 3′ ends of restriction fragments of interest. Primers used should be 20–22-mers with a Tm between 55 and 65 °C, with a maximum difference of 2 °C for each experiment. Primers can be designed using existing design tools, such as Primer3, to ensure specificity.

6. A high efficiency score obtained using CRISPR sgRNA design tools does not necessarily imply efficient dCas9 binding. Multiple candidate guides (we typically design 3 guides per target region) should be selected and tested by ChIP-qPCR.

7. Use 10× T4 DNA ligase buffer as it contains ATP; the supplied PNK buffer does not. Otherwise, supplement the buffer with a final concentration of 1 mM ATP.

8. A typical lentiviral titer from crude supernatant is 1×10^6 transduction units/mL. The viral suspension can be aliquoted in 25–50 μL aliquots and stored at −80 °C for up to 6 months or at −20 °C for up to 1 month.

9. Before setting up transduction, determine the optimum antibiotic concentration to be used for selection for each transduced cell line using a kill curve assay with six different concentrations of antibiotics.

10. Seed one well with nontransduced control cells during the antibiotic selection posttransduction and verify complete cell death.

11. For cells that are transduced with multiple viral suspensions, perform sequential selections if simultaneous double-selection is too harsh for the cells.

12. When thawing stably integrated cell lines, repeat antibiotic selection for a few passages.

13. Bicinchoninic acid (BCA) protein assays can also be used here.

14. If using PVDF membrane, activate it by immersing in 100% methanol for 10–15 s before use. Assemble the transfer sandwich such that there are no air bubbles trapped in the sandwich layers.

15. Blots can also be blocked in 5% dry milk in PBST solution for 1 h at room temperature.

16. ABA has a reported half-life of approximately 24 h in Chinese Hamster Ovary (CHO) cells [4].

17. The onset, rate, and reversibility of transcriptional changes due to CLOuD9 vary based on cell-type and target locus. We have observed activity as soon as 24 h. The irreversibility of chromatin looping typically occurs from 7 to 10 days.

18. An aliquot of DNA should be purified and assessed by gel electrophoresis. Sonication for ChIP requires a smear between 150 and 1000 bp.

19. Use ChIP-grade antibodies. Consider using a negative control, such as IgG, to assess nonspecific binding.

20. Formaldehyde concentration can be lowered to 1% if digestion efficiency is an issue.

Acknowledgments

This work was supported by the New York Stem Cell Foundation, the Doris Duke Clinical Scientist Development Award, a Stanford Dean's Fellowship and SunPharma Fellowship Award (to P.A.), and an A*STAR National Science Scholarship (to W.Q.S.). K.C.W. is a New York Stem Cell Foundation–Robertson Investigator, and the Stephen Bechtel Endowed Faculty Scholar in Pediatric Translational Medicine, Stanford Maternal & Child Health Research Institute.

References

1. Morgan SL, Mariano CM, Abel B et al (2017) Manipulation of nuclear architecture through CRISPR-mediated chromosomal looping. Nat Commun 8(1):15993 p1–15993 p9

2. Concordet J-P, Haeussler M (2018) CRISPOR: intuitive guide selection for CRISPR/Cas9 genome editing experiments and screens. Nucleic Acids Res 46(W1):W242–W245

3. Sanjana NE, Shalem O, Zhang F (2014) Improved vectors and genome-wide libraries for CRISPR screening. Nat Methods 11(8):783–784

4. Liang F-S, Ho WQ, Crabtree GR (2011) Engineering the ABA plant stress pathway for regulation of induced proximity. Sci Signal 4(164):rs2

5. Krijger PHL, Geeven G, Bianchi V et al (2020) 4C-seq from beginning to end: a detailed protocol for sample preparation and data analysis. Methods 170:17–32

Chapter 15

Acute Protein Depletion Strategies to Functionally Dissect the 3D Genome

Michela Maresca, Ning Qing Liu, and Elzo de Wit

Abstract

The organization of the genome inside the nucleus facilitates many nuclear processes. Because the nuclear genome is highly dynamic and often regulated by essential proteins, rapid depletion strategies are necessary to perform loss-of-function analyses. Fortunately, in recent years, various methods have been developed to manipulate the cellular levels of a protein directly and acutely. Here, we describe different methods that have been developed to rapidly deplete proteins from cells, with a focus on auxin inducible degron and dTAG methods, as these are most commonly used in 3D genome organization studies. We outline best practices for designing a knockin strategy, as well as generation and validation of knockin cell lines. Acute depletion strategies have been transformative for the study of the 3D genome and will be important tools for delineating the processes and factors that determine organization of the genome inside the nucleus.

Key words Genome editing, Acute protein degradation, Protein knockdown, Chromatin dynamics, Genome organization

1 Introduction

To facilitate nuclear processes such as DNA replication, DNA repair and transcription, the genome is organized inside the nucleus in a nonrandom manner. Breakthroughs in both genomics [1] and imaging methods [2, 3] have yielded unprecedented insight into the organization of the genome. Following the observation that chromosomes occupy their own nuclear subvolume, known as chromosome territories [4], it was found that they are further subdivided into active (A) and inactive (B) compartments. Compartment formation is counteracted by the activity of the cohesin complex [5–7]. Through a process called loop extrusion, the cohesin complex forms loops that are ultimately anchored by the DNA binding protein CTCF. Loop formation is a highly dynamic process

---

Michela Maresca and Ning Qing Liu contributed equally.

Tom Sexton (ed.), *Spatial Genome Organization: Methods and Protocols*,
Methods in Molecular Biology, vol. 2532, https://doi.org/10.1007/978-1-0716-2497-5_15,

that is regulated by among others, the NIPBL/MAU2 complex that is involved in the loading of cohesin [8] and the WAPL protein that is required for the offloading of cohesin [6, 7, 9]. This dynamic loading cycle is crucial for maintaining cell-type specific gene regulation [10]. For an in-depth review of nuclear organization we point the reader to one of many the excellent reviews out there for instance ref. 11 or 12.

From just a cursory glance at microscopy images of eukaryotic cells one can conclude that nuclear organization is highly heterogeneous. This was further cemented by single nucleus Hi-C experiments, that showed that despite clear segregation of A and B compartments, the constellation of these compartments are seemingly random between individual cells [13]. The cell-to-cell heterogeneity was also found at the level of cohesin-mediated loops [14]. This is in line with loop formation being a highly dynamic process, in which loops are constantly being formed and subsequently disassembled. Live cell imaging of individual loci has shown that the genome can indeed be highly dynamic [15]. Because of the dynamic nature of the 3D genome, determining the direct effects of loss-of-function mutations can only be achieved through rapid perturbation. Furthermore, many architectural proteins such as CTCF and cohesin subunits are essential precluding a straightforward loss of function analysis through CRISPR knockout. However, for nonessential genes clonal knockout lines may suffer from the induction of clonal differences. On the other hand, polyclonal CRISPR knockout or CRISPRi or siRNA knockdown cells, may suffer from incomplete perturbation and, more importantly, heterogeneous protein depletion. To study direct effects following protein depletion, methods are needed that can deplete proteins both rapidly and uniformly. Acute protein depletion methods, many based on small molecule degraders, fill this requirement and are rapidly becoming a standard tool in the analysis of genome organization. In this chapter, we will discuss various acute protein depletion techniques and their application in the study of genome architecture.

1.1 Targeted Protein Degradation to Study Early Molecular Events of 3D Genome Assembly

Acute protein degradation methods generally hijack the endogenous degradation pathways. This is accomplished through targeting the proteins of interest to the proteasome machinery, rather than through the inhibition of newly formed proteins in knockout or knockdown methods. Protein loss occurs within a very short time frame, ranging from a few minutes to hours at most. The nature of rapid protein degradation offers the opportunity to separate the primary effects of protein loss from the secondary effects caused by changes in the epigenome, transcriptome and proteome (for instance down-regulation of a transcriptional regulator). A further advantage of acute protein depletion over clonal knockout lines is that the consequences of protein loss are evaluated in the same clonal cell line, which to a great extent eliminates clonal differences irrelevant to the function of the protein. The majority of acute protein degradation methods are also reversible and can be easily titrated.

In the past few years, pioneering studies have delineated genome architecture using acute protein degradation techniques. Acute depletion of CTCF from mouse embryonic stem cells (within 24 h) revealed the crucial role of CTCF in the formation of the majority of TAD boundaries [16]. In addition to this, it was shown that the cohesin complex plays a crucial role in the formation of TADs [7, 17]. Reexpression of cohesin leads to a rapid reestablishment of TADs, enabling the quantitative assessment of loop forming mechanisms [7, 17]. Furthermore, when studying enhancer–promoter loop formation, the rapid depletion of YY1 protein from mouse embryonic stem cells (mESCs), enabled the analysis of the functional role of YY1 in mediating those types of loops [18]. Targeted degradation of the cohesin release factor WAPL and the cohesin subunit RAD21 also highlighted the functional contribution of cohesin and cohesin turnover in mediating enhancer–promoter interaction essential for the expression of cell type specific genes [10, 19]. Acute perturbation of RNA Pol I, II, and III, was used to selectively degrade the different polymerases essential for cell function, without the use of transcription inhibitors that may have pleiotropic effects. The acute protein depletion showed that interfering with active transcription only mildly affects the overall chromatin structure [20]. Finally, acute depletion of subunits of the condensin complex has shown the crucial role of this complex in the formation of the characteristic compacted and linearly organized mitotic chromosomes [21].

2 Systems for Targeted Protein Degradation

Numerous techniques have been developed to mediate protein degradation or stabilization in an acute manner. These systems often rely on the (genetic) tagging of a protein of interest with a specific protein domain, or "degron," that confers the degradation of the entire protein. However, there are also nongenetic methods for the acute depletion of proteins. Below we briefly outline some of the well-established methods for targeted protein degradation and list their advantages and disadvantages (summarized in Table 1).

2.1 Acute Depletion of Endogenous Proteins

Small molecule induced degradation systems, such as proteolysis targeting chimeras (PROTACs), are generating a lot of excitement in the field of drug discovery and development. They rely on covalent linking of two small molecules, one with affinity to the protein of interest and one to a ubiquitin ligase to induce selective

2.1.1 PROTACs

Table 1
Methods for targeted protein degradation

System	Transgenic elements	Tag size	Degrader/stabilizer	Reversible	Mechanism of action	Basal depletion	Applications
PROTACs	–	–	Bi-functional molecules (protein specific)	Yes[a]	Proteasomal degradation	–	[22–24]
TRIM-AWAY	–	–	TRIM21 + Antibody (protein specific antibody/nanobody)	–	TRIM21 mediated proteasomal degradation	–	[25]
Shield	FKBP12^{L106P}	12 kDa	Shld1[b]	Yes	Proteasomal degradation	Yes	[26]
SMASh	NS3-NS4A	34 kDa	Asunaprevir	Yes	Proteasomal degradation	–	[27]
AID	AID + TIR1	7 kDa	Indole-3-acetic acid (IAA)	Yes	TIR1 + CUL1 mediated proteasomal degradation	Yes	[7, 10, 14, 16, 17, 19–21, 28–35]
dTAG	FKBP12^{F36V}	12 kDa	dTAG	Yes	CRBN mediated (or VHL) proteasomal degradation	Yes	[10, 18, 36–39]

[a] Depends on the design of the molecule, [b] Shld1 molecule stabilizes the protein tagged with the FKBP12^{L106P} domain

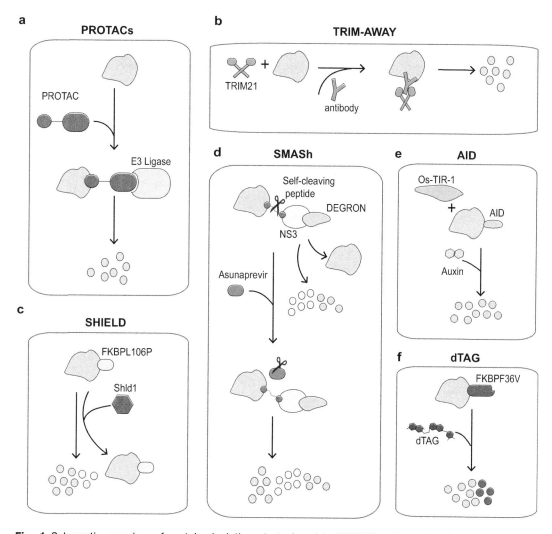

Fig. 1 Schematic overview of protein depletion strategies. (**a**): **PROTACs**. A protein of interest (gray) is recognized by bi-functional molecules (PROTAC) that bridges the protein to an E3 ligase (light blue), promoting degradation of the protein. (**b**): **TRIM-AWAY**. An antibody (red) against the protein of interest is added, which recognizes the protein. The endogenous antibody receptor and E3 ligase TRIM21 (blue) recognizes the antibody and promotes proteasomal degradation of the protein. (**c**): **SHIELD**. The protein of interest is tagged with the destabilizer domain FKBP12^{L106P} (turquoise) and is constitutively degraded. Addition of the small molecule Shld1 blocks the degradation and restores protein levels. (**d**): **SMASh**. The protein of interest is tagged with a NS3-degron domain containing a self-cleaving peptide, that cleaves off the degron domain from the protein of interest. Addition of Asunaprevir (dark blue) blocks the self-cleavage, retaining the degron and promotes protein degradation. (**e**): **AID**. The protein of interest is tagged with the AID tag (orange). Together with the expression of OsTIR1, the tagged protein is degraded upon addition of the plant hormone auxin (IAA) (green). (**f**): **dTAG**. The protein of interest is tagged with FKBP^{F36V} (dark red). Addition of dTAG molecule (purple) promotes protein degradation

protein degradation (Fig. 1a) [22, 23, 40]. The two affinity molecules are connected by a peptide linker. These so-called molecular degraders also offer the advantage of controlling the endogenous level of their protein targets. The use of degraders has greatly helped in the development of new therapeutic molecules, particularly in the field of oncology [24]. Although this system promotes a fast and selective degradation of the candidate protein, it requires the synthesis of specific affinity molecules that are specific for each protein of interest, which is challenging and time consuming, and is likely beyond the primary competences of a fundamental molecular biology lab.

2.1.2 Anti/Nanobody Mediated Degradation

An alternative approach to degrade a protein of interest is Trim-Away [25]. This method relies on the expression of the antibody receptor and ubiquitin ligase TRIM21 and the microinjection or electroporation of the antibody against the protein of interest. TRIM21 is recruited to the Fc domain of the antibody. This results in the degradation of the protein target through the proteasome pathway (Fig. 1b). By fusing a nanobody to an Fc domain, Trim-Away can be used to degrade proteins inside the nucleus. An obvious advantage is that Trim-Away does not require the generation of a genetic fusion protein. Another interesting feature is that it can be used to degrade specific isoforms of a protein or posttranslationally modified forms of a protein, for instance acetylated histones. A disadvantage is that the method relies on monoclonal antibodies or nanobodies, which need to be delivered to the cell(s). In the case of electroporation this may affect the cells, which is not an issue for the small molecule-based methods described below. Note that, although TRIM21 is often endogenously expressed, for cells that do not express TRIM21 or not highly enough, one would need to ectopically express this protein.

The nongenetic acute protein depletion methods can be particularly useful when studying proteins that have many copies in the genome such as histones or are amplified in the cell type of choice. Many copies of the gene of interest complicates the genetic tagging of each individual copy. These methods may for instance be suitable for studying common oncoproteins, such as MYC, in different cancer cell lines.

2.2 Acute Depletion Based on Endogenously Tagged Proteins

Below we describe methods that are more generalizable, but do require the fusion of a protein domain to the target. Generating endogenous protein fusions with a degradation domain, or "degron," has been greatly facilitated by CRISPR-Cas9 genome editing [41, 42]. The degron tag is used as a substrate that promotes degradation upon the addition of an external stimulus, often a small molecule. It should be noted that many of the strategies discussed below are still undergoing optimization. Ideally fusion of the degradation tag does not directly impair protein function. Only upon stimulation with specific small molecules the fusion protein is degraded by the proteasomal pathway.

2.2.1 Shield

One of the pioneer acute depletion methods employing a degron fusion is the Shield system, in which the protein of interest is fused with a FKBP12^{L106P} domain (12 kDa) [26]. The degron here is used as a destabilizing domain and is directly recognized by the proteasomal machinery and degraded. Fusion of the FKBP12^{L106P} destabilization domain to the endogenous protein leads to degradation of the target protein through the proteasomal machinery. To stop the constitutive degradation of the protein, the Shld1 agent is supplemented in the culture medium. The Shld1 molecule protects the FKBP12^{L106P} tag from being degraded and results in stabilization of the fusion protein and hence expression in the cells. Although Shield promotes a controlled and selective degradation of the protein, continuous presence of the Shld1 molecule is necessary to maintain the constitutive expression of the protein of interest (Fig. 1c).

2.2.2 Small Molecule Assisted Shutoff (SMASh-Tag)

The SMASh-tag method relies on fusing a protein of interest with the cleaving peptide NS3, a hepatitis C virus protease, together with a degron domain (total tag size: 34 kDa) [27]. After translation, the NS3 protease cleaves off itself and the degron domain from the endogenous protein (Fig. 1d). This leads to unmodified endogenous proteins and degradation of the degron domain only. To induce degradation of the endogenous protein the degron tag needs to be attached to the protein of interest. This is achieved by adding the protease inhibitor asunaprevir to the culture medium. The inhibitor selectively blocks the protease activity of the NS3, thus keeping the degron attached to the protein. After asunaprevir treatment, proteins are degraded within few hours by the proteasomal machinery. This tag system has the unique advantage of producing proteins in their native configuration, but limits the degradation only to newly synthesized proteins. Depending on the stability of the protein of interest and the proliferation rate of the cells (every cell division decreases the protein concentration) it may take some time before all proteins have been degraded.

2.2.3 Auxin Inducible Degron (AID) System

The auxin inducible system requires the fusion of the specific AID peptide (7 kDa) to a protein of interest and the expression of the *Oryza sativa* (Asian rice) receptor (OsTir1) that associates with the Skp-Cul1-F-box E3 ligase (SCF) [28]. After addition of one of the analogs of the plant hormone auxin, such as indole-3-acetic acid (IAA), the fusion protein is recognized by OsTir1 and the SCF complex is recruited, promoting targeted degradation of the protein via direct interaction with Cul1, an E3 ligase. This is followed by recruitment of the proteasomal machinery and ubiquitination and degradation of the tagged proteins (Fig. 1e). The method has been extensively used and in the majority of cases induce protein degradation within minutes. For cultured cells AID mediated degradation is reversible. Removal of IAA from the medium ("wash-off") reverts protein levels to its original level. For certain cell types, for instance embryonic stem cells that grow in colonies, we advise to trypsinize the cells following wash-off, because simple removal of the medium may result in incomplete restoration of proteins.

The main advantage of the AID system is the very fast kinetics of degradation. AID has been shown to be extremely sensitive and efficient for protein degradation. Complete depletion of the WAPL protein using the AID system takes around 30 min following IAA treatment [10]. Furthermore, the AID system can be used in all nonplant organisms. The system indeed has been widely used in eukaryotic systems, including human and mouse cell lines and *Drosophila* [29, 30]. When creating a fusion protein, it is important to consider that the AID peptide may affect expression and function of the protein independently of IAA treatment. To minimize aberrant structural configurations of the tagged protein, a smaller version of it the AID tag was designed, called mini-AID. The mini-AID tag was equally efficient in inducing protein degradation [31].

For a frequently used cell line, we advise to first integrate the OsTir1 gene at a safe harbor locus. For mouse embryonic stem cells, the *Rosa26* or *Tigre* locus has been used [16]. This parental line can then be used to create homozygous knockins at the locus encoding the protein of interest. This requires two sequential genetic manipulation steps, each demanding clonal selection, which can be time-consuming. However, the two-part system also offers the unique, but so far theoretical, advantage to generate genetically modified organisms that express *OsTIR1* gene under a tissue specific promoter, allowing conditional depletion of the protein in a cell type of interest. This approach could be useful to study the role of a specific protein during development or restricted to specific cell types.

A drawback of the AID system is the basal degradation of the fusion protein in the absence of auxin [43]. Unfortunately, to date, there is no method available that can predict the level of basal degradation for a selected protein. Notably, severe basal degradation sometimes results in difficulties to generate a homozygous knockin cell line when the protein of interest is essential for cell viability [32]. Fortunately, a number of solutions have been developed to counteract the basal degradation of AID fusion proteins. A relatively simple solution is to grow cells in the presence of the small molecule auxinole that inhibits OsTIR1 [44]. Another option is to coexpress the auxin response factor (ARF) PB1 domain [45]. Both methods have shown to block basal degradation in AID cell lines.

In addition to the above described efforts to control for basal degradation in cells tagged with the original AID system, an exciting new development is the AID2 method [32]. AID2 relies on the constitutive expression of a mutant OsTir1 (F74G). In addition, an IAA analog, 5-Ph-IAA, is used to trigger the depletion. AID2 performs favorably in terms of degradation efficiency and speed of degradation. Importantly, basal depletion of many proteins is much decreased, leading to more or less wild-type levels for selected proteins compared to the regular AID. Note, that this also enables tagging of proteins that could not be tagged with the regular AID, such as DHC1, SMC2, and CTCF in HCT116 colon carcinoma cells. AID2 also enables acute protein depletion in yeast and in living mice, although in in vivo systems one needs to consider that 5-Ph-IAA may not reach all organs equally efficiently, for instance the brain. Note, that particularly for architectural proteins such as CTCF, minute levels of protein can be enough to maintain its function in genome organization [16, 46]; therefore, how AID2 performs in 3D genome analysis remains to be assessed.

2.2.4 Degradation Tag (dTAG)

The dTAG system relies on the fusion of a candidate protein with a mutant version of the FKBP domain (FKBP12^{F36V}) (12 kDa), that can be selectively recognized by small heterobifunctional molecules (called dTAGs). The dTAG molecules are comprised of two functional moieties, an ortho-AP1867 moiety that has affinity for the FKBP12^{F36V} mutant and a thalidomide moiety. Thalidomides have the ability to directly target the cereblon (CRBN), a member of the E3 ligase complex [47]. This dual recognition promotes degradation of the tagged protein via recruitment of proteasomal machinery and protein ubiquitination (Fig. 1f). Degradation of the tagged protein is achieved by supplementing the culture medium with dTAG. The kinetics of degradation ranges from minutes to hours. As for the AID2, dTAG has been shown to be applicable in vivo with similar degradation efficiency [36], although its application is restricted to organisms that express orthologs of the CRBN protein that can bind the dTAG molecule. Similar to the AID system, degradation of the FKBP-tagged protein is reversible, where deprivation of the dTAG molecule from the medium restores the endogenous levels of the protein [18, 36].

The different dTAG molecules are denoted by the size of the linker between the two moieties, for instance, dTAG-7, dTAG-13, or dTAG-47 [36]. dTAG-13 is the most commonly used dTAG molecule and the only one that is readily available commercially (Tocris Bio). However, not all proteins are degraded equally efficiently and can even be context dependent. For instance, in HCT-116 CRBN-based dTAG degraders did not effectively degrade MED14-FKPB [37]. Different classes of dTAG molecules can be used to circumvent degradation resistance. For instance, dTAG^{V}-1 recruits the Von Hippel Lindau (VHL) E3 ligase instead of the CRBN [38], and shows improved pharmacokinetics and pharmacodynamics and displays increased degradation for certain resistant proteins.

The dTAG system requires only a single round of genetic manipulation, since it relies on endogenous E3 ligase recognition proteins. However, this may have unintended consequences, such off-target degradation of nontagged endogenous proteins in cells following dTAG treatment. Thalidomide is infamous for having disastrous consequences during human fetal development [47]. In addition, the FKBP12^{F36V} degron is larger (12 kDa) than the mini-AID degron (7 kDa), which is more likely to perturb protein structure and hence affect the function of the target protein. And although seemingly rarer than in the original AID system, the dTAG system sometimes shows basal degradation, for instance in the case of OCT4 and SOX2 [10, 39].

3   Methods: General Considerations for Targeted Protein Degradation

The generation of a degron line relies on the fusion of the degron in frame with the protein of interest. Generating a fusion protein relies on efficient strategies for manipulating the genome. Nowadays, the most commonly used knockin method is the CRISPR-Cas9-mediated homology directed repair (HDR) [48]. To do this, a single guide RNA (sgRNA) sequence for the Cas9 nuclease is designed to introduce a double strand break at a specific location in the genome. The cleaved site will be repaired by the endogenous DNA repair systems of the cells [49]. Gene fusion depends on the presence of a repair template that will be used in homologous recombination. The repair template needs to be delivered ectopically, and is designed to include the degron sequence flanked by homology arms that are homologous to the locus of interest. The most straightforward way of supplying the repair template is as a donor plasmid or a synthetic double stranded DNA fragment.

For efficient integration of the degron tag and to increase the probability of HDR, the sgRNA should cleave close to the preferred site of integration. However, successful integration of a tag into a gene of interest, strongly depends on the efficacy of the sgRNA to cleave the locus. Choosing a sgRNA that cleaves the DNA in the most efficient manner and with minimal off-target effects, is key for efficient generation of fusion proteins.

3.1   Design of a Donor Template

The design of the repair template is crucial for the site-specific integration of a degron tag to a gene of interest. Usually, a donor plasmid is designed to contain the nucleotide sequence of the degron-tag (a degron cassette), flanked by the complementary DNA sequence (referred as homology arms) 500–800 bp upstream and downstream of the integration site of the degron, to promote the integration of the degron cassette in-frame with the gene target (Fig. 2a). HDR mediated repair is usually performed with homology arms of 0.2–1 kb. However, shorter homology arms can also be

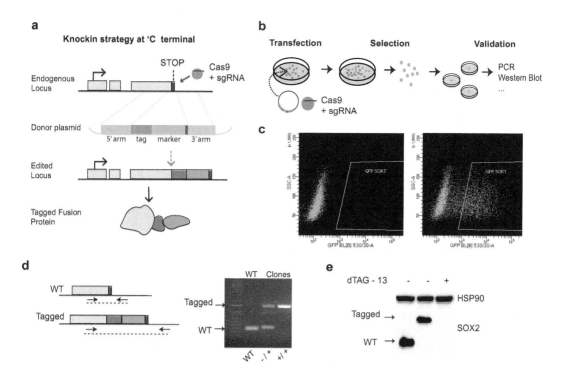

Fig. 2 Simplified overview of generation of tagged proteins. (**a**) Knockin strategy to generate fusion proteins. The 3′ end of the coding sequence of the endogenous locus (red) is cleaved by CRISPR-Cas9. A donor plasmid is designed to contain the domain tag (blue) and fluorescent marker GFP (green) sequence surrounded by homology arms (gray), used as template to repair the cleaved locus. This should result in a protein that is fused with the tag and a marker. (**b**) Schematic overview of the experimental set-up to generate edited cell lines. Cells are transfected with both donor and CRISPR-Cas9 plasmids. Edited clones are selected using FACS or antibiotic resistance and single clones are screened for the knockin and validated for correct integration. (**c**) FACS plots showing selection of single clones positive for the fluorescent marker GFP. (**d**) Clones are validated with PCR primers designed to span the endogenous sequence of the locus (left). Gel-electrophoresis of PCR product (right) showing the wild-type band (lower size), heterozygous knockin (multiple bands) and homozygous knockin (higher band) at the desired locus. (**e**) Western blot showing protein levels of SOX2 and the shift in molecular weight for the tagged protein. SOX2 protein is acutely degraded upon addition of dTAG-13 molecule

used. For instance, homology arms of 180 bp were used to target the *RAD21* gene with the AID tag [30]. The precise integration into targeted chromosome (PITCh) approach uses even shorter homology sequences (10–50 bp) and promotes integration via micro-homology mediated end joining (MMEJ) pathway [50]. PITCh has also been used to generate degron fusions [51].

Any plasmid backbone can be used for the generation of the donor plasmid. When designing homology arms spanning the endogenous PAM sequence (used in CRISPR-Cas9), the PAM sequence can be mutated to prevent reintroduction of a double-strand break after homologous recombination has occurred, and hence eliminate undesirable insertions or deletions in the engineered locus.

Construction of the donor plasmid can be done by gene synthesis or PCR amplification of the desired sequence from the genomic sequence of the parental line. Restriction enzyme-based cloning or assembly methods, such as Gibson assembly or Golden Gate cloning can be used to assemble the donor plasmid. Ordering the fragments or plasmids from commercial providers such as Integrated DNA Technologies (IDT) or Twist Biosciences can simplify the cloning procedures.

3.2 Functionality of Fusion Proteins

For the efficient generation of degron-protein chimeras, it is also important to consider which domain of the protein is most suitable to accept the donor sequence. The integration of a tag can either be at the amino terminus (or N-terminal) or at the carboxy terminus (C-terminal) of the protein. To understand which domain of the protein is the best suited option to fuse the degron to, we advise to perform a literature search to identify whether successful fusion proteins have been generated for the protein of interest. Protein databases, such as UniProt [52] can be used to identify the position of functional protein domains. Furthermore, genome-wide expression analysis, such as mRNA sequencing, can help to understand whether the gene of interest has alternative start or end sites. If either the first or last exon is skipped during transcription, generation of an N-terminal or C-terminal degron fusion, respectively, may ultimately lead to isoforms that do not contain the degron, meaning that these isoforms will not be degraded. Therefore, for genes with alternative start sites C-terminal tagging is preferred, for genes with alterative end sites, N-terminal tagging is preferred. Obviously if specific isoforms of a protein are to be targeted one can make use of this by tagging the isoform of interest.

Another issue to keep in mind is that many proteins are posttranslationally modified at their N- or C-termini. A common modification is the C-terminal covalent attachment of a lipid group, generally referred to as (iso)prenylation [53]. These posttranslational modifications protein can be essential for the physiological function of the protein. For instance, farnesylation of the carboxyterminal domain of prelamin A is required for maturation of the active form of Lamin A protein and its inclusion in the nuclear lamina [54]. Tagging the C-terminus of Lamin A might therefore lead to impairment of the nuclear lamina even without depletion of the protein.

Integration of the tag directly after the protein sequence may induce destabilization of the native configuration of the proteins. To reduce any structural instability caused by the tag, a short peptide linker sequence (such as the Gly-Ser-Gly sequence) can be used between the tag and the protein of interest.

3.3 Selection Strategies to Increase Knockin Efficiency

In mammalian cells, HDR is not a preferred pathway to repair double-strand DNA breaks. Therefore, one of the challenges in generating knockin cell lines is to improve integration efficiency of a donor sequence. One way to increase the frequency of HDR is to add inhibitors to DNA-dependent protein kinase (DNA-PK) [55]. The delivery methods of both repair template and sgRNA strongly depends on the cell line of choice. Choosing the best delivery method greatly improves the efficiency of genome editing. Transfection is preferred for cell lines, although microinjection has been used for embryos [56]. Because in-frame tagging of the protein target can be quite inefficient we advise to use a marker for the selection of successful integration events. Below we will discuss two common selectable markers.

3.3.1 Antibiotic Resistance Markers

Selection markers can be used improve selection of edited cells. A common approach is the addition of antibiotic resistance in the plasmids delivered. A few days after transfection, the specific antibiotic can be added to the cells, allowing for the selection of cells that received the plasmid. This would allow to select for transfected cells, over nontransfected, thus increasing the chances of selecting edited clones. Note that antibiotic resistance markers are often not fused in frame with the gene of interest, with the marker being expressed from its own promoter. Because of this, random integration of the repair template will also result in antibiotic resistance and the selection of clones in which the gene of interest was not tagged. Integration of an active resistance marker may also affect the expression of the proximal targeted gene. However, this can be resolved by flanking the resistance marker with loxP or FRT sites, so that they can be excised using Cre-recombinase or flippase (Flp), respectively [16, 57]. Alternatively, the resistance marker can be fused in-frame with the endogenous gene, separated by a self-cleaving P2A sequence [15].

3.3.2 Fluorescent Proteins

To increase the efficiency of selecting clonal cells, we typically fuse a fluorescent sequence, such as eGFP or mCherry, in frame with the protein of interest and the degron tag. The start codon of these fluorescent proteins can be removed to prevent the expression of these sequences driven by endogenous promoters in cells. At later stages this can also be used to track the expression of the fusion protein. To achieve this one needs to design a donor plasmid that contains the degron, a fluorescent protein and sequences with homology to the locus of interest (Fig. 2a). Fluorescent Activated Cell Sorting (FACS) can then be used to select the cells expressing the fluorescent fusion protein, which should theoretically be the cells with the integration of the donor sequence in-frame with the protein of interest. In our experience screening ~100 clones from targeted mouse embryonic stem cells results in the identification of enough clones with both homozygously and heterozygously targeted loci. To prevent out of frame integrations, additional steps can be taken. For C-terminally tagged proteins we advise, if possible, to position the 5′ end of the repair template in an intron. For N-terminally tagged proteins, placing the 3′ end of the repair template in an intron can strongly decrease the risk of introducing frameshift mutations and the formation of a truncated protein.

Fusion of the protein with a fluorescent tag enables monitoring of the expression levels of the endogenously tagged protein in live cells using FACS, which is both quantitative and sensitive (Fig. 2c). We and others used this approach to quantify endogenously WAPL [10], CTCF [7, 16, 33], and RAD21 [7, 17, 34], respectively. This enables the tracking of the degradation kinetics of the tagged protein. Fluorescent proteins can also be used for other applications such as imaging and immunoprecipitation experiments. As an alternative to fluorescent proteins, degron-tagged proteins can also be fused to a self-labeling protein domain such as the Halo tag, which can be covalently coupled to a fluorophore. Halo-tagged proteins can be used in a number of applications both in vivo and in vitro, such as FACS or immunoprecipitation experiments. Furthermore, organic fluorophores that couple to the Halo tag are bright and photostable. This property makes them extremely useful in super-resolution imaging [35].

However, the fusion of an additional protein domain might lead to further destabilization of the endogenous protein or affect its function. This can be circumvented by including a P2A sequence, which is a self-cleaving peptide that promotes ribosomal cleavage of the downstream protein domain from the protein of interest. Incorporation of the degron tag, P2A and GFP at the endogenous locus enables the measurement of the expression levels of the gene in live cells. Obviously, this abolishes the possibility of closely monitoring the kinetics of degradation in living cells. However, this application can be useful in a heterogenous population of cells, such as organoids or embryos, where not all cells express the protein of interest. For instance, when we are interested in studying the function of a lineage-specific protein in embryos, the protein can be engineered with the degron-P2A-fluorescent protein system. This system enables us to isolate a specific cell type expressing the protein of interest, because the cleaved fluorescent protein will act as a lineage marker. The fluorescent protein will remain expressed even after depletion of the protein of interest, meaning that the cells can be selected by FACS, enabling loss-of-function analysis of this protein only in the cell type(s) in which the protein was initially expressed.

3.4 Validation of Clones and Protein Functionality

Following transfection of cell lines with both donor plasmid and the Cas9 and sgRNA expressing plasmid, single cells can be sorted using an FACS-based approach or manually isolated and expanded (Fig. 2b). Validating the integration of the desired cassette at the selected locus is essential.

3.4.1 PCR Screening of Edited Clones

After delivery of the Cas9, sgRNA constructs and donor plasmids, clones are selected using FACS for the fluorescent marker or following antibody resistance. Clones are expanded and need to be validated to determine whether the editing resulted in proper endogenous targeting. The DirectPCR DNA Extraction System (Viagen Biotech Inc.) can be used to genotype the integration of the tag at the locus of interest using only a small number of cells. By choosing primers that lie within the homology arms and the tag the resulting PCR product can be used to discriminate clones that have been targeted (Fig. 2d). Since the donor plasmid can integrate randomly in the genome, an additional PCR using primers flanking the integration site needs to be performed. This PCR can also be used to determine whether the knockin is heterozygous or homozygous. A pair of PCR primers overlapping the endogenous sequence on one end and on the other end the fusion tag enables one to screen for integrations in the correct orientation. When designing a PCR genotyping approach, it is important to keep in mind that the size of the PCR product affects the genotyping approach. When using the DirectPCR lysis we recommend designing a strategy with PCR products shorter than 1 kb. For larger amplicons, isolation of pure genomic DNA is required. Because this requires many cells, this can be time-consuming for slow growing cells, precluding the rapid screening of clones.

3.4.2 Western Blotting for Functional Knockin

Stability and expression of the tagged protein might be affected by the fusion product. It is important to control for the levels of the tagged protein in physiological conditions. Determining the protein size and level on Western blot is recommended. After knockin of a tag to the protein of interest, Western blot analysis should show an increase in the molecular weight that matches the size of the endogenous protein plus the tag. An example of this is shown for the SOX2 protein tagged with $FKBP12^{F36V}$. Treatment with dTAG-13 then resulted in degradation of SOX2 (Fig. 2e). For the generation of degron cell lines, after clonal selection and validation of the protein level, it is important also to assess applicability of the degradation system. Treatment with the appropriate molecule should have an effect on protein levels within less than 24 h.

3.4.3 Sequencing for Correct Integration and Editing

After having identified the clones positive for the knockin with PCR, Sanger sequencing should be performed on the edited genomic site to control for proper repair and complete integration of the cassette into the endogenous locus. To determine that no genomic rearrangement have occurred at the locus of interest, targeted locus amplification (TLA) can be performed [58]. This has the added benefit of identifying additional integrations of the donor template at different loci in the genome.

4 Considerations for 3D Genome Applications

We have already emphasized the importance of assessing the temporal dynamics of the 3D genome and how the use of degron systems in the last few years has greatly broadened our understanding of genome function and architecture. A degron cell line can be used in a time course to track the changes in genome organization following the depletion of a protein of interest to identify transient changes. We suggest to design relatively short time course experiments, typically from few hours to maximally a few days. Cells should be seeded according to the time-frame of the experiment, considering the doubling time and density. To perform a time course experiment, a single well/plate should be seeded for every timepoint. For a 2 h depletion timepoint cells should be treated with the small molecule 2 h prior to end of the experiment. Cells of all the different timepoints should be harvested together at the end of the experiment to minimize the effects of cell density on molecular readouts (Fig. 3a).

Following cell harvest the sample can subjected to a number of different analyses to chart the 3D genome. When interested in a single gene or CTCF site, one can use 4C-seq [59] or Capture-C [60]. For pairwise genome-wide analysis the preferred method is Hi-C [1], or one of the variants. Note, that these methods generally require between one and ten million cells, although low input variants have been developed [61]. As an example, we have performed 4C-seq in RAD21 degron mESCs from a genomic site that forms a CTCF-anchored chromatin loop. Our results show a rapid resolution of the chromatin loops following RAD21 depletion (Fig. 3b). 4C and Capture-C data analysis can be performed with peakC, which can be used for visualization and peak calling [62]. For 4C analysis we recommend to sequence around one million reads per viewpoint. Further, in-depth, considerations with regard to 4C experimental design can be found in [63].

5 Future Directions

Acute protein depletion is an important tool to study the direct consequences of protein loss of function. Particularly, within the highly dynamic context of the 3D genome and gene transcription it will be an important tool to tease apart cause-and-effect relationships. It also enables the study of the functionality of different protein domains for essential proteins. For instance, by expressing a mutated version of a protein in a degron background of that same protein it is possible to determine functional domains of the protein, as was shown recently for CTCF [33]. Tagging multiple proteins with a degron enables the study of combined depletion.

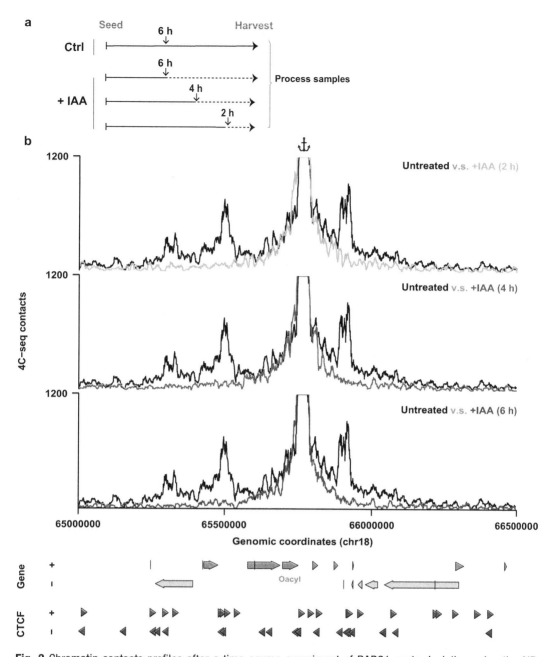

Fig. 3 Chromatin contacts profiles after a time course experiment of RAD21 acute depletion using the AID system. (**a**) Experimental design to perform a time course experiment for the degradation of tagged protein using the AID system. (**b**) (Top) High resolution 4C-seq data for the *Oacyl* locus in RAD21-AID cell lines after IAA treatment at different time points (anchor indicates the view point of *Oacyl* promoter). The black lines show the contact profile in untreated cells. The colored lines show the contact profile dynamics after 2 h (green), 4 h (purple), 6 h (blue) of IAA treatment. Bottom, annotation of genes and CTCF motif within the locus oriented for directionality (+: forward, −: reverse)

However, one can also think about tagging different proteins with different degron tags (for instance AID and dTAG). This will enable the sequential depletion of two factors to determine the causal relationship between the two proteins. The power of degrons is that it allows for the detection of rapid effects that would be missed with knockout or knockdown methods.

Because of the rapid effects following acute protein depletion it is important to measure the subject that one is interested in. For instance, a transcription factor regulates gene transcription, which ultimately results in steady state mRNA levels. We have seen, however, that methods that measure transcription rates with such as GRO-seq [64], PRO-seq [65], 4SU-seq [66], or TT-seq [67], is a better measure for the consequences of 3D genome perturbations than mRNA-seq [10]. Therefore, acute protein depletion methods need to be coupled to more direct measurements to perform truly quantitative biology in the genomics sphere.

Another interesting future avenue is the development of light induced acute depletion methods. This can either be achieved through the addition of photoactivatable small-molecules (PHO-TACs) [68] or the formation of light-induced protein dimers that can target the protein of interest for degradation, a method known as light induced protein depletion (LiPD) [69]. Although, it is still quite early days for these methods it is interesting to think about possible future applications. For instance, it would allow for protein depletion only in specific parts of the nucleus and follow the dynamics using live cell imaging after protein depletion, similar to a FRAP experiment. Alternatively, it would allow for the depletion of proteins from a specific part of an organism without requiring tissue-specific expression of effectors such as the OsTir1 receptor, and with subcellular resolution.

Clearly, exciting possibilities await the study of how the genome is organized inside the nucleus. It is clear that acute protein depletion strategies are powerful new tools that will play a crucial role in delineating how and where architectural proteins play a role in regulating the spatial genome and the role this plays in facilitating nuclear processes.

References

1. Lieberman-Aiden E, van Berkum NL, Williams L et al (2009) Comprehensive mapping of long-range interactions reveals folding principles of the human genome. Science 326:289–293

2. Boettiger AN, Bintu B, Moffitt JR et al (2016) Super-resolution imaging reveals distinct chromatin folding for different epigenetic states. Nature 529:418–422

3. Beliveau BJ, Boettiger AN, Avendaño MS et al (2015) Single-molecule super-resolution imaging of chromosomes and in situ haplotype visualization using Oligopaint FISH probes. Nat Commun 6:7147

4. Cremer T, Cremer C (2001) Chromosome territories, nuclear architecture and gene regulation in mammalian cells. Nat Rev Genet 2: 292–301

5. Schwarzer W, Abdennur N, Goloborodko A et al (2017) Two independent modes of chromatin organization revealed by cohesin removal. Nature 551:51

6. Haarhuis JHI, van der Weide RH, Blomen VA et al (2017) The Cohesin release factor WAPL restricts chromatin loop extension. Cell 169: 693–707.e14

7. Wutz G, Várnai C, Nagasaka K et al (2017) Topologically associating domains and chromatin loops depend on cohesin and are regulated by CTCF, WAPL, and PDS5 proteins. EMBO J 36:3573

8. Ciosk R, Shirayama M, Shevchenko A et al (2000) Cohesin's binding to chromosomes depends on a separate complex consisting of Scc2 and Scc4 proteins. Mol Cell 5:243

9. Kueng S, Hegemann B, Peters BH et al (2006) Wapl controls the dynamic association of cohesin with chromatin. Cell 127:955–967

10. Liu NQ, Maresca M, van den Brand T et al (2021) WAPL maintains a cohesin loading cycle to preserve cell-type-specific distal gene regulation. Nat Genet 53:100-109

11. Szabo Q, Bantignies F, Cavalli G (2019) Principles of genome folding into topologically associating domains. Sci Adv 5:eaaw1668

12. Beagan JA, Phillips-Cremins JE (2020) On the existence and functionality of topologically associating domains. Nat Genet 52:8

13. Stevens TJ, Lando D, Basu S et al (2017) 3D structures of individual mammalian genomes studied by single-cell Hi-C. Nature 544:59

14. Bintu B, Mateo LJ, Su J-H et al (2018) Super-resolution chromatin tracing reveals domains and cooperative interactions in single cells. Science 362:eaau1783

15. Alexander JM, Guan J, Li B et al (2019) Live-cell imaging reveals enhancer-dependent sox2 transcription in the absence of enhancer proximity. elife 8:e41769

16. Nora EP, Goloborodko A, Valton A-L et al (2017) Targeted degradation of CTCF decouples local insulation of chromosome domains from genomic compartmentalization. Cell 169:930–944.e22

17. Rao SSP, Huang S-C, Glenn St Hilaire B et al (2017) Cohesin loss eliminates all loop domains. Cell 171:305–320.e24

18. Weintraub AS, Li CH, Zamudio AV et al (2017) YY1 is a structural regulator of enhancer-promoter loops. Cell 171:1573

19. Thiecke MJ, Wutz G, Muhar M et al (2020) Cohesin-dependent and -independent mechanisms mediate chromosomal contacts between promoters and enhancers. Cell Rep 32:107929

20. Jiang Y, Huang J, Lun K et al (2020) Genome-wide analyses of chromatin interactions after the loss of Pol I, Pol II, and Pol III. Genome Biol 21:158

21. Gibcus JH, Samejima K, Goloborodko A et al (2018) A pathway for mitotic chromosome formation. Science 359:eaao6135

22. Sakamoto KM, Kim KB, Kumagai A et al (2001) Protacs: chimeric molecules that target proteins to the Skp1-Cullin-F box complex for ubiquitination and degradation. Proc Natl Acad Sci U S A 98:8554

23. Winter GE, Buckley DL, Paulk J et al (2015) Phthalimide conjugation as a strategy for in vivo target protein degradation. Science 348:1376

24. Sun X, Gao H, Yang Y et al (2019) Protacs: great opportunities for academia and industry. Signal Transduct Target Ther 4:64

25. Clift D, McEwan WA, Labzin LI et al (2017) A method for the acute and rapid degradation of endogenous proteins. Cell 171:1692

26. Maynard-Smith LA, Chen LC, Banaszynski LA et al (2007) A directed approach for engineering conditional protein stability using biologically silent small molecules. J Biol Chem 282: 24866

27. Chung HK, Jacobs CL, Huo Y et al (2015) Tunable and reversible drug control of protein production via a self-excising degron. Nat Chem Biol 11:713–720

28. Tan X, Calderon-Villalobos LIA, Sharon M et al (2007) Mechanism of auxin perception by the TIR1 ubiquitin ligase. Nature 446:640

29. Nishimura K, Fukagawa T, Takisawa H et al (2009) An auxin-based degron system for the rapid depletion of proteins in nonplant cells. Nat Methods 6:917–922

30. Natsume T, Kiyomitsu T, Saga Y et al (2016) Rapid protein depletion in human cells by auxin-inducible degron tagging with short homology donors. Cell Rep 15:210–218

31. Kubota T, Nishimura K, Kanemaki MT et al (2013) The Elg1 replication factor C-like complex functions in PCNA unloading during DNA replication. Mol Cell 50:273

32. Yesbolatova A, Saito Y, Kitamoto N et al (2020) The auxin-inducible degron 2 technology provides sharp degradation control in yeast, mammalian cells, and mice. Nat Commun 11:5701

33. Nora EP, Caccianini L, Fudenberg G et al (2020) Molecular basis of CTCF binding polarity in genome folding. Nat Commun 11: 5612

34. Rhodes JDP, Feldmann A, Hernández-Rodríguez B et al (2020) Cohesin disrupts

polycomb-dependent chromosome interactions in embryonic stem cells. Cell Rep 30:820

35. Gu B, Comerci CJ, McCarthy DG et al (2020) Opposing effects of cohesin and transcription on CTCF organization revealed by super-resolution imaging. Mol Cell 80:699

36. Nabet B, Roberts JM, Buckley DL et al (2018) The dTAG system for immediate and target-specific protein degradation. Nat Chem Biol 14:431–441

37. Jaeger MG, Schwalb B, Mackowiak SD et al (2020) Selective mediator dependence of cell-type-specifying transcription. Nat Genet 52:719

38. Nabet B, Ferguson FM, Seong BKA et al (2020) Rapid and direct control of target protein levels with VHL-recruiting dTAG molecules. Nat Commun 11:4687

39. Boija A, Klein IA, Sabari BR et al (2018) Transcription factors activate genes through the phase-separation capacity of their activation domains. Cell 175:1842–1855.e16

40. Lu J, Qian Y, Altieri M et al (2015) Hijacking the E3 ubiquitin ligase cereblon to efficiently target BRD4. Chem Biol 22:755

41. Jinek M, Chylinski K, Fonfara I et al (2012) A programmable dual-RNA-guided DNA endonuclease in adaptive bacterial immunity. Science 337:816–821

42. Cong L, Ran FA, Cox D et al (2013) Multiplex genome engineering using CRISPR/Cas systems. Science 339:819–823

43. Natsume T, Kanemaki MT (2017) Conditional degrons for controlling protein expression at the protein level. Annu Rev Genet 51:83

44. Yesbolatova A, Natsume T, Ichiro HK et al (2019) Generation of conditional auxin-inducible degron (AID) cells and tight control of degron-fused proteins using the degradation inhibitor auxinole. Methods 164–165:73

45. Sathyan KM, McKenna BD, Anderson WD et al (2019) An improved auxin-inducible degron system preserves native protein levels and enables rapid and specific protein depletion. Genes Dev 33:1

46. Kubo N, Ishii H, Gorkin D et al (2017) Preservation of chromatin organization after acute loss of CTCF in mouse embryonic stem cells. bioRxiv:118737

47. Ito T, Ando H, Suzuki T et al (2010) Identification of a primary target of thalidomide teratogenicity. Science 327:1345

48. Carroll D (2014) Genome engineering with targetable nucleases. Annu Rev Biochem 83:409

49. Kim H, Kim JS (2014) A guide to genome engineering with programmable nucleases. Nat Rev Genet 15:321

50. Nakade S, Tsubota T, Sakane Y et al (2014) Microhomology-mediated end-joining-dependent integration of donor DNA in cells and animals using TALENs and CRISPR/Cas9. Nat Commun 5:5560

51. Lin DW, Chung BP, Huang JW et al (2019) Microhomology-based CRISPR tagging tools for protein tracking, purification, and depletion. J Biol Chem 294:10877

52. Bateman A (2019) UniProt: a worldwide hub of protein knowledge. Nucleic Acids Res 47:D506

53. Casey PJ, Seabra MC (1996) Protein prenyltransferases. Genome Biol 4:212

54. Jung HJ, Nobumori C, Goulbourne CN et al (2013) Farnesylation of lamin B1 is important for retention of nuclear chromatin during neuronal migration. Proc Natl Acad Sci U S A 110:E1923

55. Robert F, Barbeau M, Éthier S et al (2015) Pharmacological inhibition of DNA-PK stimulates Cas9-mediated genome editing. Genome Med 7:93

56. Gu B, Posfai E, Rossant J (2018) Efficient generation of targeted large insertions by microinjection into two-cell-stage mouse embryos. Nat Biotechnol 36:632

57. Branda CS, Dymecki SM (2004) Talking about a revolution: the impact of site-specific recombinases on genetic analyses in mice. Dev Cell 6:7

58. de Vree PJP, de Wit E, Yilmaz M et al (2014) Targeted sequencing by proximity ligation for comprehensive variant detection and local haplotyping. Nat Biotechnol 32:1019

59. Splinter E, de Wit E, van de Werken HJG et al (2012) Determining long-range chromatin interactions for selected genomic sites using 4C-seq technology: from fixation to computation. Methods 58:221–230

60. Davies JOJ, Telenius JM, McGowan SJ et al (2015) Multiplexed analysis of chromosome conformation at vastly improved sensitivity. Nat Methods 13:74–80

61. Díaz N, Kruse K, Erdmann T et al (2018) Chromatin conformation analysis of primary patient tissue using a low input Hi-C method. Nat Commun 9:4938

62. Geeven G, Teunissen H, de Laat W et al (2018) peakC: a flexible, non-parametric peak calling package for 4C and Capture-C data. Nucleic Acids Res 46:e91–e91

63. Krijger PHL, Geeven G, Bianchi V et al (2020) 4C-seq from beginning to end: a detailed protocol for sample preparation and data analysis. Methods 170:17

64. Core LJ, Waterfall JJ, Lis JT (2008) Nascent RNA sequencing reveals widespread pausing and divergent initiation at human promoters. Science 322:1845

65. Mahat DB, Kwak H, Booth GT et al (2016) Base-pair-resolution genome-wide mapping of active RNA polymerases using precision nuclear run-on (PRO-seq). Nat Protoc 11:1455

66. Schwarzl T, Higgins DG, Kolch W et al (2015) Measuring transcription rate changes via time-course 4-thiouridine pulse-labelling improves transcriptional target identification. J Mol Biol 427:3368

67. Schwalb B, Michel M, Zacher B et al (2016) TT-seq maps the human transient transcriptome. Science 352:1225

68. Reynders M, Matsuura BS, Bérouti M et al (2020) PHOTACs enable optical control of protein degradation. Sci Adv 6:eaay5064

69. Deng W, Bates JA, Wei H et al (2020) Tunable light and drug induced depletion of target proteins. Nat Commun 11:304

Correction to: Versatile CRISPR-Based Method for Site-Specific Insertion of Repeat Arrays to Visualize Chromatin Loci in Living Cells

Thomas Sabaté, Christophe Zimmer, and Edouard Bertrand

Correction to:
Chapter 13 in: Tom Sexton (ed.), *Spatial Genome Organization: Methods and Protocols*, Methods in Molecular Biology, vol. 2532, https://doi.org/10.1007/978-1-0716-2497-5_13

In the original version of this book, Chapter 13 was published with few errors. This has been rectified in the updated version of this book.

The updated version of this chapter can be found at
https://doi.org/10.1007/978-1-0716-2497-5_13

Tom Sexton (ed.), *Spatial Genome Organization: Methods and Protocols*,
Methods in Molecular Biology, vol. 2532, https://doi.org/10.1007/978-1-0716-2497-5_16,
© The Author(s), under exclusive license to Springer Science+Business Media, LLC, part of Springer Nature 2022

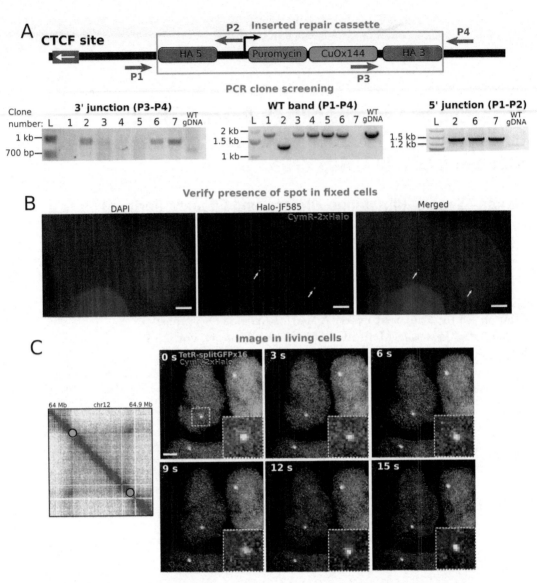

Fig. 2 Clone screening and imaging. (**a**) PCR screening to verify the insertion of the repair cassette. Top: Schematic of the inserted repair cassette in the genome. Pink arrows represent the location of screening primers at both sides of the repair cassette. Bottom: Agarose gel of 7 different clones after antibiotic selection. Left: Junction-PCR on the 3′ side (908 bp band is expected if the repair cassette is inserted). Middle: WT allele PCR (a band at 1.8 kb is expected if at least one allele is WT). Clone 7 has a homozygous insertion since no WT band is seen. Right: Junction-PCR on the 5′ side for clones that were positive on the 3′ side (a band at 1.4 kb is expected if the repair cassette is inserted). L: DNA ladder. (**b**) Fixed cell images of a PCR-screened positive clone. White arrows show the fluorescently tagged locus. Scale bar: 3 μm. (**c**) Live-imaging of loop anchors in HCT116 cells. Left: Hi-C contact frequency map of a 2 Mb region on chromosome 12 containing the two inserted operator arrays (circles) at the anchors of a 400 kb-long chromatin loop. Right: Spinning disk images of the loop anchors. Left anchor (green) was labelled by TetOx96 (inserted with the same cloning method described here for CuOx144) bound by TetR-splitGFPx16 and right anchor (red) was labeled by CuOx144 bound by a CymR-2xHalo (imaged with JF585). The inset shows a magnification of the region outlined by a white dotted line. The images are maximum intensity projections of 11 z-stacks each separated by 1 μm. One full z-stack was imaged every 3 s. Scale bar: 3 μm

INDEX

Tom Sexton (ed.), *Spatial Genome Organization: Methods and Protocols*,
Methods in Molecular Biology, vol. 2532, https://doi.org/10.1007/978-1-0716-2497-5,
© The Editor(s) (if applicable) and The Author(s), under exclusive license to Springer Science+Business Media, LLC, part of Springer
Nature 2022

Printed in Great Britain
by Amazon

52569919R00196